COMPREHENSIVE BIOCHEMISTRY

ELSEVIER SCIENCE PUBLISHERS

1 Molenwerf, P.O. Box 211, Amsterdam

ELSEVIER SCIENCE PUBLISHING Co. INC.

52, Vanderbilt Avenue, New York, N.Y. 10017

With 16 plates and 47 figures

Library of Congress Cataloging-in-Publication Data
(Revised for vol. 2)

Main entry under title:

Selected topics in the history of biochemistry.

(Comprehensive biochemistry ; v. 35-36. Section VI, A history of biochemistry)
Includes bibliographical references and index.
1. Biological chemistry–History–Addresses, essays, lectures. I. Semenza, G., 1928- . II. Series. III. Series: Comprehensive biochemistry ; 35.
QD415.F54 vol. 35 574.19'2 s 83-20491
[QP511] 574.19'2
ISBN 0-444-80507-9 (U.S. : v. 1)
ISBN 0-444-80702-0 (U.S. : v. 2)

PRINTED IN THE NETHERLANDS

COMPREHENSIVE BIOCHEMISTRY

COMPREHENSIVE BIOCHEMISTRY

SECTION I (VOLUMES 1–4)
PHYSICO-CHEMICAL AND ORGANIC ASPECTS OF BIOCHEMISTRY

SECTION II (VOLUMES 5–11)
CHEMISTRY OF BIOLOGICAL COMPOUNDS

SECTION III (VOLUMES 12–16)
BIOCHEMICAL REACTION MECHANISMS

SECTION IV (VOLUMES 17–21)
METABOLISM

SECTION V (VOLUMES 22–29)
CHEMICAL BIOLOGY

SECTION VI (VOLUMES 30–35)
A HISTORY OF BIOCHEMISTRY

COMPREHENSIVE BIOCHEMISTRY

ALBERT NEUBERGER
Chairman of Governing Body, The Lister Institute of Preventive Medicine, University of London, London (Great Britain)

LAURENS L.M. VAN DEENEN
Professor of Biochemistry, Biochemical Laboratory, Utrecht (The Netherlands)

GIORGIO SEMENZA
Laboratorium für Biochemie, ETH-Zentrum Zurich (Switzerland)

Editor of VOLUME 36

SELECTED TOPICS IN THE HISTORY OF BIOCHEMISTRY PERSONAL RECOLLECTIONS. II.

ELSEVIER SCIENCE PUBLISHERS

AMSTERDAM · OXFORD · NEW YORK

1986

GENERAL PREFACE

The Editors are keenly aware that the literature of Biochemistry is already very large, in fact so widespread that it is increasingly difficult to assemble the most pertinent material in a given area. Beyond the ordinary textbook the subject matter of the rapidly expanding knowledge of biochemistry is spread among innumerable journals, monographs, and series of reviews. The Editors believe that there is a real place for an advanced treatise in biochemistry which assembles the principal areas of the subject in a single set of books.

It would be ideal if an individual or a small group of biochemists could produce such an advanced treatise, and within the time to keep reasonably abreast of rapid advances, but this is at least difficult if not impossible. Instead, the Editors with the advice of the Advisory Board, have assembled what they consider the best possible sequence of chapters written by competent authors; they must take the responsibility for inevitable gaps of subject matter and duplication which may result from this procedure.

Most evident to the modern biochemist, apart from the body of knowledge of the chemistry and metabolism of biological substances, is the extent to which we must draw from recent concepts of physical and organic chemistry, and in turn project into the vast field of biology. Thus in the organization of Comprehensive Biochemistry, sections II, III and IV, Chemistry of Biological Compounds, Biochemical Reaction Mechanisms, and Metabolism may be considered classical biochemistry, while the first and fifth sections provide selected material on the origins and projections of the subject.

It is hoped that sub-division of the sections into bound volumes will not only be convenient, but will find favour among students concerned with specialized areas, and will permit easier future revisions of the individual volumes. Towards the latter end particularly, the Editors will welcome all comments in their effort to produce a useful and efficient source of biochemical knowledge.

M. Florkin†
E.H. Stotz

Liège/Rochester

There is a history in all men's lives.
W. Shakespeare, Henry IV, Pt. 2

History is the essence of innumerable biographies.
T. Carlyle, On History

PREFACE TO VOLUME 36

Perhaps one of the most exciting developments in biological sciences in our times has been their merging with chemistry and physics with the resulting appearance of biochemistry, biophysics, molecular biology, and related sciences. The nearly explosive development of these 'newcomers' has led to the almost unique situation that these new biological sciences have come of age at time when their founding fathers, or their scientific sons, are alive and active.

It was therefore an almost obvious idea to ask them to write, for the benefit of both students and senior scientists, personal accounts of their scientific lives. With this idea in mind I have already edited two volumes for John Wiley & Sons, who had, however, a somewhat different format.

The chapters in this and in future volumes are meant to complement, with personal recollections, the History of Biochemistry in the Comprehensive Biochemistry series (Vols. 30–33, by M. Florkin and Vol. 34 (forthcoming), by E. Schoffeniels). In fact, it is hoped that the biographical or autobiographical chapters will convey to the reader lively, albeit at times subjective, views on the scientific scene as well as the social environment in which the authors have operated and brought about new concepts and pieces of knowledge. The Editor considered it presumptuous to give the authors narrow guidelines or to suggest changes in the chapters he received; he thinks that directness and straightforwardness should be given priority over uniformity. The contributions assembled in this volume will convey the flavour of each author's particular personality; whatever the optical distortion of one chapter, it will be compensated by the views in another.

The development of today's life sciences was acted upon by serious and often tragic historical events. The Editor hopes that this message also will reach to readers, especially the young ones.

It proved an impossible task to group the contributors in a strictly logical manner whether according to subject matter, geographical area, or time. In fact, most contributions cross each of these borders. Nevertheless the Editor hopes that the reader will find these contributions as interesting as he did.

The Editor wants to express his gratitude to all individuals who made this series possible; first of all to the authors themselves, who not only wrote the texts, but also willingly collaborated in suggesting further potential contributors, thereby acting as a kind of 'Editorial Board at Large'. Thanks are due to Ms. U. Zilian who typed most of the correspondence and prepared the index of names.

Swiss Institute of Technology
Zurich, 1983

Giorgio Semenza

CONTRIBUTORS TO THIS VOLUME

H. BEINERT (Introduction for P. HEMMERICH)
Institute for Enzyme Research and Department of Biochemistry
College of Agricultural and Life Sciences
University of Wisconsin, Madison, WI 53706 (U.S.A.)

J.M. BUCHANAN
John and Dorothy Wilson Professor of Biochemistry Department of Biology,
The Massachussetts Institute of Technology, Cambridge, MA 02139 (U.S.A.)

J.T. EDSALL
Department of Chemistry and Molecular Biology, Harvard University
7 Divinity Ave., Cambridge, MA 02138 (U.S.A.)

P. HEMMERICH
Fachbereich Biologie, Universität Konstanz, D-7750 Konstanz (F.R.G.)

N.O. KAPLAN
Department of Chemistry and Cancer Center University of California San Diego,
La Jolla, CA 92093 (U.S.A.)

M. KLINGENBERG
Institute for Physical Biochemistry, University of Munich
Goethestrasse 33, 8000 Munich 2 (F.R.G.)

K. KURAHASHI
Institute for Protein Research, Osaka University 3-2, Yamadaoka, Suita,
Osaka 565 (Japan)

H.A. LARDY
Institute for Enzyme Research, University of Wisconsin
1710 University Avenue, Madison, WI 53705 (U.S.A.)

E. LEDERER
Laboratoire de Biochimie, C.N.R.S., 91190 Gif-sur-Yvette (France) and
Institut de Biochimie, Université de Paris Sud, 91405 Orsay (France)

N.W. PIRIE
Rothamstead Experimental Station, Harpenden, Herts. AL5 2JQ (U.K.)

E.C. SLATER
Laboratory of Biochemistry
B.C.P. Jansen Institute, University of Amsterdam
Plantage Muidergracht 12, 1018 TV Amsterdam (The Netherlands)

LIST OF PLATES

(Photographs reproduced with permission of authors, publishers, and/or owners)

Section VI

A HISTORY OF BIOCHEMISTRY

Vol. 30. A History of Biochemistry
Part I. Proto-Biochemistry
Part II. From Proto-Biochemistry to Biochemistry

Vol. 31. A History of Biochemistry
Part III. History of the identification of the sources of free energy in organisms

Vol. 32. A History of Biochemistry
Part IV. Early studies on biosynthesis

Vol. 33. A History of Biochemistry
Part V. The unravelling of biosynthetic pathways

Vol. 34. A History of Biochemistry
Part VI. History of molecular interpretations of physiological and biological concepts, and the origins of the conception of life as the expression of a molecular order

Vol. 35. A History of Biochemistry
Selected Topics in the History of Biochemistry
— Personal Recollections. I

Vol. 36. A History of Biochemistry
Selected Topics in the History of Biochemistry
— Personal Recollections. II

COMPREHENSIVE BIOCHEMISTRY, Vol. 35
Errata and Corrigenda

Pages xv and 103: N. William Pirie should read: N.W. Pirie

Page xvi, line 3: 135 should read: 129

Page 129: the address Vancouver, BC (VGT 243. . .) should read: Vancouver, BC (V6T 243 Canada)

Page 157: the first two formulae should read:

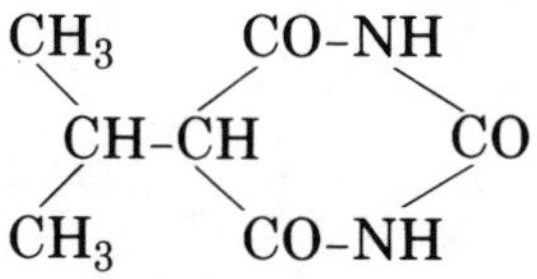

CH_3 CO–NH
CH–CH CO
CH_3 CO–NH

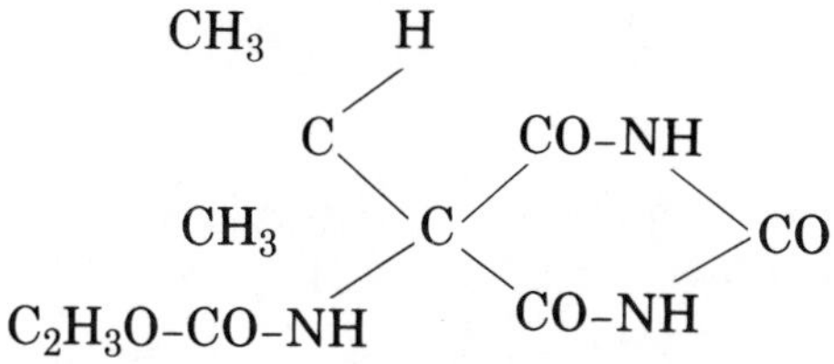

CH_3 H
C CO–NH
CH_3 C CO
C_2H_3O–CO–NH CO–NH

Page 160, line 20: thyramine should read: tyramine

Page 163, line 25: neutral should read: neural; analgetics should read: analgesics

Page 164, line 20: sluggested should read: suggested

Page 175, line 13: heat should read: heart

Page 178, line 11: Tables VI and VII should read: Tables V and VI

Page 390, ref. 38: L.V. Betlousova should read: L.V. Belousova

CONTENTS

VOLUME 36

A HISTORY OF BIOCHEMISTRY

Selected Topics in the History of Biochemistry
Personal Recollections. II.

Chapter 1. A Backward Glance
by JOHN MACHLIN BUCHANAN

Chapter 2. The Discovery of Phosphoenolpyruvate Carboxykinase. In Memoriam Merton F. Utter
by KIYOSHI KURAHASHI

Chapter 3. Jeffries Wyman and Myself: a Story of Two Interacting Lives
by JOHN T. EDSALL

Chapter 4. The BAL-labile Factor in the Respiratory Chain by E.C. Slater

Chapter 5. Experiences in Biochemistry by Nathan O. Kaplan

Chapter 6. A Half Century of Biochemistry
by Henry A. Lardy

Chapter 7. A Biochemist's View of his Struggle for Knowledge. Review of Forty Years Service to Science
by M. Klingenberg

Chapter 8. An Eventful Life Around Flavins.
by PETER HEMMERICH

Memoirs Dictated in the Last Weeks of his Life, August 1981 with an introductory note by Helmut Beinert

Chapter 9. Adventures and Research
by EDGAR LEDERER

Chapter 10. Recurrent Luck in Research
by N.W. PIRIE

G. Semenza (Ed.) Selected Topics in the History of Biochemistry: Personal Recollections (Comprehensive Biochemistry Vol. 36)

Chapter 1

A Backward Glance

JOHN MACHLIN BUCHANAN

The Department of Biology, The Massachusetts Institute of Technology, Cambridge, MA 02139 (U.S.A.)

I have recently reread a much cherished autographed volume entitled *A Trail of Research* [1] sent to me by Vincent duVigneaud over twenty years ago when our paths in research crossed briefly. This book, whose six chapters comprise the Messenger Lectures given at Columbia University in 1950, recounts his long-standing interests in sulfur metabolism dating back to his student years at the University of Illinois. At the same time he gives a vivid insight into the times when biochemistry was emerging in this country as an individual discipline. Using backgrounds in organic chemistry and nutrition, this new cadre of biological scientists, located principally at the University of Illinois and Wisconsin and at the Sheffield School of Science at Yale University, were setting into place the essential information that became the forerunner of the discipline of intermediary metabolism, one of the first major problems in the biological sciences undertaken by them.

duVigneaud was greatly influenced by two remarkable men on the faculty at the University of Illinois, William C. Rose and Howard B. Lewis, both of whom were pioneers in the field of amino acid metabolism. Rose was particularly well known for his studies on the essential dietary amino acids for the rat and the human and for the discovery of threonine. Both were involved in tracing the metabolic fates of the sulfur amino acids cystine and later methio-

Plate 1. John Machlin Buchanan.

nine, which was only discovered by J. Howard Mueller [2] as relatively recently as 1923. A principal method of experimentation at that time was feeding of suspected precursors and following the excretion of an expected product, or measuring the effect of a given compound in replacing the growth requirement of an animal kept on a regimen devoid of a particular dietary essential. As might be expected conclusive experiments of this kind were not easily planned.

Lewis accepted the chairmanship of the Department of Biological Chemistry at the University of Michigan in 1923. I first met him in the summer of 1938 when I was trying to formulate my own plans for graduate work. During the four previous years (1934–1938) I had been an undergraduate student at DePauw University in Greencastle, Indiana, majoring in chemistry and minoring in mathematics. At the beginning of my senior year I had undertaken a research problem with a young faculty member, Jesse L. Riebsomer, on the synthesis of mandelic acids [3]. During that year I became fully aware of both the thrill and frustration associated with research in organic chemistry. However, midway through the year I happened to buy and read a copy of the fourth edition of Bodansky's *Physiological Chemistry*, my first introduction to the biological sciences. When a tentative assistantship in organic chemistry at Penn State fell through because of a disastrous fire in the student laboratories, I had the opportunity to reassess my academic goals. For a while I courted the idea of applying to medical school, and even as late as the summer after graduation enrolled in two courses at Indiana University in embryology and comparative anatomy to complement my one semester of zoology at DePauw. However, the experience of this summer had convinced me that biology offered a huge store of exciting problems that could be approached with the experimental methods of the chemists. My good friend and advisor, Dr. Riebsomer, was disappointed in my decision to leave organic chemistry for a field that was new and unknown to him. Furthermore, he was unable to provide advice about possible graduate departments in biological chemistry. This advice I did obtain from a friend of my family at Eli Lilly Co. in Indianapolis. One of the schools he strongly recommended was the University of Michigan.

In approaching Dr. Lewis about the possibility of enrollment in the Fall of 1938 I was given very little encouragement about an eventual assistantship because a shift in the schedules that year at the University of Michigan Medical School had required twice the crew of graduate assistants than the usual number of appointments to teach for that year only both the first- and second-year classes.

For this reason I had to depend on our family resources to finance my first year. These financial considerations were of importance since my parents were just recovering from one of the worst economic depressions this country had experienced, and were still obligated to my brother and sister to complete their educations.

Nevertheless, the year in the Department at the University of Michigan working towards a Master's degree proved to be one of the most rewarding periods of my life. Not only did Dr. Lewis live up in full measure to his reputation as a brilliant lecturer but also succeeded in introducing topics in biochemistry to me that were to hold my interest for years to come. The topics of the departmental seminar were taken from the current biochemical literature. Papers from the Cori laboratory on the enzymes of glycogen synthesis as well as the first trickle of reports from the Columbia laboratory on the use of isotopic compounds in metabolic studies were high points of these sessions. My own reports were on the physical properties of concanavalin A and on the now outmoded Träger theory of enzyme activity contained in a long paper in German by Waldschmidt-Leitz. It seemed clear even then that these newer methods would be necessary to unravel some of the complicated reactions of intermediary metabolism.

After a rather successful first term, I approached Dr. Lewis again about the possibility of an assistantship for the next year. Since no vacancies were expected at that time and my continuation in graduate work depended on obtaining financial assistance, Dr. Lewis graciously offered me his help in finding a position at another university. He suggested that I apply among others to A. Baird Hastings at the Department of Biological Chemistry at Harvard Medical School. By the time an offer had been received from Harvard, a position did become available at Michigan. It was an extremely difficult decision for me to accept the Harvard offer because of my real fondness and admiration for Howard Lewis, whose friendship I did maintain until his death in 1954. Lewis was a member of the examiner's board for the American Medical Association for many years, and his responsibilities with this organization brought him to Boston and Philadelphia. On several occasions I was able to meet him for dinner and an evening at the theatre, which was one of his favorite forms of leisure.

I arrived at Harvard Medical School in early September, 1939, at the time when Germany was invading Poland and World War II had begun. My first year was spent on the Cambridge campus taking courses in organic chemistry, physical chemistry and biophysics. During the year John Edsall offered me an assistantship on the Cambridge campus to augment my stipend from the Medical School. I also filled in as a tutor in Biochemical Sciences for a vacancy that unexpectedly developed there. It was indeed a busy schedule and one that brought me into contact with many new areas of biology.

Glycogen synthesis and CO_2 incorporation

After one year devoted primarily to course work my research training began in September 1940 in the laboratory of Baird Hastings, who was to become a life-long friend and mentor. At the instigation of President James B. Conant of Harvard, a program on the synthesis and metabolism of lactic acid labeled with the short-lived radioactive isotope, ^{11}C, was initiated as one of the projects centered around the Harvard cyclotron. A group of chemists and biochemists including Richard Cramer, Friedrich Klemperer, Arthur Solomon, Birgit Vennesland, George Kistiakowsky and Baird Hastings was formed to synthesize carboxyl and α,β-labeled lactic acids [4] and study their conversion to glycogen in the fasted rat [5,6]. Since the half life of ^{11}C is only 20 min, it was necessary to complete the entire synthesis and metabolic reactions within a period of 3 or at the most 4 h. This limit required a highly organized effort. As a beginning graduate student and junior member of this team, my duties were to help wherever possible, for example, to provide the leg work in conveying the irradiated boron oxide target from the cyclotron to the chemistry laboratory in Cambridge. Arthur Solomon had constructed one of the first geiger counters in the Boston area and spent a great deal of time keeping it in proper working order. Birgit Vennesland was a postdoctoral biochemist, who became during the next 2 years my actual supervisor and instructor in the art of research. No one could have had a more gentle and diplomatic friend to initiate one into the insights and judgments required in the first stages of becoming an experimental scientist. In conjunction with our program with labeled compounds Birgit and I began a series of studies on the effects of inorganic ions on the synthesis of glycogen in rat and rabbit liver slices. As it turned out these in vitro studies eventually complemented our in vivo studies on glycogen synthesis from radioactive precursors. The in vitro experiments ultimately became my sole responsibility as a part of my doctoral thesis. In part of this work I was joined by Frances Nesbett, who provided excellent technical assistance. As a long-time student of inorganic chemistry as applied to physiological systems, Hastings had developed the hypothesis that complex biosynthetic reactions such as glycogen synthesis from glucose might prefer an incubation medium patterned after an intracellular rather than extracellular milieu [7]. Our first studies comparing an intracellular medium high in potassium and magnesium ions versus Ringer's solution appeared to bear out his hypothesis [8]. However, it soon became apparent that calcium ions at a concentration of 5 to 10 mM were needed in addition to magnesium and potassium ions for an optimal rate of synthesis of glycogen in both rat and rabbit liver slices [9]. Later on we became interested in the problem of glyconeogenesis from pyruvate [10]. We found that sodium and calcium ions were very necessary for total carbohydrate synthesis from pyruvate, but potassium ions were also needed for the conversion of the carbohydrate into glycogen. The addition of magnesium ions did not appear to influence the formation of total carbohydrate or its redistribution into the glycogen compartment.

Depending on the state of repair of the cyclotron during this time, experiments with radioactive substrates were carried out. In experiments performed principally by my senior colleagues the radioactivity of α,β-labeled lactic acid was found in the liver glycogen of rats to an extent approximately twice that of carboxyl-labeled lactate [5,6]. Both compounds had evidently entered a large pool of carbohydrate compounds from which the substrates for glycogen synthesis were drawn. These experiments confirmed work emanating from the Columbia laboratory [11] showing that ingested labeled substrates are diluted by preexisting body compounds during metabolism. However, the difference in the apparent conversion of the two radioactive species of lactic acid implied that the carboxyl group of lactate was lost in part during its transit of reactions ending in glycogen. Because of her background in bacterial metabolism Birgit was aware of very interesting reports by H.G. Wood and C.H. Werkman [12,13] at Iowa State College in which they had proposed from carbon balance studies that CO_2 was utilized as a substrate in the synthesis of organic acids by the propionic acid bacteria. They [14] as well as Evans and Slotin [15] had proposed that CO_2 and pyruvate react to form oxalacetate and that further products were derived from this 4-carbon acid. She suggested that [^{11}C]bicarbonate be tested as a precursor of glycogen. Not only was $^{11}CO_2$ utilized, but in approximately the amount predicted if it were to replace the carboxyl carbon of lactate that was lost [16]. In these experiments fasting rats had been fed glucose in addition to the administration of bicarbonate intraperitoneally to stimulate glycogen synthesis. The circumstances of these experimental conditions led to the following hypothesis: (i) the feeding of glucose increases the overall carbohydrate metabolic pool from which glycogen is formed; (ii) the components of this pool are brought rapidly into equilibrium with 3-carbon metabolites; and (iii) in resynthesis of hexoses approximately 1 carbon in 6 is derived from CO_2 after corrections for metabolic dilution in the carbohydrate pool. The reaction proposed for the incorporation of CO_2 via pyruvate included a step that resulted in its equilibration with a symmetrical molecule, i.e. fumarate. In the conversion of fumarate to 3-carbon sugar precursors, 1 carbon would thus be derived from the carboxyl carbon of pyruvate (or lactate) and the other from bicarbonate. When these experiments were repeated in vitro with rabbit liver slices under circumstances in which one would not expect an equilibration of the components of the carbohydrate pool, we found that CO_2 was not utilized during the conversion of glucose to glycogen, but it did provide approximately 1 carbon in 6 during glycogen synthesis from pyruvate [17]. As I will show, this preoccupation as a graduate student with the metabolism of CO_2 would figure in other research problems later on.

After the United States entered World War II in December 1941, my collaborators disappeared to other parts with only Frances Nesbett and myself left to continue work with ^{11}C until June 1942 when the Harvard cyclotron itself was committed to the War effort. Since with Arthur Solomon our geiger counter had also departed, we came to rely on an available electroscope for measurements of radioactivity.

During this short period there was still time to complete one more project on the metabolism of short-chained fatty acids that had interested me since my Michigan days [18]. The aphorism, I believe attributed to Georg Rosenfeld [18a] that 'fats burn in the flame of carbohydrates' had always intrigued me. Although many investigators had shown that feeding of even-chained fatty acids did not lead to the accumulation of liver glycogen, it was possible that this conversion could take place but at a rate too slow to be measured by conventional methods. Our approach with the stimultaneous feeding of glucose together with radioactive fatty acids seemed to provide a new approach to an old problem. Furthermore, the simplicity of the Grignard reaction permitted the synthesis of carboxyl-labeled acetic, propionic and butyric acids with the dispatch required for work with ^{11}C. In our assessment of the conversion of the labeled compounds to glycogen, correction for the incorporation of metabolic $^{11}CO_2$ was of course necessary. Although propionic acid, a known carbohydrate precursor, was clearly converted to glycogen in our experiments, the results with acetic and butyric acids were ambiguous. Nevertheless, these experiments, although inconclusive, were the forerunners of others that eventually did provide a conclusive answer.

By January 1943 my doctoral thesis was complete and for the

next 6 months I was employed on a project concerned with mustard gases as part of the large war-related program Eric Ball had undertaken. Hastings by that time was heavily involved in administrative responsibilities in Washington. Although I can remember no significant contribution I made to the project during this time, I nevertheless came to appreciate fully Eric's many fine professional and personal qualities. Again, I had the good fortune to be associated with a teacher and friend, whose counsel I greatly valued.

Pennsylvania days: fatty acid metabolism and the beginning of purine synthesis

In spite of his very busy schedule, Baird Hastings was able to arrange my first step up the academic ladder. He is known by all of his former students and colleagues for his fierce loyalty and continued support over their entire careers. A position of instructor had opened up at the University of Pennsylvania in the Department of Physiological Chemistry, whose chairman was D. Wright Wilson. Due to the foresight of Dr. Wilson and Samuel Gurin, the Pennsylvania Department had initiated a program on the preparation of $^{13}CO_2$ for use in metabolic studies. The possibility of continuing work with a stable isotope of carbon was exciting for me, particularly since problems could be undertaken that did not have a time limit imposed upon them.

After the active life in the Department at Harvard, the pace in Philadelphia was deceptively sleepy. Since the number of faculty and students as well as resources for research was small, in many cases junior faculty members became associated with departmental programs already underway with the possibility of making individual contributions. In addition to research with isotopic carbon, Otto Meyerhof, who had joined the Department in 1940, was working on glyceraldehyde-3-phosphate dehydrogenase, and William Stadie in an adjoining research institute was actively engaged in studies on the mechanism of fatty acid oxidation and the role of insulin in metabolism. The academic environment provided many possibilities for one's development as a biochemist. But at this time my vision was directed within a rather narrow range in what would rapidly become an exploding field of intermediary metabolism.

Oxidation of fatty acids and acetoacetate

Warwick Sakami had also just finished his doctoral research in the Department and together with Wilson, Gurin and myself formed a small team. Our first project was an exploration of the metabolism of ^{13}C-labeled acetoacetate in guinea pig kidney homogenates. Warwick was occupied with the synthesis of the labeled compound under Gurin's direction, and I was responsible for establishing the proper conditions for the metabolic system.

Although Breusch [19] and Wieland and Rosenthal [20] had already reported the utilization of acetoacetate for citrate synthesis, their experiments were disputed by Krebs and Eggleston [21] and by Weil-Malherbe [22], who claimed that in a heart preparation the disappearance of acetoacetate was matched by the production of β-hydroxybutyrate and that the increase in citrate could be attributed to a greater contribution from oxalacetate. The application of isotopically labeled compounds could obviously provide a decisive experiment. I had found that the inclusion of α-ketoglutarate in my kidney homogenate resulted in a stimulation of acetoacetate utilization that could not be accounted for by the production of β-hydroxybutyrate. When the experiment was performed with ^{13}C-carboxy-labeled acetoacetate, a significant excess of ^{13}C was found in the α-ketoglutarate isolated at the conclusion of the incubation [23–25]. The position of the label in the γ-carboxyl carbon indicated that further metabolic products such as fumarate and malate would have isotope located equally in either carboxyl carbon. When this analysis of in vitro experiments was applied to metabolism in vivo, we realized that conversion of carboxyl labeled dicarboxylic acids to the triose phosphates would result in a labeling pattern ending up eventually with carbons 3 or 4 of hexose containing heavy isotope. At first glance one might conclude that fatty acids or their metabolic product, acetoacetate, are converted into carbohydrate. Yet upon analysis of the reactions for the oxidation of acetoacetate by way of acetate or an acetyl derivative, precisely the opposite conclusion is reached. As shown in the following equation.

$$CH_3{}^{13}COOH + 2O_2 + HOOCCH_2COCOOH \rightarrow 2CO_2 + 2H_2O + HOOCCH_2C^{13}COOH \text{ or } HOO^{13}CCH_2COCOOH$$

According to this equation there is a complete oxidation of acetate or acetoacetate to CO_2 and H_2O without the net increase of carbohydrate precursor. Only a nonisotopic molecule of oxalacetate has been replaced with a ^{13}C-containing one, which in the intact animal would undergo equilibration with the carbohydrate pool and eventually end up in glycogen. This particular experiment made me fully aware of the pitfalls one might encounter in the interpretation of data from isotopic experiments.

Before leaving the topic of fatty acid and acetoacetate oxidation, I would like to mention briefly some experiments that contributed to a debate centering around the question of whether fatty acids were oxidized to ketone bodies by a process of multiple alternate oxidation or by β-oxidation to 2-carbon fragments with their recombination to 4-carbon compounds as had been originally proposed by McKay et al. [26]. W.C. Stadie [27] at the University of Pennsylvania, who worked principally with liver slices, favored the theory of multiple oxidation of fatty acids at alternate carbon atoms followed by splitting directly into 4-carbon fragments. On the other hand Weinhouse et al. [28] located at the Lankanau Hospital in Philadelphia had challenged this mechanism by an ingenious experiment in which they labeled octanoic acid in the carboxyl position with ^{13}C and showed that rat liver slices converted this isotopic substrate into acetoacetate with an equal distribution of ^{13}C between the carbonyl and carboxyl carbon atoms. Our interest in this debate was catalyzed by our proximity to both of these groups. One aspect that remained to be determined was the possibility that acetoacetate was formed directly from octanoic acid and then underwent fission into 2-carbon fragments, which recombined. Warwick Sakami prepared the carbonyl-labeled acetoacetate in addition to the carboxyl-labeled compound mentioned above. After incubation with rat liver slices, the reisolated acetoacetate showed essentially the same composition and distribution of ^{13}C as the initial substrates [29]. Thus, equilibration of the 4-carbon with a 2-carbon species did not occur, a result that favored the point of view of the Lankanau group. However, in repeating the experiment with carboxyl-labeled octanoic acid and using possibly a better method of decarboxylating acetoacetate with aniline citrate, we found that the distribution of ^{13}C between carbonyl and carboxyl carbons was not equal. In fact, the carboxyl carbon contained a somewhat higher ^{13}C concentration. In following up this discrepancy Crandall et al. [30] later discovered that the methyl-terminal 2-carbon fragment is handled differently by rat liver slices than are the remaining three 2-carbon units formed from octanoic acid. Their work was thus the first clue to the now well known fact that acetoacetate is formed by reaction of acetyl CoA and malonyl CoA, and that the methyl-terminal 2-carbon fragment of a newly synthesized fatty acid has a metabolic origin different from that of the other carbon atoms [31].

Precursors of uric acid

I suppose that our work on the oxidation of acetate and acetoacetate via the tricarboxylic acid cycle focused my attention on the general contributions of Hans Krebs to biochemistry. As a graduate student and during the early period of my days at the University of Pennsylvania I had spent rather long summer vacations at the home of my parents in Kalamazoo, Michigan, alternating the days between golf and reading in the library of the Upjohn Co. The latter time was well spent in researching the large volume of papers that Krebs had published in the *Biochemical Journal.* Aside from his major contributions in formulating a cyclic series of reactions for the oxidation of pyruvate and the synthesis of urea, Krebs had published some very notable papers on the metabolism of amino acids and the synthesis of purines in liver slices. In my opinion he was largely instrumental in turning the attention of enzymologists away from their preoccupation of isolating enzymes, primarily for the purpose of their characterization, to the possibility of exploring a complicated series of reactions as part of a biosynthetic process. By 1945 the technology for such a study was available because of the discovery and production of the long-life isotope of carbon, ^{14}C, and the introduction of chromatographic methods for enzyme isolation shortly thereafter.

Another interesting application of the use of the carbon isotopes was an investigation of the role of citrulline as an intermediate in the production of urea in rat liver slices from CO_2 and ammonium salts. Our success in capturing the isotopic products of labeled acetoacetate by interposing a bank of unlabeled α-ketoglutarate in kidney homogenates, prompted me to apply the same approach to examine the proposed intermediates in urea synthesis. The catalytic effect of citrulline in stimulating urea synthesis implied that it was readily transported into and, I presumed, out of the cell in liver slices. At this time Krebs' formulation of the reactions [32] had not yet been universally accepted [33]. Therefore, an experiment was undertaken in which a relatively large bank of L-citrulline was added to the medium of the liver slices during the synthesis of urea from $^{13}CO_2$. To the surprise of my collaborator, John Nodine, and myself, the reisolated citrulline contained only normal amounts of ^{13}C, although the ^{13}C concentration of the urea was approximately that of the $^{13}CO_2$. Of course, we recognized the difficulty of our experimental method in the assumption that the equilibration of intracellular citrulline formed de novo from CO_2 was exchangeable with extracellular citrulline at a rate greater than its conversion to urea intracellularly. In preparation for general publication we had submitted an abstract to the Philadelphia Medical Society in connection with the presentation by Nodine, a medical student, of the results of his research during one of the evening scientific sessions. Our decision not to communicate these controversial results in full was prompted by Philip Cohen's announcement that he had been able to accomplish urea synthesis in a cell-free system [34]. For years Phil in his good-natured style has threatened to reveal the existence of our abstract in the *Bulletin of the Philadelphia Medical Society*.

I have purposely included this anecdote on our research on urea synthesis because it had a definite bearing on the approach we would take on another major problem that was concurrently underway, namely, the study of the precursors and intermediates of purine biosynthesis in avian systems. Soon after arriving at the University of Pennsylvania I was fortunate in having very capable medical students join my laboratory for thesis problems or later as postdoctoral fellows. As a second year medical student John Sonne had approached me about a research problem. We initially decided to attempt a study of purine synthesis in avian liver slices, but quickly changed to the study of uric acid formation in the whole animal when we appreciated the quantities of purine that would be required for the separation of the individual carbon and nitrogen atoms for analytical purposes. These methods dating back to the old German chemical literature were reviewed in one of Krebs' early papers [35]. All of the carbon atoms as well as N_7 and N_9 of uric acid could be readily separated for ^{13}C or ^{15}N analysis by three methods previously described. Eventually it would be necessary to develop a fourth method [36–39] on the oxidation of uric acid with hydrogen peroxide in alkaline solution to oxonic acid and NH_3 to distinguish N_1 and N_3.

In the summer of 1946 John Sonne and I [40–43] completed a series of experiments that identified the precursors of the carbon atoms and N_7 of uric acid (Fig. 1). We were joined in part of these experiments by Adelaide Delluva, who synthesized both the α, β- and carboxyl-labeled lactate. Our initial strategy was centered

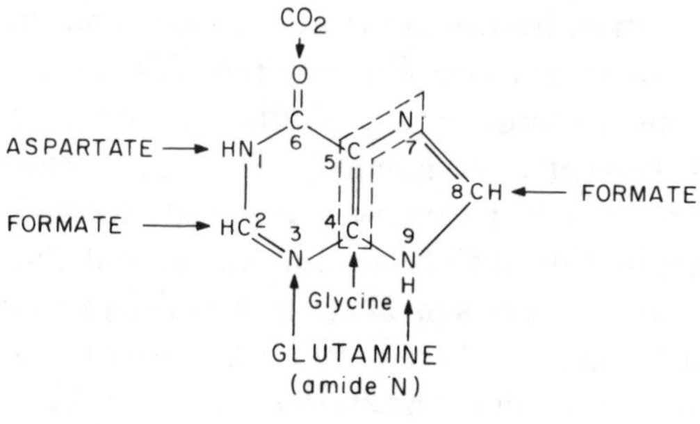

Fig. 1. Precursors of hypoxanthine. Uric acid is 2,6,8-trioxy purine.

around feeding these two compounds together with nonisotopic glucose and ammonium salts to pigeons and collection of uric acid in the excreta over a period of 12 to 20 h. The uric acid was purified by dissolving in alkali and reprecipitation. Thereafter, samples were subjected to one or another degradative procedure.

Since a 3-carbon unit derived from glucose had been postulated as the precursor of the 3-carbon backbone of uric acid, our initial working hypothesis placed the carboxyl carbon at position 6 with the α and β carbons supplying positions 5 and 4, repectively. In deference to my previous experience with the metabolism of lactic acid in our glycogen studies, a rigorous schedule for the collection of respiratory CO_2 was maintained. The results of these first two experiments definitely pointed to the direction for the next ones. It was evident that the ^{13}C concentration of C_6 of uric acid matched exactly that of the respiratory CO_2 derived from the carboxyl-labeled lactate and that carbon-4 probably contained ^{13}C derived from both this carboxyl carbon and CO_2. In a control experiment in which [^{13}C] bicarbonate was the isotopic compound administered, the ^{13}C concentration of C_6 corresponded with that of the respiratory CO_2, and the concentration in C_4 was considerably less as would be predicted if this product of CO_2 incorporation had subsequently been diluted by nonisotopic compounds prior to utilization for purine synthesis. However, CO_2 is incorporated directly into position 6 of uric acid by a reaction in which the products do not undergo further metabolic dilution. A further important result was that the carbon atoms of the two ureido groups of uric acid are not derived from CO_2 and thus differ from the carbon of urea in ureotelic animals with respect to their metabolic origins.

It was then necessary to modify our working hypothesis concerning the reactions leading to the formation of the 3-carbon backbone of uric acid. If the carbon atoms of lactate, a relatively remote precursor metabolically, were reversed, then the carboxyl carbon atom would supply position 4, the α-carbon position 5 and the β-carbon would be lost and replaced by CO_2. In 1946 the principal candidate for a 2-carbon compound fitting these specifications was glycine, whose formation had been postulated to occur by elimination of the hydroxymethyl group of the β-carbon of serine [44].

The other 2-carbon compound under consideration was acetate, but it did not meet the metabolic stipulations stated above. Nevertheless, a ^{13}C-carboxyl-labeled sample was available from Gurin's laboratory and was generously offered to me by him. I had assumed incorrectly that this particular sample had been synthesized from $^{13}CO_2$ by the Grignard reaction. When fed to pigeons it did not give rise to the appearance of ^{13}C in the carbon chain of uric acid other than that accounted for by metabolic CO_2 into C_6 and C_4. However, quite unexpectedly the two ureido carbons were generously labeled. The implied use of an organic acid in the ring closure reaction reminded me of the simple reaction of organic acids by which benzimidazole is formed by heating phenylenediamine and formic acid [45]. Moreover, Krebs and his colleagues [46,47] had reported that pigeon liver lacks xanthine oxidase and that hypoxanthine is the primary purine accumulating in the de novo synthesis in liver slices. We then synthesized [^{13}C]formate by hydrolysis of a sample of [^{13}C]cyanide that had been obtained by Gurin from the Eastman Company. When administered to pigeons, the excreted uric acid was substantially labeled in carbon atoms 2 and 8 to an equal extent. At that time reduced 1-carbon compounds were not thought to be bona fide intermediates of metabolism, particularly in the metabolism of animal tissues.

Some two years later we found it was necessary to reinvestigate our experiment with acetate [48]. Both Wright Wilson and David Sprinson [49] had independently informed me that, whereas they could confirm formate as a precursor of the purine ring, acetate carbon could not contribute in any individual role. In reviewing his records Gurin then recognized that the sample I had used earlier had been prepared by the Walden reaction [50] from the small amount of cyanide he had originally obtained from the Eastman Company. In the synthesis of other organic acids, for example lactate, we had heated the product with HgO in order to remove any contaminating formate that incidentally is formed. Had I realized the correct origin of the sample at the time, I undoubtedly would have treated it similarly and our clue to the role of formate would have been lost.

Returning now to our investigation of the other 2-carbon compound as a precursor of the carbon chain of uric acid, carboxyl-labeled glycine was synthesized and found to contribute in a major way to C_4. Although we did not prepare ^{15}N-labeled glycine, we found that nonisotopic glycine reduced the utilization of ^{15}N ammonium salts specifically for the fraction containing $N_7 + N_9$. However, subsequent to our initial report Shemin and Rittenberg [51] reported that [^{15}N]glycine was utilized in the formation of uric acid in man and Karlsson and Barker [52] showed that [2-^{13}C]glycine gave rise to the incorporation of ^{13}C in the 5 position of uric acid.

The earlier work by Barnes and Schoenheimer [53] had demonstrated that both purines and pyrimidines were formed de novo from ^{15}N-labeled ammonium salts. Örström et al. [47] had also reported that glutamine, asparagine, pyruvate and oxalacetate stimulated hypoxanthine formation in pigeon liver slices. It was thus apparent that the other three nitrogen atoms of uric acid were derived from nitrogenous sources that underwent exchange with other nitrogen-containing compounds and that the feeding of ^{15}N-labeled compounds would lead to ambiguous results. An in vitro, preferably cell-free, system was required for completion of the search for the remaining precursors of nitrogen atoms 1, 3 and 9. In view of the complications in establishing the role of citrulline in urea synthesis in rat liver slices, it now seems unlikely that use of pigeon liver slices would have provided a clear-cut solution to the question at hand.

In departing from the chronological order of experiments on the precursors of uric acid, we concluded our research in this area of the problem after a 2-year sabbatical leave on my part. During this interval G. Robert Greenberg [54], then at Western Reserve University in Cleveland, reported that he had achieved the de novo synthesis of hypoxanthine from radioactive formate and CO_2 in homogenates of pigeon liver. This important result opened the way for the eventual separation of the individual reactions of purine nucleotide synthesis, particularly since these enzymes were later shown independently by him [55] and by us [56] to be contained in the soluble fraction of the homogenate. By this time Sonne had completed his internship at the University of Pennsylvania and decided to continue his research in my laboratory as a Jane Coffin Child fellow. Using ^{15}N amino-labeled glycine, aspartate and glutamate, and glutamine labeled in the amide nitrogen, prepared by our co-worker I. Lin, Sonne measured the amount of ^{15}N incorporated into the various nitrogens of uric acid as a ratio of the incorporation of ^{14}C-labeled glycine [57,58]. Glycine contributed its nitrogen to N_7; 2 mol of the amide nitrogen of glutamine were utilized, one specifically located in N_9 and the other present in the $N_1 + N_3$ fraction. The amino groups of glutamate and aspartate likewise contributed one nitrogen to the $N_1 + N_3$ fraction. Thus, although the general outline of the contributions of the various nitrogenous precursors had been resolved, nevertheless it was still necessary to develop a chemical degradation procedure that could distinguish N_1 from N_3 and to obtain an enzymatic system that would identify either glutamate or aspartate as the contributor to the $N_1 + N_3$ fraction. The former problem was resolved by Hans Brandenberger, who jointed my laboratory during the last year it was located in Philadelphia.

John Sonne had recognized that a method for the degradation of uric acid by hydrogen peroxide in alkaline solution had the potential for splitting the purine ring into the $N_1 + N_7$ and $N_3 + N_9$ fractions. Since N_7 could be determined directly by another method, N_1 could be calculated. Brandenberger found that the previously described mechanism for the formation of oxonic acid required substantial revision.

After our laboratory moved to the Massachusetts Institute of Technology (MIT) in 1953, Bruce Levenberg, who had begun his graduate studies at Pennsylvania, and Standish Hartman, a new graduate student at MIT, collaborated to bring to a conclusion the identification of the specific nitrogenous precursors [59]. By use of the new degradation procedure they were able to establish that N_3 as well as N_9 was derived exclusively from the amide nitrogen of glutamine and that the amino nitrogen of either glutamate or aspartate contributed to N_1. As our knowledge of the properties of the enzymatic system for de novo purine nucleotide synthesis developed, we recognized that we could perform the entire synthesis of inosinic acid from its precursors in a reconstituted system in which the enzymes were supplied from an alcohol precipitate of the dialyzed, soluble pigeon liver protein. Under these circumstances only the combination of glutamine and aspartate in addition to the other substrates and factors resulted in optimal synthesis of inosinic acid. Aspartic acid was thus established as the immediate precursor of N_1 and the amino group of glutamate contributed only indirectly by transamination reactions.

Sabbatical leave in Stockholm

At one point in 1945 Hastings had asked me if I would join him in writing a comprehensive article for *Physiological Reviews* on *The use of isotopically marked carbon in the study of intermediary metabolism,* a feat never possible after that time [60]. We had worked on this paper in the summer of 1945 in Boston. One evening at dinner Baird suggested that it was time to take a sabbatical leave to further my experience in enzyme chemistry if such an arrangement could be made with Wright Wilson. The war was coming to a conclusion and the National Research Council Medical Fellowship Board, of which Hastings was then the Chairman, was again accepting applications. Hastings had suggested that I contact Hugo Theorell about spending two years in his laboratory at the Medical Nobel Institute in Stockholm. This extended period of time, which I felt necessary from a training point of view, posed a problem for Wilson, who would have to find a replacement for me during that period. I offered to resign my position, but fortunately that turned out not to be necessary since the various teaching obligations in the Department were increasing and an additional appointment was made available by the Dean. Minor J. Coon filled this position during the next year. My National Research Council fellowship began on November 1, 1946. After a 10-day voyage on the Gripsholm, I was somewhat surprised at the rather ancient quarters of the Department of Chemistry of the Caroline Institute and the adjoining Medical Nobel Institute, then located near the City Hall in the center of Stockholm. Einar Hammarsten, one of the occupants, was the patron of three brilliant protégés, Hugo

Theorell, Torbjörn Caspersson and J. Erik Jorpes. As each of the three established his own independent research program, Hammarsten allocated a part of his own space to their use. What each lacked in space and facilities was compensated for by the intimacy and camaraderie of compression. Hammarsten was a small peppery man with a slight lisp and a charming inclination to reveal more of the workings of Swedish science than most would have allowed. He had been a pioneer in nucleic acid chemistry and at the time of my arrival was constructing a mass spectrometer to study the biosynthesis of the pyrimidine nucleotides. Among others in his laboratory, Richard Abrams had arrived from the United States as a postdoctoral fellow and Peter Reichard was then a doctoral student. There could have been no greater contrast in personalities than those exhibited by Hammarsten and Theorell. Theo, as Theorell was called, was a large muscular person in spite of his lameness caused by an attack of polio during his childhood. He was immensely versatile, a master of several languages all pronounced with the same typical Swedish intonation, and a musician, who served for a long time as the president of the Stockholm Symphony Orchestra. Above all, he was a master of mime and a raconteur of many humorous stories.

As I arrived Sune Bergström was just leaving the laboratory to become a professor at Lund. Åke Åkesson was the general manager of the laboratory from whom we obtained much of the technical information needed for enzyme purification. Karl Gustav Paul, Sven Paléus and Bo Sorbo were students. Among the first visitors to the laboratory after the war were Britton Chance, Ralph Holman, Christian Anfinsen and myself from the United States, Christian de Duve from Belgium, Andreas Maehley from Switzerland and Elèmer Mihalyi from Hungary. I did not appreciate at the time what a distinguished group was collected in Theo's small laboratory. Four of the above named including Theorell himself have since become Nobel Laureates.

In regard to research, I eventually undertook a problem on the isolation of aconitase with Chris Anfinsen [61], who had been a fellow graduate student at Harvard, and for a while before his marriage had shared an apartment with me in Brookline. Theo had warned us to pick a stable enzyme to isolate with a relatively rapid method of assay since we would receive no more credit for purifying a difficult enzyme than an easy one. To our regret we failed to heed this advice. Aconitase was partially purified from pig hearts and stabilized to some extent by its substrate. However, we missed the important feature that aconitase contains ferrous iron in sulfide linkage [62]. With proper additions it becomes stabilized with a very interesting structure.

After the first year the laboratory moved to its magnificent new quarters on the outskirts of Stockholm. It was here that Chris and I nearly met with a fatal accident. We had been fractionating our enzyme preparation in the -20°C cold room with an alcohol bath cooled with dry ice. Dressed only lightly, we discovered that the latch on the solid, concrete door had failed, and realized that there was little chance that we would be discovered since the rest of the laboratory was out for lunch. After about 20 min of a combined futile experimentation with the mechanism, I was able by sheer luck to adjust the handle of the door so that it finally opened. In a gesture of relief Chris closed the door behind him as he exited the cold room. It took the repairman an hour and a half to open it again from the outside. This is the closest I have ever come in believing in providential intervention.

In the early 1950's enzyme isolation was to be revolutionized by the introduction of many new reagents and procedures that greatly surpassed the effectiveness of the standard techniques of fractionation by organic solvents, salts and electrophoresis. Thus, it is difficult to evaluate my training during my fellowship other than that I was determined upon returning home to apply my knowledge to unraveling the individual enzymatic reactions of a complex biosynthetic process, namely purine nucleotide synthesis.

During the last part of my stay in Sweden I was joined by my sister, Jean, and her son, Gordon, who provided a family life that I had missed for a long time. During that last summer I thought that I had finally mastered the rather complicated Swedish dance, the hambo. However, at one point my partner suggested that if I would twirl in the opposite, customary direction, the dance might go easier. With the cooperation of the weather, the light-filled Swedish summers are a delight, affording an unsurpassed leisure time of swimming and sailing in the magnificent Stockholm archipelago. Midsummer's night is the high point of the summer season. As I will describe in a further section, my sojourn in Sweden was one of the happiest times of my personal life and eventually led to a companionship that has lasted now 37 years.

Enzymatic synthesis of inosinic acid de novo

I am somewhat reluctant to chart the course of our research on the enzymatic synthesis of inosinic acid in any detail since this has been the topic of several reviews [63–65] and of a recent historical paper presented in one of the sessions of the American Society of Biological Chemists in connection with the fiftieth anniversary of the *Annual Reviews of Biochemistry* [66]. However, in my opinion these studies together with our work on the purine precursors constitute my principal contribution to science. I was fortunate to have in this phase of the problem the collaboration of a magnificent group of graduate students and postdoctoral fellows: Joel Flaks, Standish Hartman, Lewis Lukens, Bruce Levenberg, Edward Korn, Charles Remy, Martin Schulman, Richard Miller, Leonard Warren, William Williams, Frixos Charalampous, Samuel Love and Irving Melnick. All were highly motivated and made individual contributions reflecting their own ingenuity and resourcefulness. In another context I wish to mention my admiration for the contributions of our friendly competitors, G. Robert Greenberg and David Goldthwait, whose research meshed with and complemented ours to the distinct advantage to the progress of the program. Greenberg's report [54] on the development of a cell-free system was followed in time with other major contributions. One such contribution was the finding that inosinic acid is the first compound formed with a completed purine ring and that it is the precursor of inosine and hypoxanthine, and not vice versa [67]. These experiments demonstrated that the elements of ribose phosphate became attached to a purine precursor prior to final completion of the ring (Fig. 2). This finding was followed within a short time by

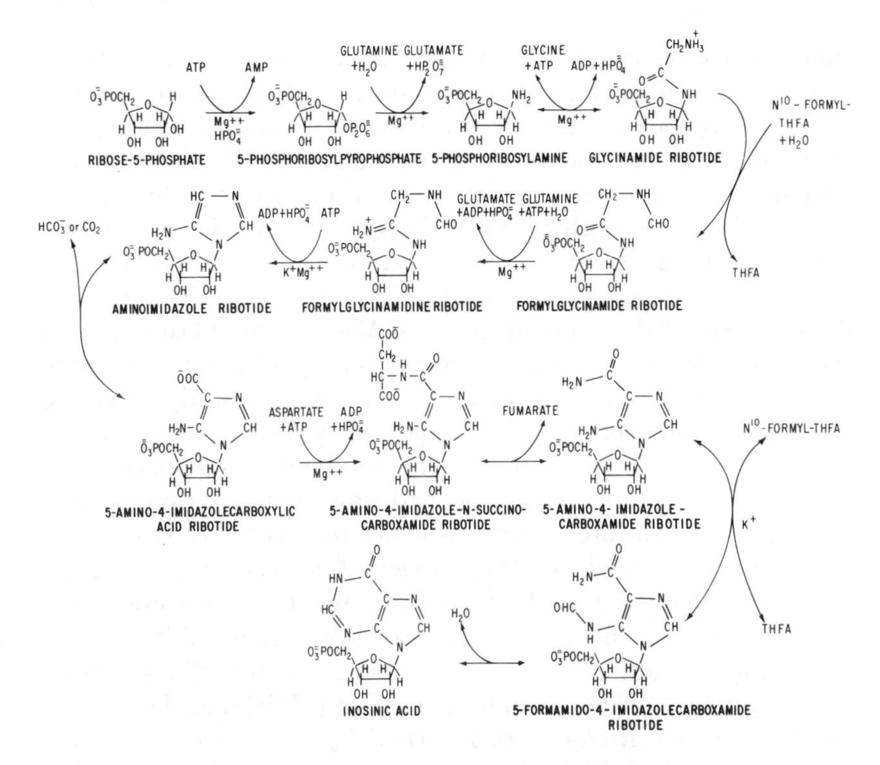

Fig. 2. Reactions in the de novo synthesis of inosinic acid from its elementary precursors.

the discovery that two phosphoribosyl compounds accumulate in pigeon liver extracts incubated with radioactive glycine or glycine and radioactive formate [68–70]. The first compound was 5′-phosphoribosyl glycinamide (GAR), the second, 5′-phosphoribosyl *N*-formyl glycinamide (FGAR). During the interval between these two reports, we had begun a study on the enzymes catalyzing the synthesis of inosinic acid from its cognate base, hypoxanthine. We had picked this enzymatic system for investigation because we thought that it might represent a model for the reactions involved in the de novo pathway. The approach that we adopted in this first series of experiments was typical of the methods used eventually throughout the entire project. In 1950 the most plausible route for inosinic acid synthesis from hypoxanthine seemed to include the formation of inosine from ribose-1-phosphate and hypoxanthine catalyzed by nucleoside phosphorylase [71], followed by the phosphorylation of the nucleoside to the nucleotide. The latter kinase reaction had not been reported, however. We had found that radioactive hypoxanthine is readily equilibrated with nonisotopic inosinic acid in the presence of an extract of pigeon liver [72]. The latter was fractionated into two components, neither of which alone could catalyze the reaction [73]. We then found that inosinic acid could be synthesized by incubation of ribose-5-phosphate, ATP and hypoxanthine with the two fractions. By purifying nucleoside phosphorylase from beef liver [74], we were able to determine that highly purified enzyme could not replace either fraction, and consequently the pathway for the formation of a nucleoside as an intermediate was not involved. We then found that Fraction I catalyzed the reaction of ribose-5-phosphate and ATP to yield a relatively heat-labile product that was converted in the presence of hypoxanthine and Fraction II to inosinic acid [75]. This ribose intermediate isolated in our laboratory [76] and independently by Kornberg and his associates [77,78] was identified as 5-phosphoribosylpyrophosphate (PRPP) by them. Fraction I became known as PRPP synthetase, Fraction II as hypoxanthine-guanine 5′-phosphoribosyl transferase [75,79]. The latter enzyme also utilizes guanine and the fraudulent purine, 6-mercaptopurine [80], as substrates. A second phosphoribosyl transferase catalyzes the reaction of PRPP with adenine [75, 79] and 5-amino-4-imidazole carboxamide [81] to their corresponding ribotides.

The reactions catalyzed by the purine phosphoribosyl transferases have been referred to commonly as the salvage pathway of nucleotide synthesis. A genetic loss of either enzyme in man [82,83] results in the failure of the purine bases to be reclaimed as nucleotides and in their eventual conversion to insoluble oxidation products, 2,8-dioxyadenine, and uric acid. The absence of the hypoxanthine-guanine phosphoribosyl transferase in tissues results in an arthritic, debilitating condition called the Lesch–Nyhan syndrome which is characterized by self-mutilation and retardation.

As hoped, PRPP was a participant in the synthesis of 5′-phosphoribosyl glycinamide from glycine and glutamine provided ATP was present [70,85]. This reaction was subsequently fractionated [86–89] into two steps in which PRPP reacts with glutamine to yield 5-phosphoribosylamine, which then is converted to 5′-phosphoribosyl glycinamide upon reaction with ATP and glycine. The enzymes catalyzing these two reactions have now been purified and designated as glutamine-PRPP amidotransferase and 5′-phosphoribosyl glycinamide synthetase, respectively.

There are, of course, two reactions in which formate is utilized in purine nucleotide biosynthesis. The first of these reactions to be studied was the ring closure of 5′-phosphoribosyl 5-amino-4-imidazole carboxamide (AICAR) with 10-formyl tetrahydrofolate to yield inosinic acid. The cognate base of the above nucleotide had been isolated from *E. coli* poisoned with sulfonamides [90] and its structure correctly determined by Shive et al [91,92]. By the late 1940's the citrovorum factor [93] had been identified as 5-formyl tetrahydrofolate [93–95], and it was correctly predicted that it or a near derivative would serve as a coenzyme for the transfer of formyl as well as other reduced 1-carbon groups [96,97]. Greenberg's laboratory [98,99] and our own [100–103] were responsible for showing that the 10-formyl derivative is the actual compound involved in the final step in inosinic acid synthesis. The overall reaction probably takes place in two steps with 5′-phosphoribosyl 5-formamido-4-imidazole carboxamide serving as an intermediate.

It is now known that 10-formyl tetrahydrofolate participates in the formulation of 5′-phosphoribosyl glycinamide. We [104,105] had originally proposed that the formyl donor was 5,10-methenyl tetrahydrofolate, but Benkovic and his colleagues [106] have now shown that the unnatural enantiomorph of 10-formyl tetrahydrofolate produced in the chemical but not enzymatic hydrolysis of the methenyl derivative is a powerful inhibitor of the reaction of the natural 10-formyl enantiomorph in the case of GAR transformylase but not AICAR transformylase. They [107] have also made the interesting observation that 4-enzyme activities related to 10 formyl tetrahydrofolate synthesis and utilization are packaged in a complex, which may be isolated intact from tissue extracts.

The middle pieces of the purine pathway soon fell into place by a succession of investigations in which individual components of the complete synthesis of inosinic acid from glycine, formate, CO_2, glutamine, aspartate and formyltetrahydrofolate were omitted in a systematic manner. For the most part the intermediates could be identified by application of a simple colorimetric test developed by Bratton and Marshall [108] in which a diazotizable amine was coupled with *N*-(1-naphthyl)-ethylenediamine to yield a colored product. By this time we had isolated FGAR in sufficient quantity to warrant its use as a substrate in routine assays. As an example of the utility of the Bratton–Marshall reagents we found that incubation of a relatively crude extract with FGAR, glutamine and ATP yielded a product (5′-phosphoribosyl 5-aminoimidazole or AIR), which upon derivatization assumed a salmon orange color. If bicarbonate and aspartate were included in the incubation mixture as well as FGAR, glutamine and ATP, AICAR, which gave a purple Bratton–Marshall product, was formed. With these simple landmarks of the overall reaction established by Levenberg [109,110], fractionation of the pigeon liver extracts was begun. The conversion of FGAR into AIR was catalyzed by two enzymes, FGAR amidotransferase, and AIR synthetase. The intermediate in the reaction was 5′-phosphoribosyl formylglycinamidine (FGAM). Both reactions required the participation of ATP [111]. When AIR was used as substrate together with ATP, bicarbonate and aspartate, AICAR was formed [112–118]. Initially the crude enzyme preparation was fractionated into two parts. An intermediate, resulting from incubation of all substrates mentioned above with Fraction I, was a highly negatively charged compound [112,114]. This intermediate failed to yield a color indicative of the presence of an arylamine if the diazotization reaction was performed at room temperature, the condition that normally had been used. However, we found that the diazotized intermediate may undergo decomposition at room temperature, but is stable at 0°C. Upon coupling at this lower temperature it yields a purple-colored product. This difference in the reaction at the two temperatures turned out to be a useful property, since it permitted us to distinguish it from AICAR whose diazotized product is stable at either temperature. The intermediate formed by Fraction I described above was 5′-phosphoribosyl 5-amino-4-imidazole succinocarboxamide (SAICAR), which contains an aspartyl residue. Upon reaction with Fraction II carbon atoms of this residue are split off as fumarate. The enzyme catalyzing this second reaction proved to be identical with adenylosuccinase [113,116–118].

Thus, only one more reaction catalyzed by Fraction I required characterization, namely the conversion of AIR into SAICAR [112,115]. Incubation of AIR with bicarbonate yielded a product, carboxy-AIR, whose Bratton–Marshall product was a cherry red. Carboxy-AIR is converted into SAICAR by reaction of aspartate and ATP. The enzymes catalyzing the overall conversion of AIR to SAICAR have been separated and partially purified. Since the carboxylation reaction does not require ATP, a biotin-containing enzyme is not involved.

This laborious progression of fractionation of crude enzyme solutions and isolation of intermediates was an early example of an approach that has been used extensively since then for other complex biochemical systems. I remember many times Theorell's admonition that a simple, accurate and rapid assay is a prerequisite for enzyme purification. Although we did use isotopes and a simple colorimetric procedure to their best advantage, nevertheless the conditions under which we established enzyme assays were anything but orthodox.

These years were not without their humorous moments. After I returned from Sweden, Martin Schulman came to the University of Pennsylvania from Berkeley as the first postdoctoral fellow ever to join my then small group. In April, as the Federation Society meetings were approaching, we had an urgent need for a few pigeons to complete an essential series of experiments. Our cages were, however, empty and there was not sufficient time to obtain more from a distant supplier. Marty and I therefore set out to Rittenhouse Square, in the center of Philadelphia carrying two gunny sacks and a bag of corn with the intention of capturing some birds. We had successfully lured two or three into our trap and were about to depart when we were approached by a policeman for an explanation of our actions. At that moment a little old lady came bearing down on the three of us waving her umbrella. Marty escaped dropping all but one lone pigeon. I necessarily stood the ground only to find out that she was heaping abuse on the policeman for stopping those fine young men from ridding her lovely park of those messy birds. She never realized how grateful we were nor the extent of her service to science.

The irreversible inactivation of enzymes by active site-directed reagents or the affinity labeling of enzymes

I have left until the end of this section reference to a program that started as a side project but eventually developed into a major effort of the laboratory. We had moved by 1953 to the Department of Biology at the Massachusetts Institute of Technology in Cambridge, MA. At the suggestion of Arthur Cope of our Chemistry Department, Dr. Alexander Moore of Parke Davis and Co. contacted me requesting that I test two of their antibiotics, L-azaserine and 6-diazo-5-oxo-L-norleucine (DON), on our enzymatic system for the de novo synthesis of inosinic acid. Skipper and his colleagues [119] had shown that injection of these compounds into tumor-bearing mice (sarcoma 180) resulted in the inhibition of the tumor growth and an inhibition of incorporation of purine precursors into nucleic acids. The compounds therefore had a potential for use in cancer chemotherapy. When Bruce Levenberg [69,120,121] applied these drugs to our enzymatic system, he found that inosinic acid synthesis from glycine was substantially inhibited and that the rate of inactivation was a function of the concentration of glutamine in the reaction mixture. The structural relationship of the two antibiotics to glutamine was thus first recognized (Fig. 3).

Although we now know that most if not all enzymes utilizing glutamine are inhibited by these antibiotics to one degree or another, by far the most sensitive enzyme is FGAR amidotransferase [122]. The nature of the inhibitory effect of these compounds proved to be abnormal in that the antibiotics were competitive

NH_2-CO-CH_2-CH_2-$CHNH_2$-COOH
GLUTAMINE

$\overset{-}{N}$=$\overset{+}{N}$=CH-CO-O-CH_2-$CHNH_2$-COOH
AZASERINE

$\overset{-}{N}$=$\overset{+}{N}$=CH-CO-CH_2-CH_2-$CHNH_2$-COOH
6-DIAZO-5-OXO NORLEUCINE

NH_2-CO-NH-CH_2-$CHNH_2$ - COOH
ALBIZZIIN

Fig. 3. Antimetabolites of glutamine.

with glutamine in their action yet eventually caused irreversible inactivation of the enzyme, either in a crude or highly purified state. If either antibiotic is incubated with enzyme in the absence of glutamine, activity is immediately lost and is not regained by addition of glutamine. This type of inhibition differed, for example, from the inhibition by malonate of succinic dehydrogenase, which is a simple function of the ratio of the concentrations of the substrate, succinate, and the inhibitor.

The characteristics of the inhibition by azaserine or DON, however, indicated that their structural similarity to glutamine permitted their interaction at the enzyme site for the natural substrate, but once in position an irreversible reaction with the enzyme occurred.

To prove this hypothesis, Richard Day synthesized azaserine containing radioactive carbon [123]. FGAR amidotransferase was purified from *Salmonella typhimurium* by Thayer French and Igor Dawid [123] and from chicken liver by Kiyoshi Mizobuchi [124]. With either type of preparation there is approximately a mole-for-mole reaction of azaserine with enzyme. The reactive residue of the enzymes is a cysteine. The amino acid sequence around this residue has been determined on the bacterial enzyme by French and Dawid [125,126] and on the liver enzyme by Shiro Ohnoki and Bor-Shye Hong [127,128]. Since it was possible that reaction of this particular cysteine was specific only for the two antibiotics studied, we carried out further studies with another glutamine analogue, L-albizziin, which is 3′-(*N*-carbamyl) amino 2′-

aminopropionic acid and bears perhaps a closer structural relationship to glutamine. Again experiments by Duane Schroeder and Joyce Allison [129] on the inhibition of FGAR amidotransferase by this compound indicated that a sulfhydryl residue was involved at the glutamine-reactive site on the enzyme.

These experiments were thus instrumental in establishing that the sulfhydryl group of a cysteinyl residue is a nucleophilic agent in the reactions of glutamine [130]. Since this time several amidotransferases have been found to contain a sulfhydryl group at the active site [122]. FGAR aminotransferase is a complex enzyme that catalyzes the formation of a carbon-to-nitrogen bond by utilizing the energy generated by the splitting of ATP to ADP and inorganic phosphate. In addition to the glutamine-reactive site mentioned above, the enzyme was shown by Kenyon and Mizobuchi [131] to have a second site for the binding and reaction of ATP and FGAR. The rates of partial reactions on one site of the enzyme are regulated by modifications on the second.

Frère and Schroeder [132] recognized that the enzyme undergoes polymerization by oxidation of nonessential cysteinyl residues but without loss of activity. From a kinetic analysis, Li [133] has proposed that the three substrates add sequentially to form a quaternary complex with the enzyme before undergoing release of products, and that glutamine adds first to the enzyme with ATP and FGAR adding randomly thereafter (i.e., by a partially compulsory order mechanism).

The importance of the experiments on the reactions of azaserine and DON on FGAR amidotransferase is that they were the first clue to the existence of this type of irreversible active-site-directed reagents. Although these compounds as well as L-albizziin are of natural origin, other active-site-directed compounds have been devised and synthesized to perform specific inhibitions of enzyme activities. For example, the chloromethylketones synthesized by Elliot Shaw [134] are very useful inhibitors of the proteases, trypsin and chymotrypsin. I take this opportunity to point out that another inhibitor of these two proteases, diisopropyl fluorophosphate, is not an active-site-directed reagent as is sometimes claimed even though it does specifically react with an activated serine on either enzyme. The inhibitor is not active-site-directed since it cannot distinguish between enzymes such as trypsin and chymotrypsin, which have such different specificities for substrates.

Methionine biosynthesis: relationship between folate and vitamin B_{12}

I have often wondered about the seemingly trivial incidents that intervene not only in one's personal but also professional life to bring about major changes in the directions of one's efforts. I believe that these impulses from the outside are, in fact, occurring continuously, but at certain times one is particularly receptive to seizing opportunities for change. I was in that particular state in 1956 when we were just concluding our major thrust in characterizing our enzymes and reactions in purine synthesis. On the basis of the interesting preliminary results with azaserine, I had committed myself to an in-depth research program on one enzyme of the purine biosynthesis pathway, namely FGAR amidotransferase. Yet I could not foresee spending a lifetime in this same specific area. Up until my move to MIT my interests and approaches in research had been taken primarily from the point of view of a chemist. I felt most comfortable working in the field of intermediary metabolism because of my training as a student and young faculty member. Yet I realized that important problems would be emerging from that vast discipline of Biology. For example, the problem of regulation of biochemical pathways was coming more and more to the forefront. At a conference sponsored by the Ciba Foundation on the Chemistry and Biology of the Purines in London in May of 1956, I met Donald D. Woods [135], who was then engaged in an interesting problem of methionine biosynthesis from homocysteine and serine by bacterial extracts of *E. coli*. One of the components of this system was a protein containing vitamin B_{12} as a prosthetic group. During the previous year I had spent a month as a visiting professor at the University of Texas at the invitation of William Shive. I was well aware of Shive's competition-inhibition analysis technique for studying metabolism in bacterial systems and of his prediction that both folate and vitamin B_{12} would be involved in methionine synthesis [96,97]. Of course, duVigneaud [1] in his research described in *A Trail of Research* describes his work on methyl synthesis in the rat and the role of folate and vitamin B_{12} as complementary agents in animal nutrition. Likewise, my good friend Thomas Jukes [136] had made several notable contributions to this problem in his capacity of director of a large productive group at Lederle Laboratories.

Therefore, when the rather innocuous question was posed by Woods [135] during the discussion period of one session of whether vitamin B_{12} might be involved in purine biosynthesis, I was certainly prepared and receptive to looking into the possibility that the reactions of reduced 1-carbon compounds of purine and methionine synthesis might be related.

Upon returning home, I obtained from Bernard Davis two methionine-requiring strains of *E. coli* that had mutations in the pathways for methyl synthesis [137]. One was denoted as 113-3, the other as 205-2. The former may also be grown in a methionine-free medium to which vitamin B_{12} is supplied. Thus, an extract of either mutant by itself was incapable of synthesizing methionine from a reaction mixture containing ATP, 5,10-methylene tetrahydrofolate, homocysteine and a reducing source, but when combined they could [138–142].

The problem then was to develop an assay so that either enzyme could be isolated and purified. By using the strategy employed so frequently in our studies on purine biosynthesis, one enzyme was purified in the presence of an excess of the other and vice versa without knowledge of the characteristics or intermediates of the reaction. However, when the reaction was carried out anaerobically, reduced flavin adenine dinucleotide was an effective cofactor in the reaction. It was possible eventually to order the sequence of the action of the two enzymes as well as the required substrates for each step. The enzyme obtained from the 113-3 mutant grown on methionine contained the enzyme lacking in the 205-2 mutant. This enzyme catalyzed the reduction by $FADH_2$ of 5,10-methylene tetrahydrofolate to a new intermediate [141–144]. In the presence of homocysteine and catalytic amounts of ATP, this intermediate was converted to methionine by an enzyme isolated from the 113-3 mutant grown on vitamin B_{12}. After purification, the latter enzyme was shown to contain vitamin B_{12} or a derivative thereof. At this time Mangum and Scrimgeour [144,145] reported that ATP functioned by reacting with methionine to yield *S*-adenosylmethionine, which was the agent acting catalytically in the reaction.

The new intermediate (I), which had an absorption maximum at 290 nm, upon exposure to oxygen was converted to a product (II) with double absorption peaks at 290 and 250 nm. Both compounds, however, could be utilized for methionine synthesis in a reducing atmosphere. Compound I had a rather special property of supporting the growth of *Lactobacillus casei* but not that of *Streptococcus faecalis* or *Leuconostoc citrovorum* [146]. Folic acid supports the growth of *L. casei* and *S. faecalis* and tetrahydrofolate supports the growth of all four microorganisms. These growth-supporting properties drew our attention to a folate compound isolated by Keresztesy and Donaldson [147]. These investigators along with Noronha and Silverman [148] had demonstrated that prefolate A is a major folate derivative of many tissues, apparently serving as a storage form of folate compounds. Keresztesy and Donalson had characterized prefolate A as being more reduced than tetrahydrofolate. As I was becoming convinced that our derivative in methionine synthesis was 5-methyltetrahydrofolate, I contacted Edward Taylor at Princeton University, an authority on pterine chemistry, who assured me that the pteridine ring could undergo no further reduction and that the extra hydrogens on prefolate A must be located in a substituent. I was thus convinced that prefolate A was 5-methyltetrahydrofolate and that the oxidized form was 5-methyl 5,6-dihydrofolate. The location of the double bond at the 7,8 position was predicted on the basis that chemical reduction of the dihydro derivative results in the reformation of a tetrahydro derivative fully capable of transferring a full complement of methyl groups to homocysteine. Had the double bond been at the 5,6 position or 6,7 position, chemical reduction would have resulted in two enantiomorphs only one of which would have been enzymatically reactive in the transfer reaction. Finally, the correctness of the structure proposed was corroborated by my former colleague,

Warwick Sakami [149], who synthesized 5-methyltetrahydrofolate by the chemical reduction of methylene tetrahydrofolate.

The mechanism of transfer of methyl groups to homocysteine via the B_{12}-containing transmethylase is still uncertain except that the methyl derivative of vitamin B_{12} is believed to be the active prosthetic group [150]. As far as is known, however, it is not this methyl group that is formed from 5-methyltetrahydrofolate and transferred to homocysteine. Quite possibly this is the function of *S*-adenosylmethionine, which does not play a stoichiometric role in the reaction.

In view of the report by J. Bremer and D.M. Greenberg [151] that *S*-adenosylmethionine can enzymatically methylate a number of sulfhydryl-containing compounds, Rosenthal and Smith [152] compared the reactivity of bacterial B_{12}-containing transmethylase to methylate 2-mercaptoethanol. Again, it was found that the methyl groups derived from 5-methyltetrahydrofolate and *S*-adenosylmethionine do not converge metabolically in the reaction.

The physiological significance of these studies is apparent when the reactions are presented in the form of a cycle (Fig. 4). In this series of reactions methylene tetrahydrofolate is formed from serine and tetrahydrofolate and is then reduced in an essentially irreversible reaction to 5-methyltetrahydrofolate catalyzed by methylene tetrahydrofolate reductase [153,154], whose activity is repressed by cells grown on methionine [154]. This compound then serves as a reservoir of methyl groups, which may be transferred to homocysteine provided an active B_{12}-containing transmethylase is available. The transfer reaction results in the liberation of tetrahydrofolate, so important in the utilization of reduced 1-carbon compounds. If through a vitamin B_{12} deficiency, tetrahydrofolate becomes completely methylated, tissues would be depleted of this necessary cofactor and a nutritional deficiency would result. Thus, 5-methyltetrahydrofolate serves as a metabolic trap or sink of folate compounds in vitamin B_{12} deficiencies. The finding by Loughlin and Elford [155] in this laboratory that pig liver contains the vitamin B_{12} methyltransferase was the final piece of the puzzle supporting the trap theory as applicable to animal tissues. The relationship of these reactions to pernicious anemia is obvious.

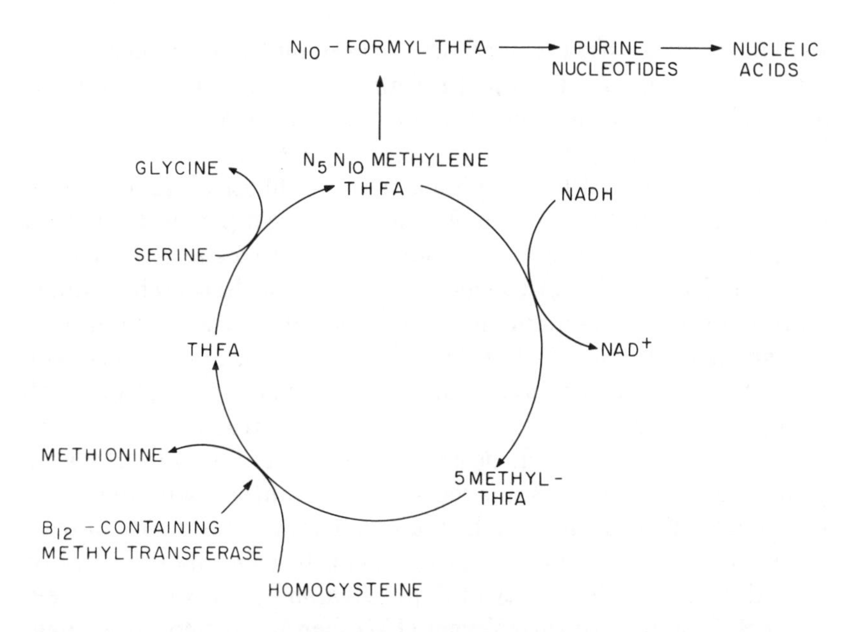

Fig. 4. Biosynthesis of methionine illustrating participation of tetrahydrofolate and vitamin B_{12} derivatives; example of a metabolic trap (the methyl-tetrahydrofolate metabolic trap).

Recently Fujii et al. [156] have performed elegant experiments with leukemia cells that confirm the metabolic trap hypothesis. Thus, after approx. 5 years (1956–1961) I was able to reply in the affirmative to the innocent question that prompted this research. Vitamin B_{12} is involved in the macroregulation of nucleotide and consequently nucleic acid synthesis, both of which are essential for cell growth and cell division. My companions in this story were Frederick Hatch, Shige Takeyama, Renata Cathou, Allan Larrabee, Richard Loughlin, Howard Elford, Spencer Rosenthal, Louis Smith, Barbara McDougall and Howard Katzen.

Bacteriophage metabolism

As mentioned in the preceding section my laboratory was at one of its turning points in 1956. I had a strong inclination to enter a new field of research that involved problems in molecular biology. This infant discipline was then just forming from a confluence of ideas and technology originating in biophysics, biochemistry, microbiology, virology and genetics.

Upon arriving in Philadelphia after my sabbatical leave in Sweden, I was invited to join the Medical Fellowship Board, the first panel of the National Science Foundation, and a subcommittee of Blood and Related Substances of the National Research Council. Through these extracurricular activities I made the acquaintance of many who became life-long friends and in some cases advisors on several such occasions. I particularly remember conversations with Rollin Hotchkiss about his work on DNA as a transforming factor for pneumococcal bacteria as a possible subject for a new research project. As interesting as this research was from the point of view of the geneticist, I did not see how a biochemist could contribute to such a difficult problem at that time. On the other hand I recalled work done at the University of Pennsylvania by my colleague, Seymour S. Cohen who, with Wyatt [157], had identified 5-hydroxymethylcytosine as replacing cytosine in the DNA formed in *E. coli* infected with the T even bacteriophage. At about this time Joel Flaks was concluding his work for his doctorate thesis with me at MIT and planned to join Cohen as a postdoctoral fellow. He brought to my attention the courses given on bacteriophage metabolism and physiology at the Cold Spring Harbor Laboratory in which we both enrolled in the summer of 1956. I had never had any experience in bacterial virology and found this summer's introduction very useful. During the next fall James Koerner from Robert Sinsheimer's laboratory and S. Varadarajan from Alexander Todd's laboratory in Cambridge, England had arrived at MIT for postdoctoral work. It was a fortunate coincidence since their scientific backgrounds complemented each other in enzyme and organic chemistry, respectively. Flaks and Cohen [158,159] had already initiated a program in the synthesis of 5-hydroxymethyl deoxycytidylic acid from methylene tetrahydrofolate and deoxycytidylic acid and were well on their way of the study of the first phage-induced enzyme to be reported, namely dCMP hydroxymethylase. We did not wish to overlap their work, yet needed an entree to the general problem. This clue was provided by an abstract of the Federation Meetings in which Kornberg, Lehman and Simms [160] reported that, although partially purified DNA polymerase from extracts of *E. coli* catalyzed the synthesis of DNA from the precursors, dCTP, dATP, dGTP, and dTTP, this capacity was greatly diminished for enzyme prepared from extracts of *E.coli* infected with T2 or T4 bacteriophage. Since, if anything, intact infected cells synthesize DNA at an accelerated rate as compared with uninfected cells, it seemed plausible that intervening enzymes programmed by the bacteriophage were preventing the incorporation of the standard set of deoxynucleoside triphosphates into T2 DNA. This hypothesis was proven correct by incubating infected and uninfected extracts together and showing an inhibition of DNA synthesis of the latter by the former. Sinsheimer [161] and Volkin [162] had meanwhile established that hydroxymethyl deoxycytidylic acid existed in glucosylated form. Koerner and Varadarajan [163] were able to prepare hydroxymethyl deoxycytidylic acid (dHMP) and its glucosylated derivative (glucosyl-dHMP) by the action of a DNase on T2DNA. Chemical phosphorylation yielded the triphosphates dHTP and glycosyl-dHTP. dHTP but not glucosyl-dHTP could support DNA synthesis from dGTP, dATP and radioactive dTTP in both infected and uninfected extracts.

The new phage-induced enzyme activity regulating the composition of the DNA synthesized was dCTPase [164–166], which catalyzes the splitting of dCTP and dCDP to dCMP and pyrophosphate or orthophosphate, respectively. It was later shown by Greenberg [167] and Warner and Barnes [168] that dUTP and dUDP are also substrates of this enzyme.

Within a short time the entire sequence of reactions accounting for the formation of T-even DNA was reported. The laboratories of Kornberg [164] and G.R. Greenberg [169] identified a new deoxynucleotide kinase that phosphorylated dHMP, dGMP, and TMP to the diphosphate. In addition, Kornberg [164] isolated a phage-specific DNA polymerase, and enzymes that glucosylated the hydroxymethyl groups in the α- and β-configurations.

Prior to the publication of the now classical paper of R.H. Epstein and R. Edgar and their associates [170] on the production of a number of mutants of bacteriophage T4, Salvador Luria of our Department of Biology was able to obtain specimens of those known to be involved in DNA synthesis. With the discovery of the set of enzymes concerned with the synthesis of phage DNA mentioned above, we began a collaborative effort aimed at the identification of the functions or activities missing in the conditional lethal *amber* mutants. The first to be established was the assigment of dCMP hydroxymethylase to gene 42 [171]. DNA polymerase was also lacking in these extracts, but the specific mutant of gene 42 studied, *am* N122, was later shown to contain a double mutation.

The genetic contribution of both the bacteriophage and the host to the production of mutant enzyme was next explored by Wiberg and by Dirksen and Hutson, respectively. Although no formation of dCMP hydroxymethylase in *E. coli* B after infection with the gene 42 mutant, am N122, could be detected, active enzyme could be demonstrated if the host were *E. coli* K-12 CR63, *E.coli* C600 or *E. coli* CS^{P} [172]. The kinetics of production of enzyme as a function of time of infection and pH differed significantly. The same findings were also made with other *amber* mutants in gene 42, am N269, am N55 and am N211. The formation of mutant enzymes is of course dependent on the presence of transfer RNAs in the permissive host capable of reading the stop codon as a codon for one of the amino acids. For example, *E. coli* CR63 contains supD ($Su\mathrm{I}^{+}$) and C600 contains SupE ($Su\mathrm{II}^{+}$), which substitute a serine or glutamine residue, respectively, at the location in the enzyme where the mutation in the bacteriophage DNA has resulted in change of code reading for the wild-type residue to a stop or terminal codon. These seemingly minor changes in the compositions of the mutant enzymes or proteins [173,174] have important effects on their stability at different pH values.

A second series of experiments were undertaken to demonstrate altered properties in the enzyme, dCMP hydroxymethylase, formed on infection of *E. coli* B with genetically altered bacteriophage T4 [175]. The properties of enzymes induced by infection with two temperature-sensitive mutants of gene 42, *ts* G25W and *ts* L13, were compared with those of the wild-type enzyme. Again striking differences in enzyme stability and activity as a function of pH, temperature, and substrate concentration were observed. These results constituted the strongest evidence possible that the structural gene for dCMP hydroxymethylase resides in the phage itself. They also served to illustrate the promise offered by phage systems to provide new correlations between genetic fine-detail mapping and enzyme properties and structure. These early papers were eventually followed by numerous publications matching individual cistrons with specific enzymes or structural proteins (see [176] for summary). For example, the structural genes for dCTPase [177], deoxynucleotide kinase [178], DNA polymerase [179,180], and DNA ligase [181] are T4, cistrons 56, 1, 43 and 30, respectively.

At approximately the same time that the mutants were being studied, we initiated experiments on the regulation of the synthesis of enzymes and proteins formed after bacteriophage infection. Flaks et al. [182] as well as Vidaver and Kosloff [183] had previously shown that irradiation of phage results in the loss of the capacity for DNA synthesis, which normally begins at 8–10 min, and hence its capacity for a productive infection. However, infection of the host with irradiated phage did result after an initial lag of 3 min in the formation of an amount of dCMP hydroxymethylase over the first 15 min comparable to that found for unirradiated phage.

These experiments thus divided the infection into pre- and post-replicative periods. From this temporal sequence the components synthesized during the prereplicative period have been designated as 'early' functions and during the post-replicative period as 'late' functions. As was found later, these periods can be further subdivided based on effects of several inhibitors of protein synthesis [184,185] and by the expression of bacterial phage products from radioactive precursors as measured by polyacrylamide gel electrophoresis and autoradiography [186,187].

A characteristic feature of the so-called early enzymes is that after a brief lag of 3 min, enzyme synthesis accelerates until approx. 10 min after infection, when further synthesis halts more or less coincidently with the beginning of DNA synthesis. However, when the production of several of the novel enzyme activities was measured over an extended period of time after infection with irradiated T2, the initial rate of formation of enzyme was in some cases not affected until higher intensities of ultraviolet light were used. Most strikingly, however, the length of time of enzyme production was extended in some cases for as long as 60 min [188,189]. By this time the amount of enzyme may reach levels 2 to 10 times greater than that resulting from a normal infection with unirradiated phage [188]. This exaggerated enzyme synthesis was related to the inhibition of DNA synthesis caused by dimer formation of adjacent thymine residues of the viral template DNA, which occurs during irradiation.

A similar extended and exaggerated synthesis of the early enzymes was observed when *E. coli B* was infected with *amber* mutants incapable of effecting DNA synthesis [171]. These parallel experiments with T4 mutants and with irradiated wild-type T2 demonstrated that the initiation of DNA synthesis was correlated with the cessation of early enzyme synthesis. On the other hand, synthesis of many of the late proteins could only occur coincidently with DNA synthesis. The regulation of early enzyme synthesis may thus represent a competition of early and late messenger RNAs for the translation machinery.

These studies then stimulated an examination of mRNA synthesis, first by measurement of general classes by the competition-hybridization technique [184,185] and later by translation of specific mRNAs and measurement of enzyme activity. The activity of DNA-directed RNA polymerase after bacteriophage infection was of course central to these studies. Ola Sköld, who obtained his doctorate degree in Sweden with Peter Reichard, joined our group for a short time. He examined the activity of DNA-directed RNA polymerase as a function of time after infection of cells that were then ruptured in the Hughes press [190]. He observed a substantial loss of RNA polymerase activity during the first 10 to 15 min when the rate of synthesis of an early enzyme was just reaching a peak and then leveling off. In these early experiments the inhibition appeared to be related to the production of a macromolecular compound of unknown action whose formation was prevented by inclusion of chloramphenicol in the medium at the time of infection. In a continuation of these studies, Arland Oleson and Jaakko Pispa [191] found that if infected cells were gently lysed, there is a transient, substantial increase (up to two-fold) of RNA polymerase activity during the first 1 to 2 min followed by a rapid decay, so that the final level of activity was about one-third to one-quarter of that at the highest transient value. Again, the inclusion of an inhibitor of protein synthesis in the initial infection medium resulted in the maintenance of the activity at the highest transient level. This effect is not related to another inhibition reportedly caused by a factor isolated from uninfected *E. coli* B, which subsequently was identified as an ATPase. In this case inclusion of an ATP-generating system in the reaction resulted in a linear response of enzyme activity to concentration [192].

As has been noted by several investigators, infection of *E. coli* B with T4 bacteriophage results in an immediate cessation of host protein synthesis, which has been ascribed in part to the action of phage-induced nucleases that hydrolyze *E. coli* DNA. However, the effect is still evident when mutant phage lacking intact cistrons for nuclease production (gene 46^-, 47^-) are used. The most compelling explanation, however, comes from the finding that immediately upon infection of *E.coli*, DNA associated with the RNA-polymerizing system in the bacterial membrane is almost entirely replaced with T4 DNA in a transcription complex [193]. The relatively rapid loss in RNA polymerase activity in the case of bacteriophage T4 but not T5 infection still remains a mystery. However, it must be in some way connected to the beginning of the synthesis of the late functions. In contrast to the T7-infecting system [194, 195], the synthesis of a phage-specific core RNA polymerase for T4 has not to my knowledge been reported [195], although ADP-ribosylation of the α subunits occurs [197] and modifying proteins such as *sigma* [198,199] and *rho* [200] factors, as well as several T4 gene products, are known [201].

The electropherograms of radioactive proteins formed from radioactive amino acids at different times after T4 infection indicated a diverse composition of early products that could be grouped into several classes, which are defined by the time of initiation and termination of the production of individual components [187]. This approach could not reveal, however, the relationship between these proteins and the synthesis of mRNAs from which they were derived. For this reason DNA-RNA competition-hybridization experiments were undertaken by Grasso [184] in this laboratory and by Salser et al. [185] to study the synthesis of early T4 mRNAs under a variety of conditions, for example, during inhibition of protein synthesis. When chloramphenicol, an agent that effectively inhibits protein synthesis in prokaryotes, is included at the time of infection of *E. coli* B with T4 bacteriophage, a class of mRNAs is formed that may represent up to 30% of the total early species. This RNA has been designated as Class I or immediate-early RNA, whose synthesis is initiated at so-called early promoters (E) on T4 DNA, which are recognized by the native *E. coli* RNA polymerase. By use of unlabeled T4 RNA formed in the presence of chloramphenicol, it has been possible to investigate the production of other later 'early' RNA species, whose production does require a short period of protein synthesis to yield T4 factors required for transcription of further segments of the T4 gene. At approx. 3 min after infection a second class, or delayed early RNA, is formed.

From their competition-hybridization experiments Salser et al. [185] have recognized species of delayed-early RNAs called quasi-late RNA, whose synthesis is more pronounced later in the prereplicative period but continue on into the postreplicative period. On the basis of in vivo [202] and in vitro [203] experiments in which protein products were recognized as radioactive bands after electrophoresis, O'Farrell and Gold have amplified the above proposal that some delayed early mRNAs are initiated at a separate or quasi-late promoter (Q). Following up on an earlier report of Milanesi et al. [204] O'Farrell and Gold have proposed that many delayed-early species arise passively by simple extensions of immediate-early transcription without the active participation of phage-specific protein factors, such as sigma and an antiterminator factor rho. The effect of inhibitors of protein synthesis such as chloramphenicol and puromycin in limiting transcription to immediate-early species is ascribed by them to a polarity effect in vivo, i.e., the requirement for coupled transcription and translation for complete expression of delayed-early cistrons.

The types of experiments described above only reveal gross classes of RNA without being able to distinguish regulatory features related to the synthesis or utilization of mRNA for a particular enzyme or protein. For this reason, George Guthrie [205], followed later by Akira Kuninaka and Kenneth Lembach [206,207], constructed a two-stage incubation system with protoplasts of *E. coli* B in which infection and transcription occurs during inhibition of protein synthesis (Step I). Then, upon adding an inhibitor of RNA synthesis, translation was performed, ideally in the absence of further RNA synthesis (Step II). Thereafter, measurement was made of individual enzyme activities. Experiments of this type had been attempted earlier with actinomycin D as the inhibitor of RNA synthesis [205], but as we were to discover ourselves, inhibition with this drug is neither sufficiently rapid nor complete to prevent some RNA synthesis during Step II, thus clouding the interpretation of the results. However, the drug, rifampicin, an inhibitor of initiation of RNA synthesis, is rapid and complete in its action. Likewise, chloramphenicol would be the drug of choice as an effective inhibitor of protein synthesis, but operationally has the disadvantage that a lapse of approx. 25 min between Step I and Step II is required to wash cells free of chloramphenicol. During this time decay of RNA would occur. Nevertheless, experiments using chloramphenicol and rifampicin, clearly established that mRNAs for dihydrofolate reductase and dCMP hydroxymethylase belong to Class I or the immediate-early category.

From an operational point of view 5-methyltryptophan is a useful inhibitor since protein synthesis can be restored by the simple addition of tryptophan to the incubation mixtures. However, 5-methyltryptophan does allow the synthesis of some Class II or delayed early mRNA species during Step I. For this reason experiments in which 5-methyltryptophan was used as an inhibitor of protein synthesis cannot be compared directly with those in which chloramphenicol was employed. Incorporation of ^{14}C-labeled amino acids into proteins was still appreciable with 5-methyltryptophan, although synthesis of active enzymes was blocked. Using 5-methyltryptophan, Lembach [207] was able to classify the early enzymes into at least three groups according to the requirement of protein synthesis for formation of the respective messengers (see also Cohen [208]).

Group 1 : dCMP hydroxymethylase, dCTPase and dihydrofolate reductase. The mRNA for these enzymes accumulates during incubation in the presence of 5-methyltryptophan in an amount sufficient to direct the synthesis in Step II of the control levels of enzyme activity observed in the absence of rifampicin.

Group 2 : thymidylate synthetase and DNA polymerase. The transcription for these early enzymes occurs during incubation with 5-methyltryptophan, but enzyme levels lower than normal are observed in the absence of further transcription with concomitant protein synthesis.

Group 3 : deoxynucleotide kinase. The mRNA for this early enzyme accumulates to a barely detectable level before protein synthesis is initiated by the addition of tryptophan.

These experiments suggest that initiation of transcription of mRNA for group 1 and group 3 enzymes occurs at different promoter sites and that mRNA for group 2 enzymes may be formed in part from either site.

The case of deoxynucleotide kinase clearly indicated that it fell into a separate class of early enzyme, possibly the quasi-late variety of delayed early enzyme. However, this designation has been modified recently [209] to distinguish genes expressed late in the early period that either do or do not require the participation of the protein product of gene 55. The former are believed to be initiated from the quasi-late promoter, Q, whereas the latter are initiated from a so-called 'middle' promoter, M. mRNA for deoxynucleotide kinase belongs to this latter class. The protein product of the *mot* gene [210] is believed to be involved in the initiation of this mRNA [211].

The unusual initiation characteristics of deoxynucleotide kinase prompted further study of this system by Shigeru Sakiyama, Peter Natale and Carrie Ireland. For example, deoxynucleotide kinase is translated in vivo from an mRNA of approx. 12–15S, which is of sufficient size to allow the synthesis of a protein subunit of 24 000 Da [212–215]. This mRNA is thus probably monocistronic. On the other hand α-glucosyltransferase mRNA may be formed as a polycistronic mRNA in vitro and in vivo [214–216]. However, in the latter case a smaller or 14.5S species has been observed, which may represent a product formed by processing of the larger 20S form, or alternatively, from a separate promoter site [215].

These investigations have permitted the estimation of the half-lives of deoxynucleotide kinase mRNA of 4.5 min at 37°C [217] and of α-glucosyltransferase of 3 min at 30°C [216]. The addition of chloramphenicol to the bacteria 3 min after infection with T4 appears to cause an overproduction of both kinase and α-glucosyltransferase mRNAs. At this time the polarity effect caused by chloramphenicol is no longer evident. Upon addition of rifampicin to this chloramphenicol-treated system at 11 or 17 min, kinase mRNA decreases at the expected rate, but α-glycosyltransferase mRNA is completely stable [217].

Even though the expression of certain genes in vivo appear to require T4 protein modifiers of *E. coli* polymerase, deoxynucleotide kinase as well as several other enzymes can be synthesized in vitro by the transcription and translation machinery of uninfected bacterial cells. Initiation must be occurring, therefore, from early promoter sites. In order to study the initiation step, an experimental system was constructed in which highly purified DNA-directed RNA polymerase was incubated briefly with various combinations of nucleoside triphosphates, followed by addition of rifampicin and the remaining nucleoside triphosphates, and transcription was continued for a prescribed period of time. After termination of transcription, translation was performed on the accumulated mRNA, followed by analysis of various enzyme activities. The level of enzyme was then a function of the initiation of mRNA synthesis in the first step. Of five enzyme systems studied deoxynucleotide kinase was unique in that initiation occurred to a substantial but not maximal degree in the presence of only one nucleoside triphosphate, namely ATP. However, upon addition of UTP initiation was as complete as usually found in the presence of all four nucleoside triphosphates [218]. Natale [219] also found that initiation could be effected by the oligonucleotides, ApApApA, UpUpUpU and ApUpU. Neither ATP nor any of the oligonucleotides, when labeled with ^{32}P, appeared in the RNA formed de novo. Natale then postulated that a particular sequence of nucleotides would be found as a part of the promoter of the kinase cistron and that the function of the oligonucleotides, including polyadenylate formed from ATP, would be to separate the DNA chains to allow positioning of RNA polymerase. Recently the nucleotide sequence of kinase gene 1 has been determined [220] and the predicted promoter region identified. The combined results of these two contributions illustrate the usefulness of functional studies in the interpretation of structural analysis.

One final interesting series of experiments with T4 initiated by Wai Mun Huang [222] in my laboratory concerned the identification of early proteins that could bind to DNA and that were, therefore, probably involved in DNA synthesis in bacterial membranes. She found that the phage-specific DNA polymerase was a primary participant in these reactions and that the binding of other proteins was in some cases interdependent and at other times not. The recent reports of Alberts [222] and Greenberg [223] have shown so elegantly the importance of these complexes in DNA function and in the activity of enzymes associated with DNA. As Greenberg has pointed out the ordered and sequential association of DNA binding proteins and enzymes may be a prominent method of metabolic regulation in phage development.

T5-bacteriophage metabolism

Infection of *E. coli* F with bacteriophage T5 is a system in which the existence of an independent immediate-early class of transcripts can be demonstrated without resort to use of inhibitors of protein synthesis. We were fortunate to have James McCorquodale spend a sabbatical year in our laboratory. He brought with him a long experience in research with T5 bacteriophage, which has an interesting discontinuity of one DNA strand. One advantage of the T5 system is that the process of infection of *E. coli* F can be divided into discrete steps [224]. After adsorption of the phage, 8% of its DNA is initially transferred (FST-DNA) into the host in the presence of protein synthesis and in a temperature-dependent manner

[225,226]. If protein synthesis is blocked at this step, the phage-bacterium complexes remain in a state of partial DNA transfer. When protein synthesis is allowed to continue, the remaining 92% of the phage DNA is transferred, but only after a 3- to 4-min period of protein synthesis. One of the proteins synthesized during this period is required for the completion of the transfer of DNA [227]. After the transfer is complete, a calcium-dependent (10^{-3}M) event occurs; this is followed by the synthesis of additional phage-specific proteins, which are required for protein synthesis. If the infection is interrupted at the stage of the first step transfer by keeping the phage-bacterium complexes at 0°C and the remaining 92% of T5 DNA removed by shearing in a Waring blender, the expression of the FST-DNA can be studied independently of subsequent metabolic events.

This system thus seemed ideal for a study of the types of proteins synthesized during the FST-reaction and their function in subsequent phage development. It was apparent that the proteins coded for by the FST-DNA were the so-called Class I proteins [228]. Three such radioactive proteins were formed from radioactive leucine and identified by polyacrylamide gel electrophoresis. The role of FST-DNA in preparing the host cells for the synthesis of several enzymes (or class II functions) was then demonstrated. Cells that had carried out the FST reaction at 37°C were capable of initiating enzyme synthesis immediately. Cells held at 0°C during this period required a 10- to 12-min delay.

In subsequent research [229–236] Richard Moyer, David Sirbasku and Jaakko Pispa clearly defined the types of proteins and RNA formed during the entire infection period and the role of calcium ions in these first initial reactions. Although the details of these experiments are too complex to be treated here, their studies served as a basis for further examination of this ideal system for exploring the initial reactions of bacteriophage infections.

Cell biology: studies with eukaryotic systems

It was clearly evident to me by 1963–64 that many of the observations made on the regulation of macromolecular synthesis in bacteria had interesting counterparts in eukaryotic systems. Aided by a grant from the Guggenheim Foundation, I was able to spend a year (1964–65) in the laboratory of Renato Dulbecco, who had just moved to the newly founded Salk Institute for Biological Studies in LaJolla, California. Dulbecco and Marguerite Vogt, his long-time collaborator, were noted for their important studies on the transformation of cells in culture by the DNA-containing polyoma virus. During that year Dulbecco, Hartwell and Vogt [237] had undertaken a systematic analysis of the effect of viral infection on DNA synthesis in the host cell. They found that 16 h after infection three enzyme activites, thymidine kinase, dCMP deaminase and DNA polymerase increased many fold as the rate of cellular DNA synthesis was stimulated. They postulated that these enzymatic changes might play some part in cellular transformation. However, it did not seem to me that such an approach would easily lead to much enlightenment on the problem of transformation, since hosts infected with nontransforming viruses also show an elevation of enzymes concerned with DNA synthesis. My own time was spent in an unsuccessful search for a factor in mouse salivary glands that might be as important for the cultivation of liver cells as nerve growth factor had been for the nutrition of nerve ganglia.

Nevertheless, my sabbatical year had an important effect on decisions to be taken upon return to my laboratory at MIT. Our work on bacteriophage metabolism was still in full swing and I could not at that time recognize a niche in the problem of viral transformation where, as a biochemist, I would make a significant contribution. Work on cells in culture would also have involved a considerable reorganization of both the goals and technical outlay of my laboratory.

However, I did maintain an interest in this area of research in case the proper opportunity should arise. As a member of the thesis committee of a graduate student working with another member of the faculty of the Biology Department, I was consulted about an interesting finding made by them concerning the relative incorporation of radioactivity from thymidine and the β-carbon of serine into DNA after fertilization of arbacia eggs. Although thymidine was utilized in the first round of cell division, serine carbon did not appear in the thymine of DNA until several cell divisions later. These interesting results implied that the unfertilized eggs lacked a key regulatory enzyme concerned with DNA synthesis that was formed after fertilization. There was the possibility that this enzyme would be involved in dUMP synthesis or in subsequent reactions leading to dTMP. As a part of his thesis Simcha Fass [238] could not find any differences in the levels of the three enzymes of the latter group, namely dihydrofolate reductase, thymidylate synthetase and serine hydroxymethylase. The problem was then dropped by him and his advisor with the stipulation that we might continue if we wished. John Noronha and Gerald Sheys [239] then undertook an examination of the enzymatic system for deoxynucleotide synthesis, namely ribonucleotide reductase. This research was complicated by the difficult and time-consuming assay of the enzyme when present in very low concentration. They were able to provide convincing experiments demonstrating that unfertilized eggs did not contain ribonucleotide reductase in any significant amount and that its formation did occur rapidly after fertilization. Since enzyme formation did not take place in the presence of inhibitors of protein synthesis, the reductase was being synthesized de novo. However, the lack of inhibition by actinomycin D led to the conclusion that it was translated from so-called 'masked' messenger RNAs.

Studies on the cell cycle

We would have pursued this interesting system further except that work with arbacia eggs in New England is very seasonal. For this reason we turned to a study of the effects of serum and purified hormonal factors on the metabolism of a line of mouse cells in culture that were first caused to rest in the Go state by serum limitation and then supplied with higher concentrations of serum, which induced their entry into S phase and cell division. We thus became involved in the complexities of regulatory mechanisms of the eukaryotic cell cycle. Drawn there in the first place by an interest in ribonucleotide reductase, we quickly appreciated from the literature [240] that this enzyme is in fact present in low concentration in resting cells and is rapidly formed relatively late in the cell cycle (12 h) at a time immediately preceding DNA synthesis or S phase.

In a problem initiated by Margaret Smith and now continued by Philip Wendler we decided rather to investigate reactions that change early in the cell cycle as a result of serum stimulation. Believing that the regulation of nucleotide synthesis might be a crucial factor in cell development, we have found that the rate of synthesis of both purine and pyrimidine nucleotides is low in resting cells and substantially increased relatively soon after hormonal supplementation [241]. The limiting reaction appears to be the synthesis of 5-phosphoribosylpyrophosphate. How this enzyme activity or metabolic system responds to hormones is still to be clarified.

Proteases as mitogens

Mitogens are thought to operate by interaction with receptors on cell walls. However, until recently identification of these receptors for specific mitogenic agents had not been established. The report of Sefton and Rubin [242] that trypsin could stimulate the proliferation of confluent chick embryo fibroblasts encouraged us to study the mitogenic effects of proteases. For example, as a proteolytic enzyme it might cleave a specific cell surface protein or receptor important in conveying the mitogenic signal.

In fact, proteases had become prime candidates not only as mitogenic but also as tumorigenic agents. Examination of cell surface proteins labeled with ^{125}I revealed that a large protein of 230 000 Da, fibronectin or LETS protein, was lost in normal chick embryo fibroblasts treated with trypsin as well as in many transformed cells. For a while, fibronectin was believed to be the cell surface signal for the regulation of cell growth [243]. This point of view was reinforced by the important studies of Edward Reich and his colleagues [244,245], who showed that chick embryo fibroblasts transformed with the Rous sarcoma virus secreted an enzyme that activated a serum factor, later identified as plasminogen, to yield plasmin, which has a potent proteolytic activity in fibrinolysis, a method for its assay. The relatively weak proteolytic activity of the cell factor, termed a plasminogen activator, was thus amplified by its reaction with plasminogen.

Lan Bo Chen, in collaboration with Nelson Teng, Bruce Zetter and Victor Ambros, was able to show that loss of fibronectin and the production of proteases, although perhaps characteristic of transformed cells, are not necessarily the central casual agents of mitogenesis in transformation. Realizing that trypsin is relatively indiscriminate in the bonds it cleaves, he obtained from David Waugh of our Department and from John Fenton highly purified thrombin, which splits a very limited number of bonds, in particular arginyl-glycyl bonds during the conversion of fibrinogen to fibrin. Chen [246] then showed that thrombin exerts a potent mitogenic action on chick embryo fibroblasts without cleavage of fibronectin [247]. However, other cell surface proteins are lost that may be of importance in the stimulation of lymphocytes and fibroblasts.

In a second series of experiments with highly purified fibrinogen obtained again from Waugh and with cells grown in plasminogen-free media and serum, Chen [248] showed that plasminogen was not a necessary factor in the expression of the transformation phenotype, pictures of which were so elegantly captured by him and Ambros [249] in scanning electron micrographs. Although these experiments were discouraging in that they detracted from then current hypotheses about cellular transformation, they clearly indicated that another approach would be required to identify the central transforming agent produced in cells during infection by tumor viruses.

Cell transformation

The above research on proteases was in progress at approximately the same time that we were concluding experiments on the translation of bacteriophage messenger RNA. During a Departmental seminar by John Coffin of neighboring Tufts Medical School on the genetic differences between transforming and transformation-defective avian retroviruses [250], it occurred to me that a comparison of the translation products labeled with [^{35}S]-methionine of the RNAs obtained from these two viruses might reveal important differences in their two electropherogram patterns. James Kamine had just joined my group, having spent a postdoctoral period in Harry Rubin's laboratory in Berkeley and was well acquainted with the technology of avian viruses. Previous attempts at finding differences in the patterns of proteins by translation of viral RNAs had not been successful because of an interfering background. However, a new method of treating extracts used in the translation reaction reported by Pelham and Jackson [251] had overcome this technical difficulty. In a series of papers [252–255] that appeared during the summer of 1977 from three laboratories including our own, transformation-specific products were discovered. What is now known to be the physiologically important product, namely a 60 000-Da protein, was obtained by Raymond Erikson, then at the Colorado Medical School in Denver and now at Harvard. With his colleague, Joan Brugge, he had developed a specific antiserum against the transforming protein, and was able to isolate it from chick cells infected with Rous sarcoma virus [255]. In a second classic paper published in 1978 Erikson demonstrated that the transforming protein was capable of phosphorylating its own specific antibody and is thus a protein kinase [256], specific as shown later by Hunter and Sefton [257] for tyrosyl residues.

These outstanding contributions by Erikson's laboratory have opened the way for a deluge of papers from many sources developing this important problem. Our own contributions have been modest by comparison. Kamine was possibly the first to bring attention to the fact that the transforming protein may associate with membranes [258]. Research by John Burr [259] since then has identified the cytoskeleton as the location of the active phosphorylated form of virally induced transforming protein in chick fibroblasts. Presumably it is through its action on the cytoskeletal and membrane elements that the morphological characteristics of the transformed cell develop.

In a more recent approach Andrew Laudano has been joined by Frank Boschelli in exploring the use of specific antisera derived from synthetic polypeptides in the precipitation of viral oncogenes or normal cellular proteins closely related in structure to these oncogenes. By comparison of homologous regions in several oncogenes, the conserved regions have been identified. Moreover, since the carboxyl terminal regions of viral *src* and its cellular counterpart differ, Laudano has been able to prepare antibodies raised against polypeptides synthesized with regard to these differences. With these useful reagents he should be able to observe the effects of the viral transforming protein on a very plausible target, namely, its cellular counterpart. These experiments thus represent one approach at determining not only the location of the transforming protein but also the physiological targets of its action.

Modification of the activity of proteins by phosphorylation is a well-established method of metabolic regulation. Recognition of the application of this principle to the transformation process has been a large step in providing a central biochemical concept for an ancient and elusive biological problem, namely the metabolism of the cancer cell.

Biology at MIT

Highly complimentary ratings given to our department in national evaluations over the last two decades have been received with some surprise and retrospection by our own faculty and Administration. The Cartter report of 1966 [260] had placed Biochemistry at MIT among its distinguished group, and in the recent review sponsored by the National Academy of Sciences in 1983 [261] our Department has been rated preeminent in three categories, biochemistry, microbiology and molecular/cellular biology. I thought, therefore, it might be of interest to relate how this Department has evolved since 1942 through the efforts of many individuals.

Upon arriving at MIT as its president in 1930, Karl Taylor Compton undertook the upgrading of the sciences to match the eminence of its engineering departments. In 1937 an internal Institute committee recommended expansion of the life sciences, then devoted principally to areas of public health, nutrition and food technology, to include new disciplines in biology and biological engineering. Francis (Frank) Schmitt of Washington University succeeded Samuel Prescott as chairman of the Department in 1941, but his plans for reorganization of the Department were delayed until the close of the war in 1945. At that time the faculty involved in applied biology was split off as a separate Department of Nutrition and Food Technology, and the faculty with interests in the basic sciences, including John Loofbourow, Bernard Gould and Irwin Sizer, remained with the reorganized Biology Department. To this group had been added three new faculty members, Richard Bear, David Waugh and Cecil Hall. This newly appointed group including Schmitt himself was interested in applying physical methods to the study of the structure of macromolecules of biological importance. The term 'Quantitative Biology' was invented to cover the area of their research and the new curriculum to be developed. Waugh's research was primarily concerned with the physical chemistry of blood and milk proteins, Bear's with X-ray diffraction and Schmitt and Hall's with the electronmicroscope. Many publications centering on the structure of collagen and the muscle and blood proteins resulted from this fruitful interaction and sometimes collaboration. A brilliant group of postgraduate medical doctors, who wished to gain experience in research in the basic sciences, joined this effort, and many have since become leaders in a discipline often referred to as "Biophysics".

With the acquisition of the new Dorrance building, James R. Killian, who succeeded to the presidency of MIT in 1948, was able to expand the faculty of the Biology Department to augment its research and teaching in biochemistry. Towards this end I was invited by Frank and President Killian to organize a Division of Biochemistry within the Biology Department under provisions that included a separate budget and the promise of three new appointments in this general area. Although this was a rather unorthodox arrangement from an administrative point of view, it assured me of the independence I required, but at the same time, most importantly, kept the biological sciences under one roof. During the first years several of our outstanding graduate students stayed on after completion of their doctoral dissertations to serve as instructors in temporary positions. However, Gene Brown was appointed to the faculty in 1954, Vernon Ingram in 1958 and Phillips Robbins in 1960 in so-called tenure-track positions. With the opening of the new Whitaker building, two more appointments in biochemistry were made by me in 1967, namely Paul Schimmel and Lisa Steiner.

Together with Gould and Sizer, the members of the Biochemistry Division represented quite a diverse spread of interests, ranging from enzymology, biosynthesis of natural products, structural (sequencing) protein chemistry, human genetics, physical biochemistry, cell wall polysaccharide chemistry and immunology. However, there were threads that linked all of these individual interests together.

By 1955 Frank relinquished the chairmanship of the Department to Irwin Sizer to become an Institute Professor, and soon thereafter Richard Bear left the Institute to become Dean of the College of Sciences and Humanities at Iowa State University. These shifts required the appointments of Cyrus Levinthal and Alexander Rich to reinforce the biophysics program.

As the infant discipline of molecular biology began to grow, it was obvious in 1956 that the Department needed leadership in the crucial areas of microbiology, virology and genetics to complement its strengths in biophysics and biochemistry.

During our summer in Cold Spring Harbor, Gene Brown and I both recognized the remarkable qualifications of Salvador Luria as a teacher and a leader in the field of bacterial physiology and genetics, particularly as related to bacteriophage metabolism. With Irwin Sizer's concurrence, we invited Luria as a visiting professor for the year 1958–59 with the intention of persuading him to head up another area of the Department. Irwin was also successful in convincing Julius Stratton, who had been acting president since 1957, to provide the resources and space for essentially a third 'division' of the Department, that is, for Microbiology. Luria accepted and was successful in recruiting Boris Magasanik and James Darnell to round out the Department in Bacterial Physiology and Animal Cell Virology, respectively.

These faculty members represented the first wave of appointments of a Department of Molecular Biology that was to expand to double this size as new space and appointments developed in the Chemistry Department, in the Cancer Center, the Whitaker College of Health Sciences, Technology and Management, and more recently in the Whitehead Institute. The Department has three Nobel Laureates, Salvador Luria, David Baltimore and Gobind Khorana. These appointments during the 1940's and 1950's set in place the central themes of the Department into interlocking spheres of research interests that are currently at the leading edge of molecular biology. My participation in administrative affairs came to an end in 1967 when the chairmanship of the Department reverted to Boris Magasanik, and eventually at the end of his term in 1977 to Gene Brown. Both have since played active and important roles in the recent growth of Biology at MIT.

In summary, the Department of Biology has prospered over these last four decades because of (1) the excellent cooperation and effective leadership of the Central Administration and the Department chairmen together with its advisory boards in supplying resources for quarters and faculty positions in the several disciplines represented in the Department; (2) the active role taken by senior members of the faculty in identifying and recruiting junior faculty within the three major disciplines and their success in fostering the careers of their younger colleagues both financially and intellectually; (3) the retention of the three disciplines within a single Department under circumstances that provided both interaction and interdependence, yet allowed for flexibility to strike out in individual directions; (4) the frequent overlapping and change of research interests so that each faculty member at one time or another may have made contributions to more than one field; (5) the good fortune of creating a Department de novo when funds from governmental and private sources were expanding.

Personal notes

There was nothing particularly unusual about my early life that indicated a predisposition towards a career in the sciences. I was born on September 29, 1917 in my maternal grandparents' home in Winamac, Indiana, although my parents permanent residence at that time was Indianapolis. My forebears were of mixed Anglo-Saxon and Germanic stock, who for the most part had migrated to this country in the late 1700's or early 1800's and settled probably first in Pennsylvania. As the country opened up they apparently moved to Ohio and later to Indiana.

During my high school years at South Bend, Indiana I had a certain attraction to languages, Latin and French, but soon came to realize that it would be difficult to devise a living based on this type of education alone. During my senior year I enrolled in a chemistry course that ultimately stimulated my interest in the sciences. Since I had graduated from high school in the depth of the depression, the award of a Rector scholarship by DePauw University made possible my higher education.

I suppose that I had been programmed at quite an early time to continue my education towards a graduate degree and to enter academic life, although my parents had stopped their education with high school diplomas. My father, Harry James Buchanan, was a salesman and my mother, Eunice Miller, had been a secretary to my paternal grandfather before her marriage. Since my father's business kept him away from home for much of the time, my mother was a frequent user of the nearby library and her more literary interests were subconsciously passed on to her children. However, my father was an ardent sportsman, so that weekends and summers were devoted to our favorite family recreation, golf.

In retrospect it seems that the most important decisions in my life have always occurred quickly with complete conviction of my choice. One such decision was to become a biochemist. In 1938 this discipline was still a virgin field with untold possibilities for novel research and integration into an academic environment. The second such decision occurred some 8 years later. I had arrived in Sweden in November 1946 at the height of their Christmas festival period when parties attempt to compensate for the winter darkness of that latitude. On the train back from one such pleasant evening I made the acquaintance of a graduate student at the University of

Stockholm. He subsequently introduced me to a friend of his, Elsa Nilsby, who was about to spend a year in the United States on a fellowship in actuarial science. Nevertheless, after a year apart we were able eventually to consolidate our plans and were married just before returning home to Philadelphia.

I was soon to learn in quite an unexpected manner the influence of the distaff member of the family. In reestablishing my laboratory at the University of Pennsylvania, a good deal of work in the evenings and weekends was required. With the exception of a short period of time during the war I had had a key to the outside doors of the building and hoped to retrieve it upon returning to Philadelphia. However, Wright Wilson seemed reluctant to comply with my request. Finally, after 6 months of leaving the doors unlatched at times when I needed to leave the laboratory for dinner, I was caught one night red-handed by the guards, who threatened to report me to no less than the Dean himself.

The next morning I approached Wilson about my problem with the guards and again asked for a key. My request was approved immediately with the explanation that he had withheld the key at the suggestion of his wife, Helene, who felt that a new bridegroom should not be spending so much of his free time in the laboratory. Knowing that Helene was an incurable romantic, I could nurse no ill feelings. Yet I could not escape the reminder that I had turned over a new leaf in my life.

Whatever compromises marriage might have entailed, they were more than compensated for by the ministrations of a partner who, for the moment at least, had unselfishly set aside her own professional goals, to become a wife, manager, hostess and mother of four children. Claire was born in 1950, Stephen in 1951, Lisa in 1955 and Peter in 1959. Claire has her PhD in limnology, Lisa and Peter are still graduate students in biology and geology, respectively, and Stephen is the business man of the family. Elsa returned to her actuarial work 10 years ago in a demanding but rewarding position.

Aside from one's own family and colleagues, there are those unusual persons who have influenced one's life to brighten and broaden significantly its horizons. Francis and Lelia Stokes were two such friends. I had met them for the first time when I had spent an Easter vacation at their home in Germantown, Pennsylvania with their son, Allen, a friend of my Harvard graduate school days. The friendship with the Stokes parents had been renewed when I moved to the University of Pennsylvania in 1943. During the year before my sabbatical leave in Stockholm I had been invited to live with them and, thus, came to witness at close range not only their devotion for each other but also their concern and hospitality for many within their extensive circle of friends. From them I was able to gain an insight into the traditions and philosophy of the Society of Friends for which Philadelphia is so widely noted.

In summarizing briefly my professional extracurricular activities I have participated in the study sections for Physiological Chemistry and the Biochemistry Training Program at the National Institutes of Health, the Research Advisory Board of the American Cancer Society, and in editorial boards of the Journal of Biological Chemistry, the Journal of the American Chemical Society, Physiological Reviews, the Federation Proceedings and the Journal of Molecular and Cellular Biochemistry. I have also served on the Councils of the Division of Biochemistry of the American Chemical Society and the American Society of Biological Chemists. In connection with the latter's Educational Affairs Committee I had the enjoyment of developing six films on protein structure and enzyme activity [262].

I am a member of the American Academy of Arts and Sciences and the National Academy of Sciences and have received honorary doctorate degrees from two of my alma maters, DePauw University and the University of Michigan. In 1951 I was the recipient of the Eli Lilly Award in Biochemistry from the American Chemical Society and in 1958 participated in the lecture series of the Harvey Society [263].

In 1967 I became the John and Dorothy Wilson Professor of Biochemistry. I have been fortunate to know my benefactors personally and to have spent with my wife several enjoyable vacations with them at their winter residence in Barbados. Mr. Wilson has been until recently Secretary to the Corporation at MIT and has had an influential role in alumni and Corporation affairs at the Institute.

In looking back over the last 40 years I cannot imagine a more gratifying profession to have chosen. Until recently, at least, resources have been relatively generous to support research, and I have had the opportunity to participate in two disciplines, biochemistry and molecular biology, at a time when they were in their ascendency. For the most part I have stayed rather close to studies on purine or pyrimidine nucleotide biosynthesis and regulation. The rapid development of the biological sciences has required, however, major shifts in one's approaches with sometimes painful but necessary reorganization of the goals of one's laboratory. Above all I have enjoyed working with a perpetually youthful group of collaborators, and witnessing their realization of personal goals and professional ambitions. After all, this is the stuff of which Universities are made.

REFERENCES

1 V. duVigneaud, A Trail of Research in Sulfur Chemistry and Metabolism and Related Fields, Cornell University Press, Ithaca, NY, 1952.
2 J.H. Mueller, J. Biol. Chem., 56 (1923) 157.
3 J.L. Riebsomer, R. Baldwin, J. Buchanan and H. Burkett, J. Am. Chem. Soc., 60 (1938) 2974.
4 R.D. Cramer and G.B. Kistiakowsky, J. Biol. Chem., 137 (1941) 549.
5 J.B. Conant, R.D. Kramer, A.B. Hastings, F.W. Klemperer, A.K. Solomon and B. Vennesland, J. Biol. Chem., 137 (1941) 557.
6 B. Vennesland, A.K. Solomon, J.M. Buchanan, R.D. Cramer and A.B. Hastings, J. Biol. Chem., 142 (1942) 371.
7 A.B. Hastings, Harvey Lectures, 36 (1940–41) 91.
8 A.B. Hastings and J.M. Buchanan, Proc. Natl. Acad. Sci. USA, 28 (1942) 478.
9 J.M. Buchanan, F.B. Nesbett and A.B. Hastings, J. Biol. Chem., 180 (1949) 435.
10 J.M. Buchanan, A.B. Hastings and F.B. Nesbett, J. Biol. Chem., 180 (1949) 447.
11 R. Schoenheimer, The Dynamic State of the Body Constituents, Harvard University Press, Cambridge, MA, 1941.
12 H.G. Wood and C.H. Werkman, Biochem. J., 30 (1936) 48.
13 H.G. Wood and C.H. Werkman, Biochem. J., 34 (1940) 129.
14 H.G. Wood, C.H. Werkman, A. Hemingway and A.O. Nier, J. Biol. Chem., 135 (1940), 789.
15 E.A. Evans, Jr. and L. Slotin, J. Biol. Chem., 136 (1940) 301.
16 A.K. Solomon, B. Vennesland, F.W. Klemperer, J.M. Buchanan and A.B. Hastings, J. Biol. Chem., 140 (1941) 171.
17 J.M. Buchanan, A.B. Hastings and F.B. Nesbett, J. Biol. Chem., 145 (1942) 715.
18 J.M. Buchanan, A.B. Hastings and F.B. Nesbett, J. Biol. Chem., 150 (1943) 413.
18a G. Rosenfeld, Ergeb. Physiol., 2 (1903) 50.
19 F.L. Breusch, Science, 92 (1943) 490.
20 H. Wieland and C. Rosenthal, Ann. Chem., 554 (1945) 241.
21 H.A. Krebs and L.V. Eggleston, Nature, 154 (1944) 209.
22 H. Weil-Malherbe, Nature, 153 (1944) 435.
23 J.M. Buchanan, W. Sakami, S. Gurin and D.W. Wilson, J. Biol. Chem., 157 (1945) 747.
24 J.M. Buchanan, W. Sakami, S. Gurin and D.W. Wilson, J. Biol. Chem., 159 (1945) 695.
25 J.M. Buchanan, W. Sakami, S. Gurin and D.W. Wilson, J. Biol. Chem., 169 (1947) 403.

26 E.M. MacKay, R.H. Barnes, H.O. Carne and A.N. Wick, J. Biol. Chem., 135 (1940) 157.
27 W.C. Stadie, Physiol. Rev., 25 (1945) 395.
28 S. Weinhouse, G. Medes and N.F. Floyd, J. Biol. Chem., 155 (1944) 143.
29 J.M. Buchanan, W. Sakami and S. Gurin, J. Biol. Chem., 169 (1947) 411.
30 D.I. Crandall, R.O. Brady and S. Gurin, J. Biol. Chem., 181 (1949) 845.
31 S. Wakil, J. Am. Chem. Soc., 80 (1958) 6465.
32 H.A. Krebs and K. Henseleit, Klin. Wschr., 11 (1932) 1137.
33 S.J. Bach and S. Williamson, Nature, 150 (1942) 575.
34 P.P. Cohen and M. Hayano, J. Biol. Chem., 166 (1946) 239.
35 N.L. Edson and H.A. Krebs, Biochem. J., 30 (1936) 732.
36 C.S. Venable, J. Am. Chem. Soc., 40 (1918) 1099.
37 F.J. Moore and R.M. Thomas, J. Am. Chem. Soc., 40 (1918) 1120.
38 H. Brandenberger, Biochim. Biophys. Acta, 37 (1954) 641.
39 S.C. Hartman and J. Fellig, J. Am. Chem. Soc., 77 (1955) 1051.
40 J.C. Sonne, J.M. Buchanan and A.M. Delluva, J. Biol. Chem., 166 (1946) 395.
41 J.M. Buchanan and J.C. Sonne, J. Biol. Chem., 166 (1946) 781.
42 J.C. Sonne, J.M. Buchanan and A.M. Delluva, J. Biol. Chem., 173 (1948) 69.
43 J.M. Buchanan, J.C. Sonne and A.M. Delluva, J. Biol. Chem., 173 (1948) 81.
44 D. Shemin, J. Biol. Chem., 162 (1946) 297.
45 E.C. Wagner and W.H. Millett, Org. Synth., 2 (1943) 65.
46 N.L. Edson, H.A. Krebs and A. Model, Biochem. J., 30 (1936) 1380.
47 Å. Örström, M. Örström and H.A. Krebs, Biochem. J., 33 (1939) 990.
48 M.P. Schulman, J.C. Sonne and J.M. Buchanan, J. Biol. Chem., 196 (1952) 499.
49 D. Elwyn and D.B. Sprinson, J. Biol. Chem., 184 (1950) 465.
50 P. Walden, Ber. Chem. Ges., 40 (1907) 3214.
51 D. Shemin and D. Rittenberg, J. Biol. Chem., 167 (1947) 875.
52 J.L. Karlsson and H.A. Barker, J. Biol. Chem., 177 (1949) 597.
53 F.W. Barnes Jr. and R. Schoenheimer, J. Biol. Chem., 151 (1943) 123.
54 G.R. Greenberg, Arch. Biochem., 19 (1948) 337.
55 G.R. Greenberg, Fed. Proc., 10 (1951) 192.
56 M.P. Schulman and J.M. Buchanan, Fed. Proc. 10 (1951) 244.
57 J.C. Sonne, I. Lin and J.M. Buchanan, J. Am. Chem. Soc., 75 (1953) 1516.
58 J.C. Sonne, I. Lin and J.M. Buchanan, J. Biol. Chem., 220 (1954) 369.
59 B. Levenberg, S.C. Hartman and J.M. Buchanan, J. Biol. Chem., 220 (1956) 379.
60 J.M. Buchanan and A.B. Hastings, Physiol. Rev., 26 (1946) 120.
61 J.M. Buchanan and C.B. Anfinsen, J. Biol. Chem., 180 (1949) 47.
62 S.R. Dickman and A.A. Cloutier, J. Biol. Chem., 188 (1950) 379.
63 J.M. Buchanan and S.C. Hartman in F.F. Nord (Ed.), Advances in Enzymology and Related Subjects of Biochemistry, Vol. 21, Interscience, New York, 1959, p. 199.

64 S.C. Hartman and J.M. Buchanan, Ergeb. Physiol., 50 (1959) 75.
65 J.M. Buchanan, J.G. Flaks, S.C. Hartman, B. Levenberg, L.N. Lukens and L. Warren, in G.E.W. Wolstenholme and C.M. Conner (Eds.), The Chemistry and Biology of Purines, Churchill, London, 1957, p. 233.
66 J.M. Buchanan, Annu. Rev. Biochem., Proc. Biochem. Symp. (1982) 57.
67 G.R. Greenberg, J. Biol. Chem., 190 (1951) 611.
68 D.A. Goldthwait, R.A. Peabody and G.R. Greenberg, J. Am. Chem. Soc., 76 (1954) 5258.
69 S.C. Hartman, B. Levenberg and J.M. Buchanan, J. Am. Chem. Soc., 77 (1955) 501.
70 S.C. Hartman, B. Levenberg and J.M. Buchanan, J. Biol. Chem., 221 (1956), 1057.
71 H.M. Kalckar, J. Biol. Chem., 167 (1947), 477.
72 M.P. Schulman and J.M. Buchanan, J. Biol. Chem., 196 (1952) 513.
73 W.J. Williams and J.M. Buchanan, J. Biol. Chem., 203 (1952) 583.
74 E.D. Korn and J.M. Buchanan, J. Biol. Chem., 217 (1955) 183.
75 E.D. Korn, C.N. Remy, H.C. Wasilejko and J.M. Buchanan, J. Biol. Chem., 217 (1955) 875.
76 C.N. Remy, W.T. Remy and J.M. Buchanan, J. Biol. Chem., 217 (1955) 885.
77 A. Kornberg, I. Lieberman and E.S. Simms, J. Am. Chem. Soc., 76 (1954) 2027.
78 A. Kornberg, I. Lieberman and E.S. Simms, J. Biol. Chem., 215 (1955) 389.
79 A. Kornberg, I. Lieberman and E.S. Simms, J. Biol. Chem., 215 (1955) 417.
80 L.N. Lukens and K.A. Herrington, Biochim. Biophys. Acta 24 (1957) 432.
81 J.G. Flaks, M.J. Erwin and J.M. Buchanan, J. Biol. Chem., 228 (1957) 201.
82 J.E. Seegmiller, F.M. Rosenbloom and W.N. Kelley, Science, 155 (1967) 1682.
83 W.N. Kelley, R.I. Levy, J.F. Henderson and J.E. Seegmiller, J. Clin. Invest., 47 (1968) 2281.
84 M. Lesch and W.L. Nyhan, Am. J. Med., 36 (1964) 561.
85 S.C. Hartman, B. Levenberg and J.M. Buchanan, J. Am. Chem. Soc., 77 (1955) 501.
86 D.A. Goldthwait, G.R. Greenberg and R.A. Peabody, Biochim. Biophys. Acta, 18 (1955) 148.
87 D.A. Goldthwait, J. Biol. Chem., 222 (1956) 1051.
88 S.C. Hartman and J.M. Buchanan, J. Biol. Chem., 233 (1958) 451.
89 S.C. Hartman and J.M. Buchanan, J. Biol. Chem., 233 (1958) 456.
90 M.R. Stetten and C.L. Fox Jr., J. Biol. Chem., 161 (1945) 333.
91 W. Shive, W.W. Ackerman, M. Gordon, M.E. Getzendaner and R.E. Eakin, J. Am. Chem. Soc., 69 (1947) 725.
92 H.E. Sauberlich and C.A. Baumann, J. Biol. Chem., 176 (1948) 165.
93 A. Pohland, E.H. Flynn, R.G. Jones and W. Shive, J. Am. Chem. Soc., 73 (1951) 3247.

94 M. May, T.J. Bardos, F.L. Barger, M. Lansford, J.M. Ravel, G.L. Sutherland and W. Shive, J. Am. Chem. Soc., 73 (1951) 3067.
95 D.B. Cosulich, B. Roth, J.M. Smith Jr., M.E. Hulquist and R.P. Parker, J. Am. Soc., 74 (1952) 3252.
96 W. Shive, Ann. N.Y. Acad. Sci., 52 (1950) 1212.
97 W. Shive, Fed. Proc., 12 (1953) 639.
98 G.R. Greenberg and L. Jaenicke, in G.E.W. Wolstenholme and C.M. Conner (Eds.), The Chemistry and Biology of the Purines, Churchill, London, 1957, p. 204.
99 G.R. Greenberg, L. Jaenicke and M. Silverman, Biochim. Biophys. Acta, 17 (1955) 589.
100 J.G. Flaks and J.M. Buchanan, J. Am. Chem. Soc., 77 (1954) 501.
101 J.G. Flaks, M.J. Erwin and J.M. Buchanan, J. Biol. Chem., 229 (1957) 603.
102 J.G. Flaks, L. Warren and J.M. Buchanan, J. Biol. Chem., 228 (1957) 215.
103 L. Warren, J.G. Flaks and J.M. Buchanan, J. Biol. Chem., 229 (1957), 627.
104 L. Warren, and J.M. Buchanan, J. Biol. Chem., 229 (1957) 613.
105 S.C. Hartman and J.M. Buchanan, J. Biol. Chem., 234 (1959) 1812.
106 G.K. Smith, P.A. Benkovic and S.J. Benkovic, Biochemistry, 20 (1981) 4034.
107 C.A. Caperilli, P.A. Benkovic, G. Chettur and S.J. Benkovic, J. Biol. Chem., 255 (1980), 1885.
108 A.C. Bratton and E.K. Marshall Jr., J. Biol. Chem., 128 (1939) 537.
109 B. Levenberg and J.M. Buchanan, J. Am. Chem. Soc., 78 (1956) 504.
110 B. Levenberg and J.M. Buchanan, J. Biol. Chem., 224 (1957) 1019.
111 I. Melnick and J.M. Buchanan, J. Biol. Chem., 225 (1957) 357.
112 L.N. Lukens and J.M. Buchanan, J. Am. Chem. Soc., 79 (1957) 1511.
113 R.W. Miller, L.N. Lukens and J.M. Buchanan, J. Am. Chem. Soc., 79 (1957) 1513.
114 L.N. Lukens and J.M. Buchanan, J. Biol. Chem., 234 (1959) 1791.
115 L.N. Lukens and J.M. Buchanan, J. Biol. Chem., 234 (1962) 1799.
116 R.W. Miller, L.N. Lukens and J.M. Buchanan, J. Biol. Chem., 234 (1962) 1806.
117 R.W. Miller and J.M. Buchanan, J. Biol. Chem., 237 (1962) 485.
118 R.W. Miller and J.M. Buchanan, J. Biol. Chem., 237 (1962) 491.
119 H.E. Skipper, L.L. Bennett Jr. and F.M. Schabel, Jr., Fed. Proc., 13 (1954) 298.
120 B. Levenberg, I. Melnick and J.M. Buchanan, J. Biol. Chem., 225 (1957) 163.
121 J.M. Buchanan, in G.E.W. Wolstenholme and C.M. O'Connor (Eds.), Amino Acids and Peptides with Antimetabolic Activity, Churchill, London, 1958, p. 75.
122 J.M. Buchanan, in E. Meister (Ed.), Advances in Enzymology and Related Areas of Molecular Biology, Vol. 39, New York, 1973, p. 91.
123 T.C. French, I.B. Dawid, R.A. Day and J.M. Buchanan, J. Biol. Chem., 238 (1963) 2171.

124 K. Mizobuchi and J.M. Buchanan, J. Biol. Chem., 243 (1968) 4842.
125 I.B. Dawid, T.C. French and J.M. Buchanan, J. Biol. Chem., 238 (1963) 2178.
126 T.C. French, I.B. Dawid and J.M. Buchanan, J. Biol. Chem., 238 (1963) 2186.
127 S. Ohnoki, B.-S. Hong and J.M. Buchanan, Biochemistry, 16 (1977) 1065.
128 S. Ohnoki, B.-S. Hong and J.M. Buchanan, Biochemistry, 16 (1977) 1070.
129 D.D. Schroeder, A.J. Allison and J.M. Buchanan, J. Biol. Chem., 244 (1969) 5856.
130 K. Mizobuchi and J.M. Buchanan, J. Biol. Chem., 243 (1968) 4853.
131 K. Mizobuchi, G. Kenyon and J.M. Buchanan, J. Biol. Chem., 243 (1968) 4863.
132 J.-M., Frère, D.D. Schroeder and J.M. Buchanan, J. Biol. Chem., 246 (1971) 4727.
133 H.-C. Li and J.M. Buchanan, J. Biol. Chem., 246 (1971) 4720.
134 E. Shaw, Physiol. Rev., 50 (1970) 244.
135 D.D. Woods, in: G.E.W. Wolstenholme and C.M. O'Connor (Eds.), The Chemistry and Biology of the Purines, Churchill, London, 1957, p. 316.
136 T.H. Jukes, Fed. Proc., 12 (1953) 633.
137 B.D. Davis and E.S. Mingioli, J. Bacteriol. 60 (1950) 17.
138 F.T. Hatch, S. Takeyama, R.E. Cathou, A.R. Larrabee and J.M. Buchanan, J. Am. Chem. Soc., 81 (1959) 6525.
139 F.T. Hatch, S. Takeyama, R.E. Cathou and J.M. Buchanan, J. Biol. Chem., 236 (1961) 1095.
140 S. Takeyama, F.T. Hatch and J.M. Buchanan, J. Biol. Chem., 236 (1961) 1102.
141 A.R. Larrabee, S. Rosenthal, R.E. Cathou and J.M. Buchanan, J. Am. Chem. Soc., 83 (1961) 4094.
142 A.R. Larrabee, S. Rosenthal, R.E. Cathou and J.M. Buchanan, J. Biol. Chem., 238 (1963) 1025.
143 S. Rosenthal and J.M. Buchanan, Acta Chem. Scand., 17 (1963) S288.
144 S.S. Kerwar, J.H. Mangum, K.G. Scrimgeour and F.M. Huennekens, Biochem. Biophys. Res. Commun., 15 (1964) 377.
145 J.H. Mangum and K.G. Scrimgeour, Fed. Proc., 21 (1962) 242.
146 V. Herbert, A.R. Larrabee and J.M. Buchanan, J. Clin. Invest. 41 (1962) 1134.
147 K.O. Donaldson and J.C. Keresztesy, J. Biol. Chem., 237 (1962) 1298.
148 J.M. Noronha and M. Silverman, in H.C. Heinrich (Ed.), Proc. of the Second European Symposium of Vitamin B_{12} and Intrinsic Factor, Enke-Verlag, Stuttgart.
149 W. Sakami and I. Ukstins, J. Biol. Chem., 236 (1961) PC 50.
150 J.R. Guest, S. Friedman, D.D. Woods and E.L. Smith, Nature, 195 (1962) 340.
151 J. Bremer and D.M. Greenberg, Biochim, Biophys. Acta, 46 (1961) 217.

152 S. Rosenthal, L.C. Smith and J.M. Buchanan, J. Biol. Chem., 240 (1965) 836.
153 R.E. Cathou and J.M. Buchanan, J. Biol. Chem., 238 (1963) 1746.
154 H.M. Katzen and J.M. Buchanan, J. Biol. Chem., 240 (1965) 825.
155 R.E. Loughlin, H.L. Elford and J.M. Buchanan, J. Biol. Chem., 239 (1964) 2888.
156 K. Fujii, T. Nagasaki and F.M. Huennekens, J. Biol. Chem., 256 (1981) 10329.
157 G.R. Wyatt and S.S. Cohen, Biochem. J., 55 (1953) 774.
158 J.G. Flaks and S.S. Cohen, Biochim. Biophys. Acta., 25 (1957) 667.
159 J.G. Flaks and S.S. Cohen, J. Biol. Chem., 234 (1959) 1501.
160 A. Kornberg, I.R. Lehman and E.S. Simms, Fed. Proc. 15 (1956) 291.
161 R.L. Sinsheimer, Science, 120 (1954) 551.
162 E. Volkin, J. Am. Chem. Soc., 76 (1954) 5892.
163 J.F. Koerner and S. Varadarajan, J. Biol. Chem., 235 (1960) 2688.
164 A. Kornberg, S.B. Zimmerman, S.R. Kornberg and J. Josse, Proc. Natl. Acad. Sci. USA, 45 (1959) 772.
165 J.F. Koerner, M.S. Smith and J.M. Buchanan, J. Am. Chem. Soc., 81 (1959) 2594.
166 J.F. Koerner, M.S. Smith and J.M. Buchanan, J. Biol. Chem., 235 (1960) 2691.
167 G.R. Greenberg, Proc. Natl. Acad. Sci. USA, 56 (1966) 1226.
168 H.R. Warner and J.E. Barnes, Proc. Natl. Acad. Sci. USA, 56 (1966) 1233.
169 R.L. Somerville, K. Ebisuzaki and G.R. Greenberg, Proc. Natl. Acad. Sci. USA, 45 (1959) 1240.
170 R.H. Epstein, A. Bolle, C.M. Steinberg, E. Boy de la Tour, R. Chevalley, R.S. Edgar, M. Susman, G.H. Denhardt and E. Lielausis, Cold Spring Harbor Symp. Quant. Biol., 28 (1963) 375.
171 J.S. Wiberg, M.-L. Dirksen, R.H. Epstein, S.E. Luria and J.M. Buchanan, Proc. Natl. Acad. Sci. USA, 48 (1962) 293.
172 M.-L. Dirksen, J. Hutson and J.M. Buchanan, Proc. Natl. Acad. Sci. USA, 50 (1963) 507.
173 D.R. Helinski and C. Yanofsky, Proc. Natl. Acad. Sci. USA, 48 (1962) 173.
174 B.D. Maling and C. Yanofsky, Proc. Natl. Acad. Sci. USA, 47 (1961) 551.
175 J.S. Wiberg and J.M. Buchanan, Proc. Natl. Acad. Sci. USA, 51 (1964) 421.
176 W.B. Wood and H.R. Revel, Bacteriol. Rev., 40 (1976) 847.
177 J.S. Wiberg, Proc. Natl. Acad. Sci. USA, 55 (1966) 614.
178 D.H. Duckworth and M.J. Bessman, J. Biol. Chem., 242 (1967) 2877.
179 A. DeWaard, A.V. Paul and I.H. Lehman, Proc. Natl. Acad. Sci. USA, 54 (1965) 1241.
180 H.R. Warner and J.E. Barnes, J. Virol., 28 (1966) 100.
181 G.C. Fareed and C.C. Richardson, Proc. Natl. Acad. Sci. USA, 58 (1967) 665.
182 J.G. Flaks, J. Lichtenstein and S.S. Cohen, J. Biol. Chem., 234 (1959) 1507.
183 G.A. Vidaver and L.M. Kozloff, J. Biol. Chem., 225 (1957) 335.

184 R.J. Grasso and J.M. Buchanan, Nature, 224 (1969) 822.
185 W. Salser, A. Bolle and R. Epstein, J. Mol. Biol., 49 (1970) 271.
186 G. Fairbanks Jr., C. Levinthal and R.R. Reeder, Biochem. Biophys. Res. Commun., 20 (1965) 393.
187 C. Levinthal, J. Hosoda and D. Shub, in J.S. Colter and W. Paranchych (Eds.), The Molecular Biology of Viruses, Academic Press, New York, 1967, p. 71.
188 M.-L. Dirksen, J.S. Wiberg, J.F. Koerner and J.M. Buchanan, Proc. Natl. Acad. Sci. USA, 60 (1960) 1425.
189 N. Delihas, Virology, 13 (1961) 242.
190 O. Sköld and J.M. Buchanan, Proc. Natl. Acad. Sci. USA, 51 (1964) 553.
191 A.E. Oleson, J.P. Pispa and J.M. Buchanan, Proc. Natl. Acad. Sci. USA, 63 (1969) 473.
192 D.J. McCorquodale, A.E. Oleson and J.M. Buchanan, in J.S. Colter and W. Paranchych (Eds.), The Molecular Biology of Viruses, Academic Press, New York, 1967, p. 31.
193 H. Bremer and M.W. Konrad, Proc. Natl. Acad. Sci. USA, 51 (1964) 801.
194 M. Chamberlin, J. McGrath and L. Waskell, Nature, 228 (1970) 227.
195 W. Summers and R.B. Siegel, Nature, 228 (1970) 1160.
196 G. Walter, W. Seifert and W. Zillig, Biochem. Biophys. Res. Commun., 30 (1968) 240.
197 A.A. Travers, Nature, 223 (1969) 1107.
198 J.W. Roberts, Nature, 224 (1969) 1168.
199 C.G. Goff and K. Weber, Cold Spring Harbor Symp. Quant. Biol., 35 (1970) 101.
200 W. Zillig, K. Zechel, D. Rabussay, M. Schachner, V.S. Sethi, P. Palm, A. Heil and W. Seifert, Cold Spring Harbor Symp. Quant. Biol., 35 (1970) 47.
201 D. Rabussay, in: C.K. Mathews, E.M. Kutter, G. Mosig and P.B. Berget (Eds.), T4 Bacteriophage, American Society for Microbiology, Washington, DC, 1983, p. 167.
202 P.Z. O'Farrell and L.M. Gold, J. Biol. Chem., 248 (1973) 5502.
203 P.Z. O'Farrell and L.M. Gold, J. Biol. Chem., 248 (1973) 5512.
204 G. Milanesi, E.N. Brody, O. Grau and E.P. Geiduschek, Proc. Natl. Acad. Sci. USA, 66 (1970) 181.
205 G. Guthrie and J.M. Buchanan, Fed. Proc., 25 (1966) 864.
206 K.J. Lembach, A. Kuninaka and J.M. Buchanan, Proc. Natl. Acad. Sci. USA, 62 (1969) 446.
207 K.J. Lembach and J.M. Buchanan, J. Biol. Chem., 245 (1970) 1575.
208 P.S. Cohen, Virology, 41 (1970) 453.
209 E. Brody, D. Rabussay and D.H. Hall, in C.K. Mathews, E.M. Kutter, G. Mosig and P.B. Berget (Eds.), T4 Bacteriophage, American Society of Microbiology, Washington, D.C. 1983, p.174.
210 T. Mattson, J. Richardson and D. Goodin, Nature, 250 (1974) 48.
211 C. Thermos and E. Brody, Cited in [209].
212 S. Sakiyama and J.M. Buchanan, Proc. Natl. Acad. Sci. USA, 68 (1971) 1376.

213 S. Sakiyama and J.M. Buchanan, J. Biol. Chem., 248 (1973) 3150.
214 P.J. Natale, J.M. Buchanan, Proc. Natl. Acad. Sci. USA, 69 (1972) 2513.
215 P.J. Natale, C. Ireland and J.M. Buchanan, Biochem. Biophys. Res. Commun., 66 (1975) 1287.
216 E.T. Young, II and G. Van Houwe, J. Mol. Biol. 51 (1970) 605.
217 S. Sakiyama and J.M. Buchanan, J. Biol. Chem., 247 (1972) 7806.
218 P.J. Natale and J.M. Buchanan, Proc. Natl. Acad. Sci. USA, 71 (1974) 422.
219 P.J. Natale and J.M. Buchanan, J. Biol. Chem., 252 (1977), 2304.
220 J. Abelson, Personal Communication.
221 W.-M. Huang and J.M. Buchanan, Proc. Natl. Acad. Sci. USA, 71 (1974) 2226.
222 B.M. Alberts, C.G. Morris, D. Mace, N. Sinha, M. Bittner and L. Moran, in M.M. Goulian, P.C. Hanawalt and C.F. Fox (Eds.), DNA synthesis and its regulation, Benjamin, Menlo Park, CA, 1975.
223 P.K. Tomich, C.-S. Chiu, M.G. Wovcha and G.R. Greenberg, J. Biol. Chem., 249 (1974) 7613.
224 Y.T. Lanni, Virology, 10 (1960), 514.
225 D.J. McCorquodale and Y.T. Lanni, J. Mol. Biol. 10 (1964) 10.
226 Y.T. Lanni, D.J. McCorquodale and C.M. Wilson, J. Mol. Biol., 10 (1964) 19.
227 Y.T. Lanni, Proc. Natl. Acad. Sci. USA, 53 (1965) 969.
228 D.J. McCorquodale and J.M. Buchanan, J. Biol. Chem., 243 (1968) 2550.
229 R.W. Moyer and J.M. Buchanan, Proc. Natl. Acad. Sci. USA, 64 (1969) 1249.
230 R.W. Moyer and J.M. Buchanan, J. Biol. Biol. Chem., 245 (1970) 5897.
231 R.W. Moyer and J.M. Buchanan, J. Biol. Chem., 245 (1970) 5904.
232 D.A. Sirbasku and J.M. Buchanan, J. Biol. Chem., 245 (1970) 2679.
233 D.A. Sirbasku and J.M. Buchanan, J. Biol. Chem., 245 (1970) 2693.
234 J.P. Pispa, D.A. Sirbasku and J.M. Buchanan, J. Biol. Chem., 246 (1971) 1658.
235 D.A. Sirbasku and J.M. Buchanan, J. Biol. Chem., 246 (1971) 1665.
236 J.P. Pispa and J.M. Buchanan, Biochim. Biophys. Acta, 247 (1971) 181.
237 R. Dulbecco, L.H. Hartwell and M. Vogt, Proc. Natl. Acad. Sci. USA, 53 (1965) 403.
238 S. Fass, Ph.D. Thesis, Massachusetts Institute of Technology, 1969.
239 J.M. Noronha, G.M. Sheys and J.M. Buchanan, Proc. Natl. Acad. Sci. USA, 69 (1972) 2006.
240 B.A. Nordensjold, L. Skoog, N.C. Brown and P. Reichard, J. Biol. Chem., 245 (1970) 5360.
241 M.L. Smith and J.M. Buchanan, J. Cell. Physiol., 101 (1979) 293.
242 B.M. Sefton and H. Rubin, Nature, 227 (1970) 843.
243 R.O. Hynes, Proc. Natl. Acad. Sci. USA, 70 (1973) 3170.
244 J.C. Unkeless, A. Tobia, L. Ossowski, J.P. Quigley, D.B. Rifkin and E. Reich, J. Exp. Med., 137 (1973) 85.

245 L. Ossowski, J.P. Quigley and E. Reich, J. Biol. Chem., 249 (1974) 4312.
246 L.B. Chen and J.M. Buchanan, Proc. Natl. Acad. Sci. USA, 72 (1975) 131.
247 N.N.H. Teng and L.B. Chen, Proc. Natl. Acad. Sci. USA, 72 (1975) 413.
248 L.B. Chen and J.M. Buchanan, Proc. Natl. Acad. Sci. USA, 72 (1975) 1132.
249 V.R. Ambros, L.B. Chen and J.M. Buchanan, Proc. Natl. Acad. Sci. USA, 72 (1975) 3144.
250 J.M. Coffin and M.A. Billeter, J. Mol. Biol., 100 (1976) 293.
251 R.B. Pelham and J. Jackson, Eur. J. Biochem., 67 (1976) 247.
252 J. Kamine and J.M. Buchanan, Proc. Natl. Acad. Sci. USA, 74 (1977) 2011.
253 J. Kamine, J.G. Burr and J.M. Buchanan, Proc. Natl. Acad. Sci. USA, 75 (1978) 366.
254 K. Beemon and T. Hunter, Proc. Natl. Acad. Sci. USA, 74 (1977) 3302.
255 J.S. Brugge and R.L. Erikson, Nature, 269 (1977) 346.
256 M.S. Collett and R.L. Erikson, Proc. Natl. Acad. Sci. USA, 75 (1978) 2021.
257 T. Hunter and B.M. Sefton, Proc. Natl. Acad. Sci. USA, 77 (1980) 1311.
258 J. Kamine, and J.M. Buchanan, Proc. Natl. Acad. Sci. USA, 75 (1978) 4399.
259 J.G. Burr, G. Dreyfuss, S. Penman and J.M. Buchanan, Proc. Natl. Acad. Sci. USA, 77 (1980) 3484.
260 A.M. Cartter, An Assessment of Quality in Graduate Education, American Council on Education, Washington, DC, 1966.
261 L.V. Jones, G. Lindzey and P.E. Coggeshall (Eds.), An Assessment of Research - Doctorate Programs in the United States: Biological Sciences, National Academy Press, Washington, DC, 1982.
262 J.M. Buchanan, Federation Proc., 33 (1974) 1998.
263 J.M. Buchanan, Harvey Lectures, 54 (1958-59) 104.

G. Semenza (Ed.) Selected Topics in the History of Biochemistry: Personal Recollections (Comprehensive Biochemistry Vol. 36) © *1985 Elsevier Science Publishers*

Chapter 2

The Discovery of Phosphoenolpyruvate Carboxykinase

In Memoriam Merton F. Utter

KIYOSHI KURAHASHI

Institute for Protein Research, Osaka University, 3-2, Yamadaoka, Suita, Osaka 565 (Japan)

Introduction

Whenever I read biographical sketches written by eminent scientists reflecting their work done in the past, I think of an older, prominent and usually retired professor. Since I am neither old nor retired, I should not undertake such a task. The sad and untimely death of Dr. Merton F. Utter on November 28, 1980, my first mentor in the United States, to whom I owe a great deal for my later career as a biochemist, shook my resolution. When I received an invitation from the Editor to write a chapter in this volume, I decided to write of the exciting time I spent at Western Reserve University (now called Case Western Reserve University) with Drs. Utter, H.G. Wood and other excellent teachers and stimulating colleagues.

I finished my university education at the end of World War II in Japan. I majored in experimental morphology under the guidance of Prof. Yo. K. Okada in the Department of Zoology at Tokyo Imperial University and entered into the graduate school to study enzymology under Prof. Tokusuke Goda. As the chaos of postwar

Plate 2. Kiyoshi Kurahashi.

Tokyo prevented me from doing much meaningful research, I took a stint as a laboratory technician in a U.S. Army Hospital.

It was my first exposure to American affluence. Loads of graduated pipettes and other glassware were found in drawers and cabinets. In the University we were supplied with only one of each kind of flask and volumetric pipette. In place of the Department's Dubosque colorimeter, I found a Coleman spectrophotometer to run blood-sugar analyses. The hospital library was very useful: there I found new issues of Journal of Biological Chemistry. American scientific journals were precious in those days and only available to the public at the American Cultural Center. Many Japanese scientists queued up and devoured articles. Copying machines had not been invented. I was on duty working in the hospital during the night, but during the daytime at the University I continued my thesis work on the effect of insulin, adrenal cortex extracts and growth hormone (actually extracts of pituitary glands) on glucose utilization by tissue slices and minces. An old Warburg manometer was a treasure in our laboratory and I used it for measurement of Q_{O_2}, $Q_M^{O_2}$, and $Q_M^{N_2}$ with glucose as a substrate. I myself raised rats by feeding only grains and cabbages which were salvaged from my family rations. Quite a few rats developed a kind of sarcoid, of which I still don't know the cause. For preparations of ATP from rabbit muscle we chose a cold winter day and kept all the windows of the laboratory open. A make-shift swinging-bucket refrigerated centrifuge had dry ice all around, which we had to obtain by a 2-h trip to a factory. Due to frequent suspension of the electricity and water supplies, sometimes we prepared distilled water by distillation of well water carried in a bucket. The water in a distillation flask was heated with a Bunsen burner and the vapor in the condenser was cooled by letting the well water run through a cooling coil from a bucket placed up high on a shelf. Curiously we still felt satisfied in that we were accomplishing something, though practically we weren't getting anywhere under these circumstances. Exposure to American journals of biochemistry kindled my desire to go to the United States, and I sent out inquiries and applications to several universities, seeking an opportunity to pursue biochemical researches in an advanced laboratory. Through introduction by

Plate 3. Merton F. Utter and Kiyoshi Kurahashi in 1955.

one of my friends, Richard W. Watts, a U.S. Army doctor and a graduate of Western Reserve University, I was admitted to the Department of Biochemistry of the above University as a graduate student in 1950. It was not easy to go abroad then under the conditions of occupied Japan, but with the help of many well-meaning people, Japanese and American, especially Jeffries Wyman, a professor of Harvard University stationed in Japan and Tokyo College Women's Club, I managed to leave Japan in early 1951.

On arrival at Vancouver, Canada, I was detained for a week in the custody of the captain on board the ship on which I had sailed. I was about to be deported back to Japan because of being a citizen of a hostile country when Utter intervened and I was permitted to proceed to Cleveland under parole.

I arrived in Cleveland in February of 1951 and was assigned to work on oxalacetate (OAA) carboxylase* of pigeon liver. To my great surprise, I learnt that there was quite a controversy about the Wood-Werkman reaction. I describe here briefly the development of the knowledge of CO_2 fixation into OAA before I began work on the problem.

OAA carboxylase in heterotrophic bacteria

In 1935 Wood and Werkman demonstrated for the first time that heterotrophic organisms can utilize CO_2 for synthetic purposes [1,2]. They postulated that succinic acid formed in fermentations by the propionic acid bacteria might be derived from OAA which was formed by the condensation of CO_2 with a C_3-compound, possibly pyruvic acid as shown in Reaction 1 [3].

$$CO_2 + CH_3COCOOH \rightarrow COOHCH_2COCOOH \qquad (1)$$

* A generic name; OAA carboxylase was used here to represent an enzyme(s) which was considered to synthesize OAA from CO_2 and a 3-carbon compound(s) including pyruvate. This name was also used sometimes synonymously with OAA decarboxylase before 1950.

They further observed that phosphate was necessary for optimal CO_2 utilization and succinic acid formation [4,5] and suggested that a phosphate ester, probably phosphoenolpyruvate (PEP) was involved in the reaction, because the high energy content of the phosphorylated compound might provide the source of energy for CO_2 assimilation*.

Supporting evidence for Reaction 1 was presented by Wood et al. [7,8] in 1940 who showed that all the $^{13}CO_2$ fixed into succinic acid during the fermentation of various carbon sources by *Escherichia coli* and *Propionibacterium pentosaceum* was present in the carboxyl group. Simultaneously Carson and Ruben [9] obtained evidence for incorporation of [^{11}C]bicarbonate in propionic and succinic acids during the fermentation of glycerol by propionic acid bacteria. Nishina et al. [10] also showed by the use of $^{11}CO_2$ that malic acid and fumaric acid were synthesized from pyruvate and CO_2 by *B. coli*.

Seemingly more direct evidence for Reaction 1 was presented by Krampitz et al. [11]. They used the CO_2–OAA exchange reaction which became a standard method for the study of OAA carboxylase and found that $^{13}CO_2$ was incorporated into the β-carboxyl of OAA when it was incubated with acetone-treated cells of *Micrococcus lysodeikticus* in the presence of a pool of OAA as shown in Reaction 3.

$$\mathrm{COOHCH_2COCOOH} + {}^{13}\mathrm{CO_2} \xrightleftharpoons{\mathrm{Mg^{2+}\ or\ Mn^{2+}}} {}^{13}\mathrm{COOHCH_2COCOOH} + {}^{12}\mathrm{CO_2} \tag{3}$$

This reaction was thought to be catalyzed by OAA decarboxylase present in the bacteria and to indicate the occurrence of the Wood-Werkman reaction (Reaction 1). However, the decarboxylase

---

* Twenty years later Siu and Wood [6] discovered an enzyme, PEP carboxyphosphotransferase, in *Propionibacteria* which catalyzes a primary reaction of CO_2 assimilation in this organism as shown in Reaction 2.

$$\underset{\text{(PEP)}}{\mathrm{CH_2{=}C(OPO_3H_2)COOH}} + \mathrm{CO_2} + \mathrm{H_3PO_4} \rightleftharpoons \underset{\text{(OAA)}}{\mathrm{COOHCH_2COCOOH}} + \mathrm{H_4P_2O_7} \tag{2}$$

extensively purified by Herbert [12] was proved to carry out Reaction 3 as well as decarboxylation of OAA (Reaction 4), but it could not synthesize OAA from CO_2 and pyruvate (Reaction 1) [12,13].

$$COOHCH_2COCOOH \xrightarrow{Mn^{2+}} CH_3COCOOH + CO_2 \qquad (4)$$

The first demonstration of the synthesis of a small amount of OAA (or a compound closely related to it) from pyruvate and CO_2 by *E. coli* extracts was reported by Kalnitsky and Werkman in 1944 [14]. Kaltenbach and Kalnitsky [15] extended this work and confirmed the above results by using a very high concentration of $^{14}CO_2$, but in 1951 the mechanism still remained obscure.

OAA carboxylase in plants

The first demonstration of CO_2 fixation by plant tissues in a manner similar to that of hetrotrophic organisms was reported by Ruben and Kamen [16] in 1940. They reported that ground barley root can fix $^{11}CO_2$, although the nature of the organic compounds which contained the ^{11}C was not known. In 1947, Gollub and Vennesland [17] reported that OAA decarboxylase from parsley root could carry out a $^{14}CO_2$–OAA exchange reaction similar to that of the enzyme of *M. lysodeikticus* [11]. The first clear-cut demonstration of OAA synthesis by CO_2 fixation in plants was reported by Bandurski and Greiner [18] in 1953 at about the same time as our discovery of ITP-dependent OAA carboxylase (now called PEP carboxykinase) [19]. They discovered an enzyme, PEP carboxylase, in spinach leaves which catalyzes the following reaction.

$$\underset{\text{(PEP)}}{CH_2{=}C(OPO_3H_2)COOH} + CO_2 + H_2O \xrightarrow{Mg^{2+}} \underset{\text{(OAA)}}{COOHCH_2COCOOH} + H_3PO_4 \qquad (5)$$

OAA carboxylase in animal tissues

While Wood and Werkman were extending their work on CO_2 fixation in microorganisms, Evans and Slotin [20] reported in 1940 the incorporation of $^{11}CO_2$ in α-ketoglutarate by pigeon liver minces and suggested that CO_2 combines directly with pyruvic acid to yield OAA. Simultaneously Krebs and Eggleston [21] presented similar but indirect evidence for the synthesis of OAA. In the following year Wood et al. [22,23] and Evans and Slotin [24] showed that α-ketoglutarate formed from $^{13}CO_2$ or $^{11}CO_2$ contained the labeled-carbon in the α-carboxyl group adjacent to the carbonyl group. In the following year Evans et al. [25,26] obtained cell-free extracts which carried out CO_2 fixation into citric-acid-cycle intermediates in the presence of pyruvate and fumarate. The extracts contained also a heat-labile factor which enhanced decarboxylation of OAA in the presence of Mn^{2+} at pH 5.5 and they called the enzyme OAA carboxylase.* Thus, the Wood–Werkman reaction that was originally met with cold skepticism by biochemists [27–29] came to be recognized as a reaction more widely occurring in nature.

An entirely independent supporting evidence for CO_2 fixation in animals came from the study of carbohydrate metabolism by the use of $^{11}CO_2$ by Solomon et al. [30] in 1941. They injected [^{11}C]-bicarbonate into fasted rats after feeding unlabeled lactic acid and found ^{11}C in the deposited liver glycogen. Since the pyruvate kinase reaction had been considered to be irreversible and Kalckar [31] had reported a formation of PEP during oxidation of malate or fumarate by cat or rabbit kidney preparations, Solomon et al. [30] considered that the synthesis of PEP might arise through a dicarboxylic acid cycle according to the Lipmann's postulate[32] as shown in the following diagram (Fig. 5).

Lipmann postulated that PEP would be derived by decarboxylation of phosphooxalacetate, which itself could be formed by dehydrogenation of an adduct of fumarate and phosphate. The hypothetical intermediates, phosphoenoloxalacetate or oxalacetylphos-

---

* Judging from its pH optimum and no requirement for nucleoside polyphosphates, this decarboxylase may have been the same as that intrinsic to malic enzyme as will be described in the following section.

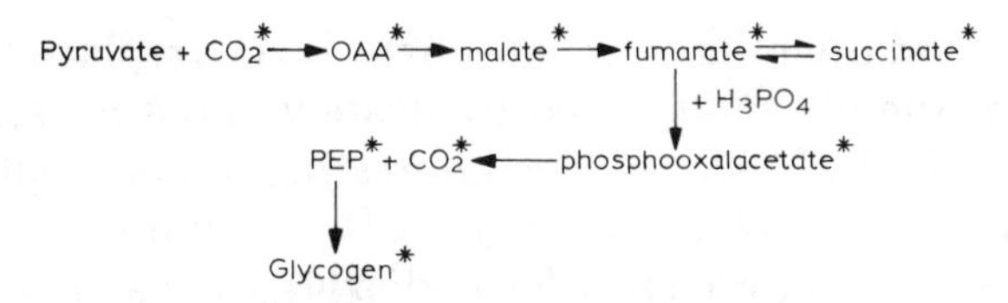

Fig. 5. Pathway of CO_2 incorporation into glycogen postulated by Solomon et al. [30] and Vennesland et al. [33].

phate were later considered possible intermediates in the synthesis of OAA from PEP and CO_2, but the existence of such compounds has never been established [18,34,35]. Leloir and Muñoz[36] confirmed the formation of PEP during oxidation of fumarate by a particulate preparation of guinea pig liver and further suggested that PEP might be converted to a phosphorylated C_4 compound. However, when Lardy and Ziegler [37] presented evidence that the pyruvate kinase reaction was at least slightly reversible, the concept of the formation of PEP from dicarboxylic acids was largely abandoned [38–41].

In 1945, Wood et al [42] tried to demonstrate the CO_2 exchange reaction with OAA using cell-free extracts of pigeon liver, but failed. Shortly afterwards Utter and Wood [43] found that ATP was essential for the exchange reaction.

$$\mathrm{COOHCH_2COCOOH} + {}^{14}\mathrm{CO_2} \xrightleftharpoons{\mathrm{ATP,\ Mn^{2+}}} {}^{14}\mathrm{COOHCH_2COCOOH} + {}^{12}\mathrm{CO_2} \quad (6)$$

They presented the following scheme as a possible explanation for the participation of ATP in the exchange reaction.

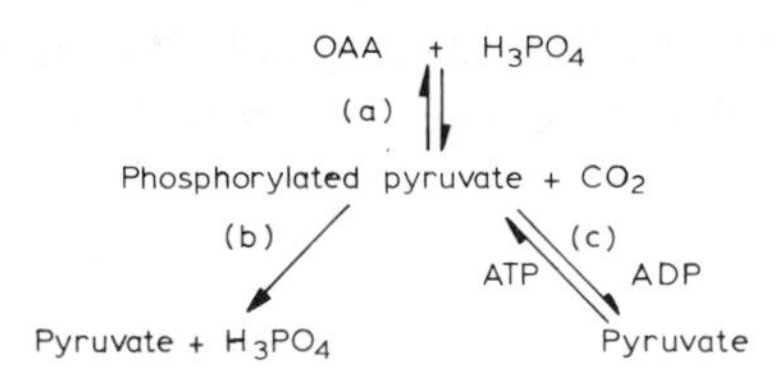

Fig. 6. A mechanism postulated by Utter and Wood [43] for fixation of CO_2 into OAA by pigeon liver OAA carboxylase.

Since PEP and thiamine pyrophosphate did not replace ATP, they stated that the phosphorylated pyruvate was not necessarily the well-known PEP. However, it is interesting to note that the synthetic reaction of OAA represented by Reaction (a) in Fig. 6 is the same as the one catalyzed by PEP carboxylase (Reaction 5) discovered by Bandurski and Greiner [18] and that, if one moves the point of the participation of ATP and ADP one step up, it represents exactly the reaction of PEP carboxykinase which Utter and I discovered in 1953 [19] as shown in Reaction 7, though the active nucleotides turned out to be inosinic and guanylic polyphosphates in later experiments.

$$CH_2{=}C(OPO_3H_2)COOH + CO_2 + ADP\ (IDP) \xrightleftharpoons{Mn^{2+}} COOHCH_2COCOOH + ATP\ (ITP) \quad (7)$$

Malic enzyme

While the mechanism of CO_2 fixation in pigeon liver had remained still rather obscure, in 1947 Ochoa et al. [44,45] reported the presence of an enzyme in pigeon liver extracts which catalyzes the reversible oxidative decarboxylation of L-malic acid to pyruvic acid and CO_2 and they purified the enzyme.

$$COOHCH_2CHOHCOOH + NADP^+ \xrightleftharpoons{Mn^{2+}} CH_2COCOOH + CO_2 + NADPH + H^+ \quad (8)$$

Although Reaction 8 is readily reversible, its equilibrium position was reported to be far to the right.* However, the net synthesis of malate from pyruvate and CO_2 was demonstrated easily when the reaction was shifted in the direction of CO_2 fixation by coupling with the Glc-6-P dehydrogenase system as shown in Reactions 9, 10 and 11 [45,47].

* Later it was found that the equilibrium of the malic enzyme reaction is actually toward the synthesis of malate by Harary et al. [46].

$$\text{Glc-6-P} + \text{NADP}^+ \rightarrow \text{6-P-gluconate} + \text{NADPH} + \text{H}^+ \quad (9)$$

$$\text{pyruvate} + CO_2 + \text{NADPH} + \text{H}^+ \xrightarrow{Mn^{2+}} \text{L-malate} + \text{NADP}^+ \quad (10)$$

$$\text{Sum: Glc-6-P} + \text{pyruvate} + CO_2 \xrightarrow{\text{NADP}, Mn^{2+}} \text{6-P-gluconate} + \text{L-malate} \quad (11)$$

In addition, this purified enzyme was shown to decarboxylate OAA at pH 4.5 in the presence of Mn^{2+} and the activity was enhanced by a catalytic amount of NADP [45,48].

$$COOHCH_2COCOOH \xrightarrow{Mn^{2+}, \text{NADP}} CO_2 + CH_3COCOOH \quad (12)$$

This intrinsic property of malic enzyme to decarboxylate OAA also introduced confusion into the picture of an apparently simple reaction of OAA carboxylase as will be described in the following section.

The relationship of OAA carboxylase and malic enzyme in pigeon liver

The findings that two reactions for the synthesis of dicarboxylic acids are catalyzed by OAA carboxylase and the malic enzyme present in pigeon liver raised the question whether one of the two reactions is a primary fixation reaction of CO_2 to form a dicarboxylic acid or whether both occur simultaneously. Vennesland et al. [49] tried first to answer this question. They found that NADP but not ATP stimulates decarboxylation of OAA and that ATP but not NADP enhances the exchange reaction between $^{14}CO_2$ and the β-carboxyl carbon atom of OAA and suggested that the two reactions catalyzed by OAA carboxylase and malic enzyme might be independent.

Utter continued his studies on OAA carboxylase using labeled CO_2 and carboxyl-labeled pyruvate and reported that the fixation of pyruvate into OAA (Reaction 13) was not correlated with the fixation of CO_2 (Reaction 6) [50,51].

$$CH_3CO^{14}COOH + COOHCH_2COCOOH \xrightleftharpoons{Mn^{2+}} CH_3CO^{12}COOH + COOHCH_2CO^{14}COOH \quad (13)$$

He further established that malate is not a precursor of OAA in CO_2 fixation catalyzed by ATP and likewise that OAA is not a precursor of malate in the NADP-catalyzed fixation [52]. On the basis of the constancy of the total fixed radioactive CO_2 in OAA and malate irrespective of the relative concentration of ATP and NADP, Utter postulated that the primary step in the two CO_2 fixation reactions leads to a precursor that may either react with NADP to form malate or with ATP to form OAA as shown in Fig. 7.

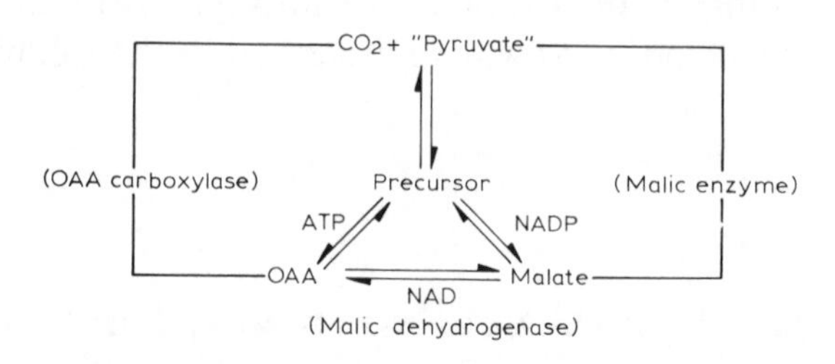

Fig. 7. Possible interrelationship between OAA carboxylase and malic enzyme catalyzed reactions as postulated by Utter [52].

Utter observed also inhomogeneity of the product, labeled OAA, that suggested that the precursor in Fig. 7 might be a labile derivative of OAA.

On the other hand, using purified malic enzyme, Ochoa and his associates presented evidence that at neutral pH malic enzyme is neither an OAA carboxylase nor a malic dehydrogenase, and free OAA cannot be an intermediate in the reversible oxidative decarboxylation of malate [48]. They [45] and Herbert [12] also showed that malic enzyme cannot be replaced by a mixture of malic dehydrogenase and OAA carboxylase of *M. lysodeikticus,* either with NAD or NADP. Neither was the synthesis of malate achieved by a mixture of the enzymes with NADH, pyruvate and CO_2 as initial reactants, whether in the absence or presence of ATP [45]. Since the experiments on fixation of CO_2 in OAA had so far been performed only with slightly purified pigeon liver preparations, Ochoa [45] pointed out that it would be desirable to carry out such experiments with purified OAA carboxylase. However, in 1950 Veiga Salles et al. [53] demonstrated that CO_2 fixation into a pool of OAA could be accomplished by a coupled reaction of malic enzyme and malic dehydrogenase in the presence of a catalytic amount of malate as shown in Reactions 14 and 15.

$$\underset{\text{(L-malate)}}{C^{*}OOHCH_2CHOHCOOH} + NADP^{+} \underset{\text{(malic enzyme)}}{\overset{Mn^{2+}}{\rightleftharpoons}} \underset{\text{(pyruvate)}}{CH_3COCOOH} + C^{*}O_2 + NADPH + H^{+} \quad (14)$$

$$\underset{\text{(OAA)}}{C^{*}OOHCH_2COCOOH} + NADPH + H^{+} \underset{\text{(malic dehydrogenase)}}{\rightleftharpoons} \underset{\text{(L-malate)}}{C^{*}OOHCH_2CHOHCOOH} + NADP^{+} \quad (15)$$

$$\text{Sum: L-malate} + \text{OAA} \underset{\text{(dismutation)}}{\overset{Mn^{2+}\text{, NADP}}{\rightleftharpoons}} \text{pyruvate} + CO_2 + \text{L-malate} \quad (16)$$

Labeled CO_2 would first be incorporated into malate (Reaction 14) which would be oxidized to OAA by Reaction 15. It was shown that malic dehydrogenase can react with NADP. They suggested that the net overall reaction would be a reversible decarboxylation of OAA (Reaction 17), that is, the Wood and Werkman reaction.

$$\text{OAA} \underset{\text{(malic enzyme + malic dehydrogenase)}}{\overset{Mn^{2+}\text{, NADP, L-malate}}{\rightleftharpoons}} \text{pyruvate} + CO_2 \quad (17)$$

Thus, the reaction catalyzed by malic enzyme was considered to be the primary CO_2 fixation reaction and the role of OAA carboxylase in the synthesis of dicarboxylic acids became questionable again. At one time or another this reaction was called the ‘wouldn’t work’ reaction [54,55] and the above observation of Utter, the ‘Utter confusion’ [55]. Utter himself stated in his talk at the Brookhaven Symposia in Biology in 1950 as follows [56]:

“I find myself in the somewhat uncomfortable position of making complex an apparently simple reaction which was accepted almost ten years ago as understood and as such has been incorporated into numerous reviews and text books. I refer to the so-called 'Wood and Werkman' reaction, first hypothesized by these authors [3] in 1938 as shown below: OAA $\rightleftharpoons CO_2$ + pyruvate."

Although evidence had been accumulating as described above for the CO_2 fixation into OAA by pigeon liver extracts [13,42,43,49–52] *M. lysodeikticus* [11,12,51,57] and *E. coli* [14,15], because the protein entity responsible for the reaction had not been isolated there remained ambiguity concerning the occurrence of the reaction in nature. Even as late as in 1952, Ochoa and Stern [58] stated in the *Annual Review of Biochemistry* as follows.

"In the opinion of the reviewers there is not sufficient evidence at present to assume the existence of two separate pathways, via malate and OAA, respectively, for the synthesis of dicarboxylic acids from pyruvate and CO_2, since only the 'malic' enzyme pathway effects a net synthesis of free dicarboxylic acids."

In the popular text book by Fruton and Simmonds [59] published in 1953 we find a statement that

"although the reversal of the enzymatic decarboxylation by the 'Wood-Werkman' reaction was considered for a time to be a major metabolic pathway for the fixation of CO_2, more recent studies by Ochoa suggest that a somewhat different mechanism for the formation of OAA from pyruvic acid and CO_2 may be involved."

In a way they were correct, because our discovery of PEP carboxykinase [19] did not yet solve the problem of OAA synthesis by the Wood-Werkman reaction, because the substrate for the synthesis of OAA was PEP but not pyruvate and though the reaction was freely reversible, it was assigned the role of replenishing PEP from tricarboxylic acid cycle metabolites [34,40,60], which are formed by malic enzyme. The elucidation of the enzymatic mechanism of the Wood–Werkman reaction had to wait another 8 years until Utter and Keech [61,62] discovered CO_2 fixation into OAA by pyruvate carboxylase requiring ATP and a catalytic amount of acetyl-CoA (Reaction 18).

$$\text{pyruvate} + HCO_3^- + \text{ATP} \xrightleftharpoons{Me^{2+},\ \text{acetyl-CoA}} \text{OAA} + \text{ADP} + P_i \qquad (18)$$

At approximately the same time Woronick and Johnson [63]

reported a similar reaction but without requirement for acetyl-CoA in *Aspergillus*. In retrospect, the above complex scheme (Fig. 7) presented in 1950 by Utter for CO_2 fixation in dicarboxylic acids in pigeon liver was inevitable, because he tried to explain three independent reactions, now known to be catalyzed by PEP carboxykinase, pyruvate carboxylase and malic enzyme, as a unified interrelated reaction. Since then four enzymes which catalyze CO_2 fixation in OAA have been discovered. They are PEP carboxykinase (Reaction 7), PEP carboxylase (Reaction 5), PEP carboxytransphosphorylase (Reaction 2) and pyruvate carboxylase (Reaction 18) [64,65].

The discovery of PEP carboxykinase

Upon my arrival at Western Reserve University, Utter explained to me the above hypothetical scheme (Fig. 7) reconciling the two reactions carried out by OAA carboxylase and malic enzyme, but in spite of my learning of practical English in Japan from association with many English speaking people, it was still beyond my capability to understand the complicated roundabout concept in native English, especially Utter's rapid conversation. What I conceived was that there were two reactions for CO_2 fixation into dicarboxylic acids. Certainly I myself had never doubted the presence of OAA carboxylase, because I naively believed that a reaction named after the discoverer must have been a well-established concrete fact. Then, the problem seemed to be reduced to a simple one that the two enzymes should be purified and separated. Before I left Japan, I had experience in the isolation and purification of ATP and some enzymes under the very primitive conditions I have described. So with a cold room and all sorts of modern equipment, like Beckman spectrophotometers and a Spinco Model L preparative ultracentrifuge, such work seemed to me very agreeable. An engineer from Spinco stood aghast on hearing that we were using it for spinning down ammonium sulfate precipitates. An ordinary high-speed centrifuge was not available to handle a large volume of materials.

It had already been shown by Utter and Wood that not only

pigeon liver but also rat and chicken liver possessed the ATP-dependent activity of CO_2 exchange with OAA (Reaction 6) [43]. After a few trials with pigeon liver, we decided to use chicken liver, since a large amount of acetone powders were needed for purification. We checked the activity of OAA carboxylase in liver of different kind of chickens and found the Barred Rock a satisfactory source. We also developed a simplified quick method for measurement of $^{14}CO_2$ incorporation into a pool of OAA. In contrast to the generally accepted concept, we found that OAA was not so very labile and we just plated an aliquot of the acidified reaction mixture on a paper disk, dried it under vacuum in cold and counted the radioactivity on the disk. This method facilitated our purification of OAA carboxylase.

I started to purify the enzyme by heat inactivation, ammonium sulfate precipitation and fractionation with calcium phosphate gel. First I concentrated upon the assay of CO_2 exchange with OAA, but in August of 1951 and after 15–20-fold purification was achieved I also began to check malic enzyme activity in all fractions. Utter was still very much concerned with the pyruvate exchange with OAA (Reaction 13) and insisted that we check that activity. We soon found that all three enzyme activities (CO_2 exchange with OAA, pyruvate exchange with OAA and malic enzyme activities) were separated into different fractions. In the progress report to the U.S. Atomic Energy Commission submitted in September 1951, Utter stated that enzyme or enzymes responsible for the fixation of CO_2 might be separated from the enzyme which incorporated pyruvate into OAA. The combination of the two fractions enabled both pyruvate and CO_2 to be fixed into a pool of OAA. A hypothetical mechanism embodying the known facts at that time is shown below, the bracketed substances indicating hypothetical intermediates of unknown nature (see Fig. 8).

ATP was needed for the fixation of CO_2 but not of pyruvate. The results were interpreted to be explained by assuming that activated [CO_2] and [pyruvate] were the substances actually undergoing condensation. In contrast to Fig. 7 the dotted line indicated the uncertainty whether or not malate was formed from a common intermediate specified as [OAA].

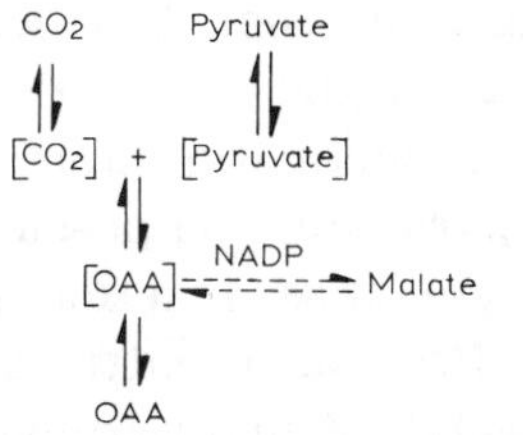

CO_2 Pyruvate

$[CO_2]$ + [Pyruvate]

NADP

[OAA] Malate

OAA

Fig. 8. A hypothetical mechanism of incorporation of CO_2 and pyruvate into OAA as postulated in September of 1951.

Since I had not participated in the research on OAA carboxylation and had not encountered the complicated and puzzling results such as inhomogeneity of the OAA formed during the CO_2 fixation reaction [52] or the lack of correlation between the fixation of pyruvate and that of CO_2 in OAA [50,51], the above described separation of OAA carboxylase and malic enzyme into two fractions free of each other seemed very straightforward. I simply took it as an indication that there were two separate enzyme entities responsible for CO_2 fixation into malate and OAA, dependent on NADP and ATP, respectively, which act independently. Once these two enzymes were separated, to solve the problem of OAA synthesis seemed easy. We needed only to find an actual CO_2 acceptor in the synthesis of OAA. As reviewed above, Utter considered that the acceptor of CO_2 for the synthesis of OAA was not pyruvate itself.

Up to this point we had used only the CO_2-OAA exchange reaction as an assay of the enzyme, but when we separated the OAA carboxylase and malic enzyme, we began to check their activities for decarboxylation of OAA, since malic enzyme had OAA decarboxylase activity that was stimulated by a catalytic amount of NADP and had a pH optimum near 4.5 (Reaction 12) [45,48]. The partially purified OAA carboxylase devoid of malic enzyme activity still had OAA decarboxylase activity in a pH range of 5.8–7.8 which is dependent on ATP [19], indicating that liver contains at least two different OAA decarboxylases. At this point we studied the fate of ATP during decarboxylation of OAA. What we found was that no inorganic phosphate was formed during the reaction and that the ratio of moles of OAA decarboxylated to the moles of ATP added was as high as 5, suggesting that ATP was acting catalytically, which was shown not to be the case previously [51,56], or it was being regenerated. Since the final product of decarboxylation was still considered to be pyruvate, one possible way of regenerating ATP was via the pyruvate kinase reaction, PEP being an intermediate of the reaction. Previously, Utter and Chenoweth [50,51] tried to exchange carboxyl-labeled pyruvate with OAA (Reaction 13). They found that the fixation of pyruvate in OAA occurred with certain enzyme preparations but was not correlated with the fixation of CO_2 and that ATP had little effect.* In retrospect, the strange part of all these experiments is that no attempts were made to use pyruvate as an acceptor of CO_2 with omission of a pool of OAA. PEP was also tested off and on by Utter and Wood [43] and Utter [13], but again it was tested only in the CO_2–OAA exchange reaction to see whether or not it could replace ATP. Actually Utter observed that PEP plus AMP could replace ATP [13], but it was incorrectly interpreted as evidence that PEP was not an intermediate in the carboxylase reaction, but only acting as a phosphate donor to form ATP from AMP through the myokinase and pyruvate kinase reactions.

With purified OAA carboxylase on hand, we attempted to determine whether PEP could act as a CO_2 acceptor in the absence of OAA. We happened to have a tiny amount of the silver salt of PEP, a kind gift of Dr. E. Baer. In a series of experiments for CO_2-OAA exchange, we included two flasks in which OAA was replaced with PEP or pyruvate. It was quite exciting to see some counts were fixed in the reaction with PEP as a substrate. The amount incorpo-

* The pyruvate carboxylase reaction (Reaction 18) was formulated to consist of two reactions as follows [66, cf.65].

$$\text{E-biotin} + \text{ATP} + HCO_3^- \xrightleftharpoons{Me^{2+},\ \text{acetyl-CoA}} \text{E-biotin}{\sim}CO_2 + \text{ADP} + P_i \quad (19)$$

$$\text{E-biotin}{\sim}CO_2 + \text{pyruvate} \rightleftharpoons \text{E-biotin} + \text{OAA} \quad (20)$$

According to this scheme, pyruvate exchange with OAA does not require ATP. The ATP-independent pyruvate exchange reaction with OAA [50,51] that Utter pursued persistently while working on OAA carboxylase may well have been Reaction 20.

rated was only a few counts above the background, but it was consistent in the triplicates with increasing amounts of the enzyme. In retrospect, our experiments were very wild. We were putting in 800 000 counts of bicarbonate in a Warburg flask. The efficiency of the counting in those days was only 10%. After the reaction we just added acid to get rid of the unreacted $^{14}CO_2$ into the air. Nevertheless the incorporation of a few counts above the background out of 800 000 counts opened the way for a fruitful investigation of PEP carboxykinase. Improving the conditions by replacing ATP with ADP and further with ITP coupled with the hexokinase system to generate IDP, we soon synthesized a sizable quantity of OAA from PEP and CO_2.

In the progress report submitted to the Atomic Energy Commission in August of 1952, the following data in Table I were included. Even with this result Utter did not fail to include a sentence that although PEP apparently gave rise to OAA synthesis, it was inadvisable to state that phosphopyruvate was a direct reactant on the basis of the present evidence, showing his extreme prudence.

Another important finding on OAA carboxylase made at about the same time was the discovery of ITP as an active cofactor in addition to ATP. Earlier Utter and Wood tried to replace ATP by other cofactors, such as NADP, NAD, pyridoxal phosphate, biotin, thiamine pyrophosphate, AMP and PEP [13,51]. Only a combination of PEP and AMP were found active, but all other compounds were unable to replace ATP as described above. We tested again

TABLE I

Net synthesis of OAA by chicken liver OAA carboxylase

Reactants added	OAA synthesized μmol
PEP, ADP	0.060
PEP	0.004
Pyruvate, ATP, ADP	0.011

Incubation time, 5 min. Starting materials: 50 μmol of $NaH^{14}CO_3$, 4 μmol of PEP, 2 μmol of ADP, 40 μmol of pyruvate, and 2 μmol of ATP in a 1-ml volume.

phosphate compounds including ITP and IMP which became available commercially, using purified OAA carboxylase. To our great surprise, ITP was found to be three to four times more active than ATP [19,67]. Our work on the reversible carboxylation of PEP to OAA dependent on ADP or IDP catalyzed by a new enzyme PEP carboxykinase (Reaction 7) was more or less completed by April of 1953, in 2 years after I entered the graduate school, and the full accounts of our work were submitted in September of 1953 to the *Journal of Biological Chemistry* and published in a series of papers in 1954 [34,67,68].

One significant aspect of our findings was that they shed light on the earlier work of Kalckar [31] that PEP accumulated during the oxidation of malate by a kidney preparation. The PEP carboxykinase reaction coupled with the malic enzyme and malic dehydrogenase systems was considered as a solution to overcome the energy barrier of the pyruvate kinase reaction [34,40], to replenish PEP in gluconeogenesis, but the actual source of OAA turned out not to be malic enzyme but pyruvate carboxylase found by Utter and Keech [61,62] in chicken liver in 1960.

The studies described above had been carried out before my prethesis committee was organized and could not be credited as a partial fulfilment of the requirements for the degree of Doctor of Philosophy. I started to study the nucleotide specificity of PEP carboxykinase and the distribution of the enzyme in various tissues and organisms to fulfil the requirements.

The finding that ITP was more active than ATP in the PEP carboxykinase reaction was the first case where nucleoside triphosphates other than ATP were shown to act as better phosphate donors than ATP in transphosphorylation reactions. The only system where ITP had been reported to act as a better substrate than ATP was the dephosphorylation reaction by myosin studied by Kleinzeller [69]. Further purification of PEP carboxykinase from chicken liver mitochondria to free it of nucleoside diphosphokinase and of commercial ATP to remove contaminating nucleotides revealed that ATP was not active at all in the carboxykinase system [70,71]. Since the occurrence of ITP or IDP in nature had never been demonstrated except in frog muscle [72], this finding raised the following questions: (1) how is ITP or IDP synthesized; (2) is ITP or IDP in all tissues which contain PEP carboxykinase; (3) are there other cofactors than ITP or IDP in nature for the enzyme? However, in December 1954 I received a sample of GDP from Dr. J.L. Strominger, and it was found more active than IDP in the PEP carboxykinase reaction [70,71]. Sanadi et al. [73] reported in 1954 that GDP was the phosphate acceptor in the substrate-level phosphorylation connected with α-ketoglutarate oxidation and other investigators reported the natural occurrence of GTP in liver and muscle [74,75]. We also found the active principle for the PEP carboxykinase reaction present in commercial ATP was GDP. These findings lessened the necessity for the search for the synthetic pathway or the natural occurrence of inosinic polyphosphates, though Webster [76] reported that a preparation of rabbit myofibril deaminated ADP to form IDP and I confirmed his results by converting ^{14}C-labeled ATP to IDP via ADP (unpublished results).

A survey of the distribution of PEP carboxykinase indicated that gluconeogenic tissues, such as liver and kidney contained this enzyme [77]. On the other hand, all other animal tissues, bacteria and plants tested failed to show any indication of the presence of the enzyme dependent on ITP/IDP. However, I found that baker's yeast catalyzed an appreciable exchange of CO_2 with OAA in the presence of ATP, but not with ITP as shown in Table II. This experiment was done just a few days before my departure to Bethesda in 1955 to work at the National Institutes of Health and was never published. Later studies by others proved that PEP car-

TABLE II

CO_2 exchange reaction with OAA by baker's yeast extracts

Nucleotides added μmol	CO_2 fixed μmol $\times$ 10^2/min/mg protein
None	0.04
ITP, 1	0.30
ATP, 1	3.3

OAA, 0.06 M; $NaH^{14}CO_3$, 0.05 M, 8 $\times$ 10^5 cpm. Total volume, 1 ml.

boxykinase of microbial sources utilizes adenylic polyphosphates rather than inosinic or guanylic polyphosphates [78, cf.64].

Appendix

After I finished my graduate studies at Western Reserve University, I received a postdoctoral fellowship of the Jane Coffin Childs Memorial Fund and moved to the National Institute of Arthritis and Metabolic Diseases, the National Institutes of Health, to work with Dr. Herman M. Kalckar, my second wise and inspiring mentor in the United States. I was again very fortunate to be a member of the Laboratory of Biochemistry and Metabolism headed by Dr. Bernard L. Horecker. The laboratory and the seminar group were full of very active and enthusiastic researchers, carrying on Arthur Kornberg's tradition.

At first I started to work on galactosemia with Kalckar but soon switched to study galactose metabolism in *E. coli* in collaboration with Joshua and Esther Lederberg. They had been accumulating many mutants of lactose and galactose nonfermenters of *E. coli*. We checked the pathway for the dissimilation of galactose in *E. coli* and found that it is the same as in yeasts and animals as described by Leloir [79]. By analyzing the enzymes of galactose metabolism in Lederberg's mutants grown in the presence of galactose, we found that they were classified into galactokinase-less, galactose-1-phosphate uridylyltransferase-less, and triple mutants lacking the above two enzymes and UDPgalactose-4-epimerase [80–82]. The galactokinase-less mutants had still the activities of galactose-1-phosphate uridylyltransferase and UDPgalactose-4-epimerase. This finding was contrary to the prevailing concept of sequential induction of inducible enzymes. When I reported this at an informal meeting on enzyme induction held at New York University in 1957, J. Monod was quite skeptical of our results. H. de Robichon-Szulmajster* who worked in our group obtained a similar result in

* She died untimely in 1974. She was a really cherished colleague with outstanding capability. I still recollect our collaboration with deep emotion.

yeasts [81,83] and it was established that inducible enzymes are not necessarily induced sequentially * [81–85].

After Kalckar left the National Institutes of Health, I worked in the section headed by Dr. Osamu Hayaishi and then by Dr. Simon Black in the same institute. There in 1957 I initiated a research on the biosynthesis of gramicidin J which later turned out to be the same as gramicidin S [87], an antibiotic oligopeptide produced by *Bacillus brevis*. At that time the mechanism of protein biosynthesis was not yet clear. The involvement of amino acids activating enzymes, soluble RNAs, and particulate fractions had been known, but messenger RNAs were not yet discovered. The difficulties of the studies of protein biosynthesis lay in that the actual structure of the protein being synthesized was not well defined. The measure of protein biosynthesis relied only on the incorporation of labeled amino acids into hot trichloroacetic acid precipitable materials. The biosynthesis of gramicidin S was undertaken in order to see whether there was any correlation between the mechanism of the biosynthesis of oligopeptides and that of protein. In other words, does the biosynthesis of gramicidin S follow the pattern of stepwise formation of peptides as is the case of biosynthesis of glutathione or does it follow the template mechanism through activated amino acids, as had been proposed for protein synthesis?

In late 1960 I received a call from Prof. Shiro Akabori, Director of the newly established Institute for Protein Research, and returned to Japan after a 10 years' stay in the United States. I continued the work on the enzymes of galactose metabolism and the biosynthesis of gramicidin S and tyrocidines. The biosynthetic mechanism of antibiotic oligopeptides turned out to be different from that of protein biosynthesis. They are synthesized by the multienzyme thiotemplate mechanism using multienzyme proteins as templates [cf. 88,89]. However, we found recently that a larger antibiotic peptide produced by *Bacillus subtilis* ATCC6633, subtilin, is synthesized by the usual protein synthetic mechanism and is processed through several intermediates [90]. There are two different mechanisms for

* It is naturally to be expected in the light of the operon concept of inducible enzymes advanced by Jacob and Monod in 1961 [86].

the syntheses of antibiotic peptides produced by Bacilli [91,92]. Though I wrote at the beginning I am still young, I shall soon be faced with mandatory retirement from this institute. I am hoping that I can solve this interesting and complicated processing mechanism of subtilin before then.

Acknowledgements

During the course of the studies on PEP carboxykinase Irwin A. Rose and Ronald J. Pennington participated in various phases of the work and we owe thanks to Dolores Zaparzony, Helen Gordon and Audrey Deal for technical assistance. I shoud also like to express my deep appreciation to many colleagues and collaborators in my other research activities for their help. To my deep regret a list of their names could not be appended because of limitations of space.

I am very grateful to Drs. Harland G. Wood, Richard W. Hanson and Michiyuki Yamada for giving me helpful advice in the preparation of this manuscript.

REFERENCES

1 H.G. Wood and C.H. Werkman, J. Bacteriol., 30 (1935) 332.
2 H.G. Wood and C.H. Werkman, Biochem. J., 30 (1936) 48-53.
3 H.G. Wood and C.H. Werkman, Biochem. J., 32 (1938) 1262-1271.
4 H.G. Wood and C.H. Werkman, Biochem. J., 34 (1940) 7-14.
5 H.G. Wood and C.H. Werkman, Biochem. J., 34 (1940) 129-138.
6 P.M.L. Siu and H.G. Wood, J. Biol. Chem., 237 (1962) 3044-3051.
7 H.G. Wood, C.H. Werkman, A. Hemingway, and A.O. Nier, J. Biol. Chem., 135 (1940) 789-790.
8 H.G. Wood, C.H. Werkman, A. Hemingway and A.O. Nier, J. Biol. Chem., 139 (1941) 365-376.
9 S.F. Carson and S. Ruben, Proc. Natl. Acad. Sci. USA, 26 (1940) 422-426.
10 Y. Nishina, S. Ando and H. Nakayama, Sci. Papers Inst. Phys. Chem. Res. Tokyo, 38 (1941) 341-346.
11 L.O. Krampitz, H.G. Wood and C.H. Werkman, J. Biol. Chem., 147 (1943) 243-253.
12 D. Herbert, Symp. Soc. Exp. Biol., 5 (1951) 52-71.
13 M.F. Utter, in W.D. McElroy and B. Glass (Eds.), Phosphorus Metabolism, Vol. 1, Johns Hopkins Press, Baltimore, 1951, pp. 646-656.
14 O. Kalnitsky and C.H. Werkman, Arch. Biochem. Biophys., 4 (1944) 25-40.
15 J.P. Kaltenbach and G. Kalnitsky, J. Biol. Chem., 192 (1951) 629-639.
16 S. Ruben and M.D. Kamen, Proc. Natl. Acad. Sci. USA, 26 (1940) 418-422.
17 N.C. Gollub and B. Vennesland, J. Biol. Chem., 169 (1947) 233-234.
18 R.S. Bandurski and C.H. Greiner, J. Biol. Chem., 204 (1953) 781-786.
19 M.F. Utter and K. Kurahashi, J. Am. Chem. Soc., 75 (1953) 758.
20 N.A. Evans Jr. and L. Slotin, J. Biol. Chem., 136 (1940) 301-302.
21 H.A. Krebs and L.V. Eggleston, Biochem. J., 34 (1940) 1383-1395.
22 H.G. Wood, C.H. Werkman, A. Hemingway and A.O. Nier, J. Biol. Chem., 139 (1941) 483-484.
23 H.G. Wood, C.H. Werkman, A. Hemingway and A.O. Nier, J. Biol. Chem., 142 (1942) 31-45.
24 E.A. Evans Jr. and L. Slotin, J. Biol. Chem., 141 (1941) 439-450.
25 E.A. Evans Jr., L. Slotin and B. Vennesland, J. Biol. Chem., 143 (1942) 565.
26 E.A. Evans, Jr., B. Vennesland and L. Slotin, J. Biol. Chem., 147 (1943) 771-784.
27 H.G. Wood in: J.F. Woessner Jr. and F. Huijing, (Eds.), Miami Winter Symposia, The Molecular Basis of Biological Transport, Vol. 3, Academic Press, New York, 1972, pp. 1-54.
28 Letter of H.G. Wood to J. Edsall and H.A. Krebs, Mol. Cell. Biochem., 5 (1974) 91-94.
29 H.G. Wood in: G. Semenza (Ed.), Of Oxygen, Fuels, and Living Matter, Part 2, Wiley, New York, 1982, pp. 173-200.
30 A.K. Solomon, B. Vennesland, F.W. Klemperer, J.M. Buchanan and A.B. Hastings, J. Biol. Chem., 140 (1941) 171-182.

31 H.M. Kalckar, Biochem. J., 33 (1939) 631-641.
32 F. Lipmann, Adv. Enzymol., 1 (1941) 99-162.
33 B. Vennesland, A.K. Solomon, J.M. Buchanan, R.D. Cramer and A.B. Hastings, J. Biol. Chem., 142 (1942) 371-377.
34 M.F. Utter and K. Kurahashi, J. Biol. Chem., 207 (1954) 821-841.
35 T.T. Tchen, F.A. Loewus and B. Vennesland, J. Biol. Chem., 213 (1955) 547-555.
36 L.F. Leloir and J.M. Muñoz, J. Biol. Chem., 153 (1944) 53-60.
37 H.A. Lardy and J.A. Ziegler, J. Biol. Chem., 159 (1945) 343-351.
38 Y.J. Topper and A.B. Hastings, J. Biol. Chem., 175 (1949) 1255-1264.
39 H.A. Krebs, Symp. Soc. Exp. Biol., 5 (1951) 1-8.
40 H.A. Krebs, Bull. Johns Hopkins Hosp., 95 (1954) 19-33.
41 J.S. Fruton and S. Simmonds, General Biochemistry, Wiley, New York, 1953, 457-458.
42 H.G. Wood, B. Vennesland and E.A. Evans, Jr., J. Biol. Chem., 159 (1945) 153-158.
43 M.F. Utter and H.G. Wood, J. Biol. Chem., 164 (1946) 455-476.
44 S. Ochoa, A. Mehler and A. Kornberg, J. Biol. Chem., 167 (1947) 871-872.
45 S. Ochoa, A.H. Mehler and A. Kornberg, J. Biol. Chem., 174 (1948) 979-1000.
46 I. Harary, S.R. Korey and S. Ochoa, J. Biol. Chem., 203 (1953) 595-604.
47 S. Ochoa, J.B. Veiga Salles and P.J. Ortiz, J. Biol. Chem., 187 (1950) 863-874.
48 J.B. Veiga Salles and S. Ochoa, J. Biol. Chem., 187 (1950) 849-861.
49 B. Vennesland, E.A. Evans Jr. and K.I. Altman, J. Biol. Chem., 171 (1947) 675-686.
50 M.F. Utter and M.T. Chenoweth, Fed. Proc., 8 (1949) 261-262.
51 M.F. Utter and H.G. Wood, Adv. Enzymol., 12 (1951), 41-151.
52 M.F. Utter, J. Biol. Chem., 188 (1951) 847-863.
53 J.B. Veiga Salles, I. Harary, R.F. Banfi and S. Ochoa, Nature, 165 (1950) 675-676.
54 H.A. Krebs, Mol. Cell. Biochem., 5 (1974) 79-82.
55 H.G. Wood, TIBS, 6 (1981) 1-2.
56 M.F. Utter, Brookhaven Symp. Biol., 3 (1950) 37-55.
57 I.R. McManus, J. Biol. Chem., 188 (1951) 729-740.
58 S. Ochoa and J.R. Stern, Annu. Rev. Biochem., 21 (1952) 547-602.
59 J.S. Fruton and S. Simmonds, General Biochemistry, Wiley, New York, 1953, 442-443.
60 M.F. Utter, Ann. N.Y. Acad. Sci., 72 (1959) 451-461.
61 M.F. Utter and D.B. Keech, J. Biol. Chem., 235 (1960) PC 17-18.
62 M.F. Utter, D.B. Keech and M.C. Scrutton, Adv. Enzyme Reg., 2 (1964) 49-68.
63 C.L. Woronick and M.J. Johnson, J. Biol. Chem., 235 (1960) 9-15.

64 M.F. Utter and H.M. Kolenbrander, in: P.D. Boyer, (Ed.) Enzymes, Vol. 6, Academic Press, New York, 1972, pp. 117-168.
65 M.C. Scrutton and M.R. Young, in: P.D. Boyer (Ed.), Enzymes, Vol. 6, Academic Press, New York, 1972, pp. 1-35.
66 M.C. Scrutton, D.B. Keech and M.F. Utter, J. Biol. Chem., 240 (1965) 574-581.
67 M.F. Utter, K. Kurahashi and I.A. Rose, J. Biol. Chem., 207 (1954) 803-819.
68 M.F. Utter and K. Kurahashi, J. Biol. Chem., 207 (1954) 787-802.
69 A. Kleinzeller, Biochem. J., 36 (1942) 729-736.
70 K. Kurahashi and M.F. Utter, Fed. Proc., 14 (1955) 240.
71 K. Kurahashi, R.J. Pennington and M.F. Utter, J. Biol. Chem., 226 (1957) 1059-1075.
72 K. Lohmann, Biochem. Z., 254 (1932) 381-397.
73 D.R. Sanadi, D.M. Gibson and P. Ayengar, Biochim. Biophys. Acta, 14 (1954) 434-436.
74 H. Schmitz, R.B. Hurlbert and V.R. Potter, J. Biol. Chem., 209 (1954) 41-54.
75 R. Bergkvist and A. Deutsch, Acta Chem. Scand., 8 (1954) 1889-1897.
76 H.L. Webster, Nature, 172 (1953) 453-454.
77 K. Kurahashi, Oxalacetate carboxylase, Ph.D. thesis, Western Reserve Univ., Cleveland, 1955, pp. 114-125.
78 J. Canata and A.O.M. Stoppani, Biochim. Biophys. Acta, 32 (1959) 284-285.
79 L.F. Leloir, Proc. 3rd Int. Cong. Biochem., Brussels, 1956, pp. 154-162.
80 K. Kurahashi, Science, 125 (1957) 114-116.
81 H.M. Kalckar, H. de Robichon-Szulmajster and K. Kurahashi, Proc. Int. Symp. Enzyme Chem. Tokyo, Kyoto (1958) 52-56.
82 H.M. Kalckar, K. Kurahashi and E. Jordan, Proc. Natl. Acad. Sci. USA, 45 (1959) 1776-1786.
83 H. de Robichon-Szulmajster, Science, 127 (1958) 28-29.
84 G. Buttin, J. Mol. Biol., 7 (1963) 164-182.
85 H.C.P. Wu and H.M. Kalckar, Proc. Natl. Acad. Sci. USA, 55 (1966) 622-629.
86 F. Jacob and J. Monod, J. Mol. Biol., 3 (1961) 318-356.
87 K. Kurahashi, J. Biochem., 56 (1964) 101-102.
88 K. Kurahashi, Annu. Rev. Biochem., 43 (1974) 445-459.
89 K. Kurahashi, in: J.W. Corcoran, (Ed.), Antibiotics, Vol. IV, Biosynthesis, Springer-Verlag, Berlin, 1981, pp. 325-352.
90 C. Nishio, S. Komura and K. Kurahashi, Biochem. Biophys. Res. Commun., 116 (1983) 751-758.
91 K. Kurahashi and C. Nishio, in E. Haber (ED.), The Cell Membrane, Plenum, New York, 1984, pp. 55-66.
92 K. Kurahashi, C. Nishio, K. Babasaki, J. Kudoh and T. Ikeuchi, in S. Ebashi (Ed.), Cellular Regulation and Malignant Growth, Jap. Sci. Soc. Press, Tokyo, 1985, pp. 177-186.

G. Semenza (Ed.) Selected Topics in the History of Biochemistry: Personal Recollections (Comprehensive Biochemistry Vol. 36) ©1985 Elsevier Science Publishers

Chapter 3

Jeffries Wyman and Myself: a Story of Two Interacting Lives

JOHN T. EDSALL

Department of Biochemistry and Molecular Biology, Harvard University, 7 Divinity Avenue, Cambridge, MA 02138 (U.S.A.)

Early life and student days at Harvard

Jeffries Wyman and I, in our life and work, have been deeply involved together for over 60 years. We both entered Harvard College in 1919, and soon became acquainted. By our third year, that relation had developed into a close friendship, which has lasted ever since. It has involved our common interests, not only in science, but in art, literature, travel, and long walks in many kinds of country. Only once did we publish a paper together (in 1936) but much of our shared thinking found its way into our book on *Biophysical Chemistry* (1958); and the constant interplay of thought between us goes on unabated. For about a dozen years at Harvard we shared the teaching of a course in biophysical chemistry, and though in later years we have lived on opposite sides of the Atlantic we have remained constantly in touch by visits and letters. Here I attempt to tell something of his life and achievements, especially in science, but with some more personal glimpses. Naturally it will involve some of my personal history as well.*

* I have written two autobiographical articles elsewhere [1,2]. Inevitably there will be some repetition here, but the emphasis here will be on Jeffries rather than myself. I will occasionally refer to those other accounts for some matters that are only briefly mentioned here.

Plate 4. John T. Edsall, about 1968.

Both of us came of families long settled on the American side of the Atlantic. Jeffries' ancestors have been New Englanders from a long way back, and the same is also true for the ancestors of my mother, Margaret Tileston. My father's ancestors, however, have lived in New Jersey and New York, beginning with Samuel Edsall, who immigrated from England in 1650, and became a prominent citizen, though his life was for a time in peril on a charge of treason (later dismissed) because of his service on the Council of the controversial Governor of the New York colony, Jacob Leisler. Among my paternal ancestors also are the DeKays, a Huguenot family who briefly settled in The Netherlands but then migrated to America around 1700. My father, David Linn Edsall, grew up in the high hilly country of Sussex County, NJ; country which I came to know and love as a boy, when I spent several summers there. David Edsall became a leader in American medicine, and a pioneer in medical research, first at the University of Pennsylvania and later at the Massachusetts General Hospital and Harvard Medical School, where he served as a powerfully influential Dean (1918–1935) [2a]. My mother was a women of broad intellectual interest and wide cultivation, beloved by her family and intimate friends. I had every encouragement to take an interest in literature, science, art, and political affairs. Tragically my mother died of pneumonia in 1912, when I was 10 years old, and my younger brother Geoffrey (later a notable immunologist and a leader in public health work) was only 4. My stepmother, whom my father married in 1915, was a woman of much ability and broad cultivation; but the relations between her and the rest of the family became difficult, and the marriage finally ended in divorce. We had moved from Philadelphia to Boston in 1912; my mother was delighted to return to New England, where she had grown up, but her death occurred soon after that. Later we lived in Milton, and in Cambridge, MA, which has been my home for most of my life, both before and after my marriage to Margaret Dunham in 1929. I entered Harvard College 10 years before that, in the class of 1923, at the same time as Jeffries Wyman.

The Jeffries Wyman of today is the third, in successive generations, to bear that name. His great-grandfather, Dr. Rufus Wyman,

Plate 5. Jeffries Wyman (1901- . . . .) Taken in 1971 at National Institutes of Health.

was a psychiatrist who was a pioneer, in the early 19th century, in the humane treatment of the mentally ill, at the McLean Asylum in Somerville, MA (now the McLean Hospital in Belmont). His son, the first Jeffries Wyman (1814–1874), became an eminent scientist. He was a great comparative anatomist, and probably the foremost American physical anthropologist of his time, becoming the first director of the Peabody Museum of American Archeology and Ethnology at Harvard. He was one of the founding members of the National Academy of Sciences in 1863 [3]. He was remarkable, not only for his scientific achievements but for his personal character and influence on his students. One of these was William James, who later wrote:

"His extraordinary effect on all who knew him is to be accounted for by the one word, character. Never was a man so absolutely without detractors. The quality which everyone first thinks of in him is his extraordinary modesty . . . his unfailing geniality and serviceableness, his readiness to confer with and listen to younger men . . ."

James also spoke of Wyman's

"disinterestedness and single-minded love of the truth",

rating him above the more famous Agassiz, both as scientist and teacher [4].

His son, the second Jeffries Wyman, started his career in the West, with the Chicago, Burlington and Quincy railroad. After his marriage he returned to Boston, as an officer of the Bell Telephone Company, then in its infancy. His son Jeffries, of whom I write here, was born in 1901.

When he entered Harvard College in 1919, Jeffries initially chose philosophy as his major field, and it left an enduring impression. His philosophical bent is often apparent in the rigor and subtlety of his scientific thinking, and in his endeavor to discern general principles in specific problems. He graduated with highest honors in philosophy, and high honors in biology. His enduring interest in mathematics and physics also developed during those years, and was to influence profoundly the character of his future work in biophysical chemistry. P.W. Bridgman's famous course in advanced thermodynamics was an important and enduring influence, and the rigor and subtlety of thought that was characteristic of Bridgman is present in his writings also.

Jeffries was much involved in informal discussions with a small group of undergraduates, of which I was not a member. Two important members, whom I was to know well not long after, were Alfred Mirsky and M.L. Anson. Robert Oppenheimer, who had entered college in 1922, with his immense range of intellectual interests and his almost feverish mental activity, also later was drawn into the group. I came to know Robert also, for we were both members of the Harvard Liberal Club, and I was editor of its newly founded small publication. He supplied its name *The Gadfly*, taken from the remarks of Socrates about himself, in Plato's *Apology*. The magazine survived through three issues, and then faded into oblivion.

My own interets in college were varied. My work centered in chemistry, with a fair amount of mathematics and physics, but relatively little biology. Outstanding among my undergraduate teachers in science were E.P. Kohler in advanced organic chemistry and Lawrence J. Henderson in Biochemistry. Kohler developed his subject with great artistry, evoking the central problems of the subject, as they were seen at that time, in a way that called for constant thinking and close attention by the students. He inspired us to look searchingly at fundamental questions concerning the structure and reactivity of organic molecules. In the course of his discussion the nature of the work that had led to the answers to such questions — often incomplete and tentative answers — gradually emerged. It was an esthetic pleasure, and an intellectual discipline, to follow Kohler's closely woven discourse. As I learned later, he destroyed his lecture notes at the end of each year, and started again from scratch the next year; a practice that partly accounts for the high quality of his lectures.

Henderson was quite different. His lectures were not as elegant, or as carefully prepared, as Kohler's; but he had broad perspectives on fundamental biology and chemistry. In the course of his career he was biochemist, physiologist, historian and philosopher of science, and sociologist. He had for several years attended Josiah Royce's philosophy seminar, where he first presented in draft form his book on *The Fitness of the Environment*, which opened my eyes to a broader vision of nature and of life in general. Problems of biological organization and regulation were central in Henderson's thinking, and were outstandingly exemplified in his work on blood as a physicochemical system. His undergraduate biochemistry course involved no laboratory work, but was a great intellectual stimulus. I studied History of Science under him and under the great scholar George Sarton, whose breadth of learning in the field vastly exceeded Henderson's. Henderson lectured to us in the Fall term, Sarton in the Spring; and both enjoyed discussion, and interchange of ideas with students. Henderson, who was a close friend of my father, remained a significant personal influence for Jeffries and me, as will appear later in this story. His vision of blood as an interacting system of components, in which a change in any one must produce changes in all the others, has obviously a fundamental relation to the Wyman linkage functions.

After our graduation, Jeffries spent a year (1923–24) in the Harvard Graduate School, chiefly to take advanced courses in physics and chemistry, while I became a first year student at Harvard Medical School. Anatomy and histology had little appeal to me, but the courses in physiology (with Walter B. Cannon) and biochemistry (with Otto Folin) were far more important. Folin's biochemistry course was totally different from Henderson's: plenty of laboratory work, largely in colorimetric methods of analysis for important biochemical constituents — methods largely devised by Folin himself. We used the old Duboscq colorimeter, comparing by eye the color in the unknown solution against a standard, and moving a plunger up and down until the colors in the two solutions matched. Photoelectric colorimetry was still many years in the future. We had lectures and laboratory guidance also from Cyrus Fiske, also a master of analytical techniques. A few years later Fiske, with Y. Subbarow, who came to Harvard from India, was to turn his method of phosphorus analysis to powerful use, in two discoveries of outstanding importance: of phosphocretine and of ATP in muscle

(ATP was independently discovered by Karl Lohmann in Meyerhof's laboratory).

Of crucial importance to me was a small piece of research, under the guidance of Alfred C. Redfield, on the strength of contraction of the heart muscles of turtles in the absence of oxygen, and their recovery when oxygen was supplied. I have written of this elsewhere [1]. Redfield's wise guidance greatly influenced my future career, and started my strong interest in muscle biochemistry.

Studies abroad; Graz, Cambridge, London

In June 1924 Jeffries and I went abroad together to work in the newly established Department of Biochemistry in Cambridge, England, headed by Sir Frederick Hopkins ("Hoppy"). We crossed the Atlantic in a slow steamer, The "Winifredian", along with some 700 cattle and 200 human passengers. Two good friends of ours were aboard, and the journey, which took 10 days, was a festive affair. On landing we visited Cambridge only briefly, to plan for our activities there when we would return in early October, and then headed for Germany and Austria. In those days a good reading knowledge of scientific German was indispensable for young scientists. Although both of us had studied German in school, we badly needed to improve our reading knowledge and also to be able to speak the language reasonably well. After traveling up the Rhine by boat and visiting Salzburg, we settled in Graz, Jeffries living with one family and I with another. We talked in German only with our respective families and read only German scientific and general literature during the summer. From then on, we felt at ease in reading German scientific literature, and could talk in German freely, though imperfectly. (24 years later, in 1948, on returning to Germany after the war, I found I was able to give lectures in German, without reading from a prepared text; the audiences evidently understood me, since their questions after the talks were much to the point.)

We carried letters of introduction to three professors at the University of Graz, all of whom received us kindly. Otto Loewi, in particular, saw us frequently, inviting us to his house on several occasions, regaling us with entertaining stories of his earlier days, and taking us for a week into the Tyrolean Alps, where in fact Jeffries and I did considerably more climbing than Loewi did. He also demonstrated for us his famous "Vagusstoff" experiment, which led to his later Nobel Prize: he stimulated the vagus nerve of a frog, until the heart stopped beating, then transferred some of the liquid around its heart to the heart of another frog; the active substance (later shown by H.H. Dale to be acetylcholine) then bringing the beating of the second heart to a stop also. Some have reported being unable to repeat that experiment; I can testify, however, that in Loewi's hands, and with his frogs, it worked.

Returning to England, we settled in Cambridge early in October, 1924, both of us living in St. John's College. The Sir William Dunn Institute of Biochemistry, where Hopkins was in charge, had opened its doors only a few months before. Hopkins, who had moved from London to Cambridge about the turn of the century, had at last, after years of delay, achieved an independent department of biochemistry, and an adequate building to contain it. For the first time he could offer an advanced course (Part II) in biochemistry; previously, biochemistry had been taught only as one aspect of the Part II physiology course. Hopkins himself has related the struggles of his earlier years [5] and Kohler [6] has described how Hopkins, with the powerful support of Sir Walter Fletcher and the Medical Research Council, achieved the creation of the Sir William Dunn Institute. In 1974 there was a commemorative gathering to celebrate the 50th anniversary of its opening, with talks by Sir Frank Young, Sir Rudolph Peters, Malcolm Dixon, Dorothy Needham and Robin Hill. These talks have been preserved in an unpublished booklet together with some cartoons and verses from that early time. Needham [7] has given a valuable picture of Hopkins's life and influence, and so has Pirie [7a].

The new biochemical institute, because of the magnetic attraction that Hopkins exerted, was almost completely occupied within a few months. For Hopkins the administrative burdens must have been almost overwhelming, and he had little time in those days to apply his genius for experimental laboratory work. He gave some inspiring lectures, imparting to us his broad view of the nature and scope of general biochemistry, and he was invariably kind and helpful on the occasions when a student might have a chance to talk with him personally; but such occasions were rare.

Second to Hopkins was the Reader, the unforgettable J.B.S. Haldane, with his large and powerful frame, his booming voice (some called him "the Bull of Bashan"), his all-encompassing intellectual curiosity and range of learning. He was lecturing on enzymes in that fall term; and his well known book *Enzymes*, which finally appeared in 1930, developed from these lectures. Later he was to turn his major field of interest from biochemistry to genetics, and to leave Cambridge for the John Innes Institute and later for University College, London. Certainly he was a major figure in the Biochemistry Department while I was there [8,9].

Much was happening in Cambridge biochemistry and physiology at that time. G.S. Adair, in his beautiful studies of the osmotic pressure of hemoglobin, established for the first time the true molecular weight of hemoglobin, showing that the molecule was four times as large as practically all earlier workers had supposed. Adair formulated the basic equations for the binding of oxygen and other ligands by hemoglobin, in terms of four successive binding constants. By a suitable choice of these constants it was possible to fit the sigmoid form of the experimental oxygen binding curves, which had first been found (1904) by Christian Bohr and his collaborators in Copenhagen, and later extensively studied by Joseph Barcroft and his colleagues in Cambridge. This was a major step forward; one could describe the cooperative interactions in hemoglobin in terms of four association constants, increasing in strength as binding proceeded, even though the underlying mechanism was still obscure. Adair's fundamental studies were to lay part of the foundation for much of Jeffries Wyman's work in later years [10–12].

Another major advance in hemoglobin chemistry was also proceeding at that time in Cambridge. Hamilton Hartridge, already famous for his brilliant ingenuity in experimental design, enlisted a younger colleague, F.J.W. Roughton, in developing the continuous flow method for studying very rapid reactions. Together they applied it to measure the rate constants for the reactions of hemoglobin with O_2 and CO. This was a tremendous achievement: it extended the range of measurable rate constants by about three orders of magnitude, compared to any previous technique.

Roughton was an instructor that year (1924–25) in the Part II biochemistry class, and he gave the students the opportunity to use the apparatus under his guidance. We worked in pairs; Jeffries and I were one such pair. There were two very large jars, one filled with oxyhemoglobin solution and the other with a dilute solution of hydrosulfite (dithionite) to remove the oxygen as it came off the hemoglobin. The liquids flowed from these two jars into the Hartridge mixing chamber, where mixing was almost instantaneous. Then the mixed solution moved down the observation tube, in turbulent flow. By spectroscopic measurements, at different points along the tube, we could follow the rate of the release of oxygen from hemoglobin. The oxygen released was immediately removed by the dithionite, so there was no back reaction. Actually we achieved fairly good measurements, with reasonable rate constants, rather to Roughton's surprise.

For Roughton this was only the beginning of the outstanding studies on blood and hemoglobin, which he pursued for almost half a century until his death in 1972 [13]. In time this brought him into close relation with both Jeffries and myself. He and Jeffries often interpreted their experiments differently; sometimes the discussion of their disagreements became rather heated, though the differences became reconciled in time. Roughton was a frequent visitor to the United States; he was in close touch with L.J. Henderson and D.B. Dill at the Harvard Fatigue Laboratory, and worked there during the war years. In later visits he frequently stayed at our house, and for some 10 years, from 1960 on, my own research dealt with carbonic anhydrase, the enzyme he had discovered (with N.U. Meldrum) in 1933, and had so extensively studied thereafter.

Mirsky and Anson, who had graduated from Harvard a year ahead of Jeffries and me, had been doing exciting work in Joseph Barcroft's laboratory on hemoglobin derivatives, which was to lead to their later important studies on reversibility of protein denaturation. Their interests were wide, and talk with them was immensely stimulating. Alfred Mirsky went on to a distinguished career; his interests shifted later to protein biosynthesis and DNA [14]. Nearly 20 years later "Tim" Anson and I were to become the first editors of *Advances in Protein Chemistry* and we worked on it together until his death in 1968 [15].

One of the greatest Cambridge scientists, David Keilin, already known as an eminent parasitologist, was rediscovering and extending the almost forgotten work of C.A. MacMunn on those cellular heme proteins that Keilin rechristened cytochromes. This was only the beginning of the magnificent researches that established Keilin as one of the world's most eminent biochemists. He was immensely kind and sympathetic to young students; when I was doing a small piece of research on phosphates in frog muscle, he showed me how to extend the work to insect muscle, and dissect out the muscles of bees for analysis. He took delight in demonstrating the spectral absorption bands of reduced and oxidized cytochrome in the muscles, using a small hand spectroscope designed by Hartridge. Always, on visits to Cambridge in later years, both Jeffries and I would come to see Keilin in his laboratory; he was a never failing source of inspiration [16,17a].

Another great man in Cambridge was W.B. (Sir William) Hardy, whose pioneer work on the physical chemistry of proteins was to be an inspiration to Jeffries and me. Hardy, who had begun as a histologist, was always pushing on into new fields of activity; when we knew him, he was working on the physics of surface phenomena. He was a man of great vitality and zest in life.

Cambridge in 1925 was an extraordinary center for work on hemoglobin and other heme proteins, with Barcroft, Adair, Hartridge, Roughton, Anson and Mirsky, and Keilin. I doubt whether, at that time, Jeffries had any inkling of the fact that hemoglobin would later come to be the major subject of his own research for nearly 40 years. However, that remarkable constellation of workers in Cambridge certainly had an enduring influence on him.

By the end of one term in Cambridge, however, Jeffries realized that the work on the dynamics of muscle, at A.V. Hill's laboratory in London, was of more interest to him than the biochemical studies in Hopkins's department. Hill was happy to accept him, and he moved to London about the end of 1924. Hill's laboratory at University College [18] was his scientific home for nearly 3 years, until he received his Ph. D., working on the relations of work and heat in tortoise muscles undergoing isotonic and isometric contractions, and subject to the effects of stretch and sudden release [19].

The training that he got in Hill's laboratory was of immense importance; the demands for rigorous and critical thinking, the need for careful planning of experiments and for the utmost skill and care in their execution — all this left its enduring mark. As Jeffries wrote to Hill, many years later, transposing from Lewis Carroll:

"...and the muscular strength that you gave to my *mind* has lasted the rest of my life."

His work with Hill had nothing to do with hemoglobin; but later Jeffries was to revive Hill's long-forgotten equation of 1910 for the binding of oxygen to hemoglobin, and would show its previously unsuspected power in the analysis of cooperative interactions in ligand binding.

In spite of his move to London, Jeffries and I remained in close touch, making frequent visits between Cambridge and London. In the following academic year (1925–26) Robert Oppenheimer, whom we had both known as an undergraduate at Harvard, arrived in Cambridge to work at the Cavendish Laboratory. We already knew that he was a phenomenal person, with immense intellectual power and intense interests in literature, philosophy, and other subjects that extended far beyond science. It was about the time of his arrival in Cambridge that Heisenberg's first great paper on quantum mechanics appeared, to be followed by Schrödinger's wave mechanics a few months later. Dirac, who was there in Cambridge at St. John's College (as I was) was still unknown to the world, but would soon be famous. It was an immensely exciting time for Robert Oppenheimer, who mastered all these new developments as soon as they appeared, and conveyed something of their meaning to Jeffries and me. In spite of the fact that he was passing through a grave personal psychological crisis that year, he pursued his work in science, and his interest in other subjects, with intensity. During the University's spring vacation in 1926, he and Jeffries and I paid a visit to Corsica, with long walking trips in the magnificent mountain country and by the sea*.

One fellow student at St. John's College, whom we both vividly remember, was Gregory Bateson, cultivated, charming, humorous, still uncertain as to the path in life that he would follow. Son of the eminent geneticist William Bateson, he was to be known in future for his work in anthropology with his wife, Margaret Mead, and still later for his independent explorations of human nature and its possibilities.

My return to Harvard Medical School: the Department of Physical Chemistry

I came back to Harvard Medical School in the late summer of 1926. Having studied pathology and bacteriology, as well as biochemistry, during my 2 years abroad, I began clinical work as a third year student. What concerns me here, however, was that I could spend some of my time in a laboratory on a biochemical problem. Though the clinical programs were very demanding, Tuesday and Thursday afternoons were kept free for each student to spend as he saw fit. I consulted Alfred Redfield again for advice, remarking on my interest in muscle biochemistry and physiology, which had begun with the work on tortoise heart in his laboratory, and continued with a small study on phosphates in muscle in Hopkins' laboratory. Redfield considered, and said

> "I think that the muscle proteins represent the most neglected aspect of muscle biochemistry. Edwin Cohn, in the Department of Physical Chemistry upstairs, has started a study of muscle proteins, but the student who worked on it is leaving, and Cohn is looking for someone to carry on. Why don't you talk to him?"

---

* For an account of Oppenheimer's year in Cambridge, including our trip in Corsica, see [20].

That conversation steered me into the laboratory that was to be my home for the next quarter century, and was to influence Jeffries greatly when he returned from England a year later. Cohn welcomed me in, and set me to work on "muscle globulin", as we called it in those days. The Department of Physical Chemistry was an unusual place; it was set up for research in what we later came to call biophysical chemistry. There was no department like it in any other medical school, as far as I know. It was originally planned for L.J. Henderson, to keep him from leaving Harvard, when Johns Hopkins had tried to lure him away in 1919. Cohn had taken his Ph. D. with Henderson, who chose him to run the laboratory. As Cohn matured and showed promise of becoming a leader in protein chemistry, Henderson's interest in blood research shifted to the Fatigue Laboratory, which had been set up for him at the Harvard Business School. Thus Cohn became in a few years the actual (and later the official) head of the Department.

At the beginning of his career Cohn had decided to devote himself to the study of proteins. After getting his doctorate working both with Henderson and with Frank Lillie of the University of Chicago, he went to New Haven to work with T.B. Osborne, the foremost American protein chemist of his day. Later, after the First World War, he spent 6 months in Copenhagen with S.P.L. Sørensen. Like these great teachers he viewed proteins as definite chemical compounds with well defined structures, and rejected the views of some of the contemporary colloid chemists, who held proteins to be variable aggregates of ill-defined composition and structure [21].

Cohn was deeply concerned with the factors governing the solubility of proteins in relation to their structure, and with their properties as acids and bases. The equipment in his laboratory was relatively simple: racks of flasks for Kjeldahl nitrogen analyses, a hydrogen electrode and potentiometer for pH titrations (glass electrodes did not come in for another decade), a couple of bucket centrifuges, each of which could handle two liters of material in a single run, and the usual laboratory equipment of the time, including jars of 10 liters capacity or more, for the large-scale protein preparations. A key facility was the cold room, essential for keeping proteins stable and avoiding bacterial growth, and for some of the preparative work. Laboratory cold rooms were rather rare in those days; there was none in Hopkins' Institute, though there was a warm room for bacterial metabolism studies. As the work in Cohn's laboratory grew, he added more cold rooms. There was no ultracentrifuge, however, until 1938; Svedberg's laboratory in Uppsala had almost a monopoly in that field until the later 1930s.

The solubility work on protein fractionation and crystallization involved primarily the classical methods of salting-out at high salt concentration, and salting-in of globulins that were almost insoluble in water, but dissolved on addition of small concentrations of salt. The classical work of Hofmeister [22] had defined the role of various cations and anions in salting out, but in essentially qualitative terms. Also Hofmeister, although he had been well aware of the effects of added acid or alkali on protein solubility, did his work 20 years before Sørensen established the pH scale. Cohn recognized the importance of the electric charge on protein molecules in influencing their thermodynamic activity, and hence their solubility. A young biochemist in the laboratory, Arda Alden Green, was doing beautiful studies of the solubility of horse hemoglobin in various salts, as a function of salt concentration, pH and temperature [23]. They remain as classics in the field. Also she was collaborating with Ronald M. Ferry in an elegant study on the oxygen binding of hemoglobin, as a function of oxygen pressure and of pH, which was important for both Linus Pauling and Jeffries Wyman, as we shall see [24]. Later she was to work with Carl and Gerty Cori, where her great skill in purifying and crystallizing proteins played an essential part in the crystallization of glycogen phosphorylase. Sidney Colowick has given an excellent appreciation of her career [25].

There were weekly seminars in the laboratory, and Cohn's interests attracted others from different departments. A major participant in the discussion was George Scatchard, physical chemist from the Massachusetts Institute of Technology. Scatchard and Cohn had become, and remained, intimate friends, and in an important sense, collaborators, although they never published a paper together. Scatchard had a much deeper knowledge than Cohn of fundamental physical chemistry. He was unusual among the physical chemists of his time in taking a deep interest in the problems of biochemists and biomedical researchers. His close collaboration with the laboratory at the Medical School continued, and even intensified, during the years of the Second World War and after. He could be a searching critic of sloppy thinking. His papers often made hard reading for most of us, but he was most patient and helpful in explaining difficult points in personal discussion [26,27].

The work of Debye and Hückel [28], on the influence of interionic forces on the thermodynamic activities of electrolytes, had immensely clarified the understanding of electrolyte solutions; and it was directly relevant to solubility studies on proteins, especially at low ionic strength. How it should be applied to such large and complicated molecules was a matter of much debate in and outside the seminars. Also of great importance to us was the reinterpretation of the ionization constants of amino acids and peptides — and hence, by implication, of proteins — by Adams [29], and later in a more comprehensive study, by Bjerrum [30]. The chemists of the early 20th century had assumed that simple isoelectric amino acids in solution must have the structure $H_2N{\cdot}CH(R){-}COOH$; thus the pK value near 2.3 was assigned to the amino, and the pK near 9.7 to the carboxyl group. Adams [29] challenged this view, pointing out, by comparison with the pK values of fatty acids and amines, that there were compelling reasons for reversing the assignment, and isoelectric amino acids therefore must carry a positively charged amino and a negatively charged carboxyl group. Bjerrum, who was evidently unaware of the work of Adams, gave essentially the same analysis but in more detail. He concluded that the ratio of this **zwitterion** (or dipolar ion) form to the uncharged form, in aqueous solution, must be in favor of the former by at least 200 to one; our own later measurements [31] indicated that the ratio was of the order of 10^5. Bjerrum also noted that the very high melting points of amino acid crystals suggested the presence of strong electrostatic forces between the molecules.

In the seminars, we soon convinced ourselves that Adams and Bjerrum were right, but it took considerable time to realize all the implications of the presence of these electrically charged groups on the properties of these molecules, and on their interactions with water and other molecules around them.

All the discussion and thinking that went on led Edwin Cohn to a major change in the emphasis of research in the laboratory. Instead of concentrating almost entirely on proteins, he decided to place the chief emphasis on the physical chemistry of amino acids, peptides, and related compounds, of known structure, as simple models related to the proteins. This period actually lasted for about a decade, from about 1929 to nearly 1940. Work on proteins did continue during that time, but the major emphasis was on the smaller compounds. The results of these intensive studies were indeed far reaching; they opened up new vistas for a small group of us, who constantly interchanged ideas and plans for new experimental tests during that decade. Jeffries, on his return from England in 1927, was to play a key role in all this, and I now return to him, and to another relationship quite apart from the laboratory, which was to be a major personal influence for both of us.

Alfred North Whitehead and Evelyn Whitehead: personal influence and inspiration

One great and inspiring influence in those years on both Jeffries and me lay outside the laboratory or biochemical research; yet it counted so much for us both that it would be wrong to omit it. Alfred North Whitehead, after his great achievements in mathematical logic and philosophy at the Universities of Cambridge and London, came to retiring age in London in 1924. It was, I believe, L.J. Henderson (who had a hand in so many of these things) who saw an opportunity for Harvard, and readily persuaded President Lowell to offer Whitehead a professorship, which brought him to Harvard, still at the height of his powers, late in 1924. Jeffries and I did not meet the Whiteheads until our return from England, but their effect on both of us was profound. As fas as we were concerned this was unrelated to Whitehead's formal teaching; neither of us ever attended one of his classes. Nor had it primarily to do with reading his books, though some of his writings did indeed convey new and valuable insights to us, while others (to me at least) remained somewhat baffling and obscure. For us and many others, Whitehead's influence was primarily direct and personal. He and his wife Evelyn drew old and young alike to their apartment overlooking the Charles River; anyone who wished was welcome to drop in on a Sunday evening, sometimes on other evenings as well, and join in the conversation, the best that I have ever known. Often it went on till midnight, sometimes into the small hours, and it ranged over everything — politics, religion, history, philosophy, science, personal recollections; anything might turn up, and lead to fresh and unexpected vistas. The Whiteheads loved young people; we knew that we were really welcome and appreciated, and it warmed our hearts. Whitehead's appearance was one of great benignity, which indeed was a reflection of his nature; his sense of humor was keen, but the sharpness and boldness of his mind could lead to sudden vistas of unconventional thought, expressed in a deceptively gentle tone. Evelyn Whitehead too was most exceptional — part Irish, part French and Spanish, she grew up in Brittany and had to master the English language after her marriage. She cared deeply about people, with a great capacity for love and appreciation; but for a few individuals, she was also capable of intense dislike. Fortunately both Jeffries and I were among those she valued.

It was a diverse group who came to those evenings at the Whiteheads, and it varied from week to week. Mostly they were young, like us; one who was to become famous later was Philip Johnson, obviously brilliant and dynamic, but in those days restless and uncertain of his future plans, before he finally emerged as one of the world's most notable architects. Among the older generation was Felix Frankfurter, later to become a Justice of the Supreme Court, then still a professor in the Harvard Law School. In 1948, after Whitehead's death at 86, Frankfurter wrote of him in a letter to the New York Times:

"Professor Whitehead had a benign and beautiful presence, a voice and diction that made music of English speech, humor that lighted up dark places, humility that made the foolish wiser and evoked the wisdom of the taciturn. For twenty years

Professor Whitehead exercised this great and radiating influence. He did so at Harvard because he was there. He did so beyond because he was what he was. People came to Harvard because he was there."

In the realm of ideas, we learned from Whitehead to think in terms of a world of events, rather than static unchanging objects; some events of long duration appear as fixed objects, but they are events none the less. At least one of his maxims is good for any investigator to remember:

"Seek simplicity, and distrust it."

Properties and significance of dipolar ions: Jeffries Wyman and the study of dielectric constants

Jeffries returned from England in the summer of 1927, to an appointment in the Harvard Zoology Department, after completing his work for the Ph. D. with A.V. Hill at University College. I had already been working for a year at Harvard Medical School on muscle proteins, under the guidance of Edwin Cohn; in what time I could spare from work as a medical student. On his return, Jeffries also spent a large part of his time in Cohn's laboratory, during the year 1927–28, and devoted himself to work on the viscosities of protein solutions, work that he never published. I was then in my last year as a medical student, and like a number of my classmates was given permission to spend most of my time working on a special project. In my case, of course, that was the muscle protein work. So Jeffries and I were constantly in the laboratory together, and were also involved in the frequent seminars where Cohn and Scatchard and the rest of us were discussing proteins, amino acids and peptides and their electrostatic interactions in the light of the interionic attraction theory of Debye and Hückel [28] and the evidence from the work of Adams [29] and Bjerrum [30] that, at neutral pH, the amino groups would be positively, and the carboxyl groups negatively, charged. This new conception, as we came to realize, carried far reaching implications for proteins and their interactions. A protein, near pH 7, would have its surface studded with positive and negative charges. Even if the net charge added up to zero, there would be strong electrical interactions between the protein and the neighboring water molecules and ions, which would be important for protein behavior. Even with zero net charge, the protein might also have a large net dipole moment, so that it would orient itself in an electric field.

As I have said, Edwin Cohn had decided to shift the direction of work in the laboratory toward study of amino acids, peptides, and related molecules, as simple models, of known structure, to provide guidance in the study of proteins; and for about a decade (1929–1939) these problems deeply involved a small group of us. Apart from Cohn and Scatchard, who were the senior members of this informal group, I must speak of at least three younger members besides Jeffries and myself. One was T.L. McMeekin [32], a skilled organic chemist with a biochemical background, who collaborated closely with Jeffries in his work on dipolar ions, and by his choice of molecules to be synthesized and studied played a central part in almost every aspect of the work.

Another was Jesse P. Greenstein, fresh from a Ph. D. at Brown University, also a skilled synthetic organic chemist, who went for a year to Max Bergmann's laboratory in Dresden to get advanced experience in the synthesis of peptides. With his great ability and immense energy, he contributed both to the organic and the physical chemistry of the peptides, and a few years later did important work on protein denaturation and sulfhydryl groups. Later he was to become head of the Biochemical Division of the National Cancer Institute at the National Institutes of Health. He became an outstanding leader in cancer biochemistry, until his sudden death in 1959, at the age of 56 [33].

A third, the youngest and the most brilliant of the group in theoretical chemistry, was John G. Kirkwood, then a young chemist at M.I.T., closely associated with Scatchard. Scatchard and Kirkwood together were the first to produce a theory of the interaction of ions and dipolar ions, using a sort of "dumbbell" model of a dipolar ion — a positive and a negative charge, at opposite ends of a rigid rod-like link. Kirkwood, with his great mathematical powers, soon produced better models, and his calculations soon proved immensely valuable in stimulating experimental work and guiding its interpretation [34]. He went on to do outstanding work in both experimental and theoretical chemistry at Cornell, at CalTech, and finally at Yale, dying from cancer at the age of 52, while working heroically to almost the last moment [35].

These were the people who were most intimately involved with the developments I am about to describe. Others, such as J.L. Oncley and J.D. Ferry, were also to play a major role a few years later. Later Dr. Cohn and I were to attempt to put the story together in a book, including essential chapters by Scatchard, Kirkwood, Oncley, and one or two others, a book to which I shall have occasion to refer again [36].

I must mention briefly the outcome of my work on the muscle protein that we then called myosin (in the light of later knowledge, it would now be classed as a form of actomyosin). The arrival, in 1928, of Alexander von Muralt in Cohn's laboratory transformed the situation. With Alex's discovery that the "myosin" solutions became birefringent on flow in a velocity gradient, we realized that we had long asymmetric particles, largely responsible for the birefringence of the muscle fibre, and we worked on the system, in great excitement, for two years [37]. I have told more about that story elsewhere [1], and Alex and I together wrote, 50 years later, an account of how it was, as we remembered it [38]. I will merely note here how much Alex, who had taken a Ph. D. in physics before going into physiology, was responsible for educating me in optics and in the physics of macromolecules. The friendship that we then established remains as strong as ever.

Jeffries Wyman and the dielectric constants of dipolar ion solutions

Simple dipolar ions in solution, such as the amino acids and peptides, should be dipoles of high electric moment. Even with the rather imperfect knowledge of interatomic distances and bond angles that we had around 1930, it was pretty easy to calculate that the positive charge on the amino group, and the negative charge on the carboxyl, should be about 3 Å (3×10^{-8} cm) apart. Since the numerical value of each charge is 4.8×10^{-10} electrostatic units (esu), the moment should be at least roughly equal to $(4.8 \times 10^{-10}) \times (3 \times 10^{-8}) = 14.4 \times 10^{-18}$ esu, or 14.4 Debye units (D). Values for numerous other molecules, studied by the methods Debye had developed, gave much smaller moments: e.g. 1.85 D for water, 2.7 D for acetone, and 4.8 D for urea [39]. His procedure involved determination of the dielectric constants of systems containing these molecules at low concentration, either in the gas phase or dissolved in a non-polar solvent. From the dielectric constant one could calculate the molar polarization (P). In the simplest case, that of molecules in a dilute vapor phase, the relation became simply:

$$P = \frac{\epsilon - 1}{3}\left(\frac{M}{\rho}\right) = \frac{4\,\pi\,N}{3}\left(\alpha_0 + \frac{\mu^2}{3\,kT}\right) \tag{1}$$

Here ϵ is the dielectric constant, μ the permanent dipole moment of the molecules, α_0 is the induced moment due to the electrical distortion of the molecule by the applied field, M/ρ is the molar volume (molecular weight divided by density) N is Avogadro's number and k is Boltzmann's constant. The permanent moment μ is what concerns us; the polarization due to this arises from the orientation of the dipoles by the field. Thermal motion tends to disorient them; therefore the orientation, at a given field strength, decreases as the temperature rises. The μ value can be obtained by making measurements of P as a function of temperature, and plotting P against $1/T$. Measurements of polar molecules in non-polar solvents follow a similar procedure; the α_0 value can be obtained from measurements of the refractive index of the system at high frequencies.

The most direct evidence for the existence of amino acids and peptides as dipolar ions would be to determine their dipole moments. To do this, dielectric constant measurements were clearly indispensable; and this is what Jeffries undertook to do. Since 1927 he had been a member of the Harvard Biology Department, at the University in Cambridge, 4 miles from the Medical School, and remained so for the next 24 years. He continued to work in close touch with the group at the Medical School. Indeed I also spent part of my time in Cambridge, as Instructor and Tutor in Biochemical Sciences, where I worked with undergraduates to help them in developing an integrated understanding of physics and chemistry in relation to biology.

Jeffries was faced with formidable problems in planning to measure and interpret the dielectric constants of amino acids and peptides. Debye's approach was inapplicable for such highly polar molecules as amino acids and peptides; their vapor pressures, at ordinary temperatures, were vanishingly small, and they were soluble only in highly polar solvents like water, to which Debye's theory was inapplicable.

Moreover, when Jeffries started his work, there was no adequate theory to relate the dielectric constants of such highly polar systems to the dipole moments of the molecules present. Yet it still seemed probable that dielectric constant measurements would give significant results. This indeed proved to be true; the results that Jeffries was to obtain even surpassed our expectations in their importance. First, however, he had to overcome various experimental difficulties. The methods then available for measuring dielectric constants worked well for liquids of low conductivity; but amino acid solutions in water had a much higher conductivity even than pure water, which itself was more conducting than most liquids. It was necessary to develop a new method that could give highly accurate results for such solutions.

With some guidance from Prof. G.W. Pierce of the Harvard Engineering School, Jeffries soon developed a highly accurate method for his dielectric constant studies [40]. The pure amino acids and peptides that he needed for the work were for the most part not available commercially; they came from McMeekin and Greenstein [41], both of whom played key roles in many of the studies on amino acids and peptides.

The results of the dielectric constant measurements were simple and dramatic. Water itself has a very high dielectric constant (78.5 at 25°C); the vast majority of organic molecules, when added to water, lower the dielectric constant of the resulting solution. In contrast, amino acids and peptides raised it remarkably. Moreover, the dielectric constant (ϵ) was a linear function of the molar concentration (c) of the solute, in aqueous solutions, over the whole range of concentration studied; up to 2.5 M for glycine solutions, for instance. Taking ϵ_0 as the dielectric constant of water, the data for each substance could be accurately described by the equation:

$$\epsilon - \epsilon_0 = \delta c \qquad (2)$$

where δ, the dielectric increment, was approx. 23 for α-amino acids, 71 for dipeptides, around 115 for tripeptides, and became progressively higher as the peptide chain was lengthened. The fundamental simple empirical relations emerged clearly in the first paper by Wyman and McMeekin [42], and all the later studies fitted the same pattern. Table I lists some of the δ values that Jeffries obtained [43]. The highest δ value of all was for lysylglutamic acid, which carried two positive and two negative charges at its isoelectric point [44].

When these measurements were made there was still no theory of such polar liquids to relate the dielectric increments to the dipole moments of the molecules involved. Qualitatively it was obvious that the high δ values reflected high dipole moments, that must increase with increasing separation of the charged groups. For the α-amino acids, the calculation given above indicated that the dipole moment should be close to 14–15 D, but for amino acids and peptides with wider charge separation there could be various conformations, involving rotations around single bonds in the chain. In a study on betaines, Jeffries got δ values on the three benzbetaines, $(H_3C)_3N^+ \cdot C_6H_4 \cdot COO^-$. In these compounds the charged groups were fairly rigidly attached to the benzene ring, and dipole distances could be approximately calculated from the molecular geometry [45].

Jeffries took the first step toward a theoretical interpretation by finding a rough empirical linear relation, for some 140 different polar liquids, between the dielectric constant of the liquid and its polarization (P) in an electric field, from dipole moment data [46]. This stimulated Lars Onsager [47] to produce the first adequate theory of the dielectric properties of polar liquids. A still more realistic approach to the situation in polar liquids came from Kirkwood, three years later [48].

Kirkwood related the molar polarization, in such a polar medium, to the product of two different, but closely related, electric moments. One moment (μ) is that of an individual molecule in its polar surroundings; the other ($\bar{\mu}$) is the total moment of a spherical region in the liquid, with the designated molecule at its center: $\bar{\mu}$ depends, not only on the moment of this particular molecule, but on the mean orientations of all the surrounding solvent molecules, as determined by the interactions with the molecule at the center. The radius of this sphere need be taken as only a few molecular diameters; essentially all the orientations that contribute to $\bar{\mu}$ lie in this small region. The two vectors, μ and $\bar{\mu}$, are not necessarily parallel; they may deviate from one another by an angle θ. In general the significant quantity in Kirkwood's theory is the product $\mu \bar{\mu} \cos \theta$, or in vector notation $\boldsymbol{\mu} \cdot \bar{\boldsymbol{\mu}}$. Jeffries gave his most comprehensive and detailed discussion of dielectric constants and dipole moments in Chapter 6 of our book on *Biophysical Chemistry* [49]; a deep and subtle analysis.

The use of the Kirkwood theory does, however, lead to an approximate and very simple relation between the dielectric increment, δ, and the product $\boldsymbol{\mu} \cdot \bar{\boldsymbol{\mu}}$ given by Kirkwood [50]:

$$3.3\, \delta^{1/2} = (\mu \cdot \bar{\mu})^{1/2} = \mu_{\mathrm{m}} \tag{3}$$

The product $(\mu \cdot \bar{\mu})^{1/2}$ can be regarded as a sort of mean moment (μ_{m}) characteristic of the molecule in question, not as an isolated unit, but as it becomes involved in mutual interactions with other polar molecules clustered in its neighborhood. Table I includes values of this quantity for a variety of amino acids, peptides, and betaines that Jeffries studied. These are taken from his discussion, which leads to a slightly more complicated formula than the simple one given above; but the resulting values for μ_{m} are much the same on either basis. The precise values listed are not so important as their general magnitudes, which are of a higher order than those of ordinary polar molecules. Even urea, one of the few molecules, other than dipolar ions, that gives a positive δ value (+ 4.8) in water solution has still much less than half the electric moment of an α-amino acid.

TABLE I

Some dielectric increments (δ) and mean dipole moments (μ_m) for amino acids, peptides and betaines in water at 25°C

Values of δ are taken from Wyman's review [43]; the values listed here are from his own measurements, obtained in collaboration with McMeekin, Greenstein, and Edsall. Wyman also lists many values of δ from other investigators, notably from G. Devoto in Italy, which generally accord well with his. Values of μ_m are from Chapter 6 of Edsall and Wyman [49], p. 372, but are rounded off here to the nearest integer in Debye units.

Substance	$\delta = d\epsilon/dc$	$\mu_m = \sqrt{\mu \cdot \bar{\mu}}$/Debye Units
Glycine	22.6	17
Alanine	23.2	17
Leucine	25	18
β-Alanine	34.6	21
γ-Aminobutyric acid	51.0	25
δ-Aminovaleric acid	63	28
ϵ-Aminocaproic acid	77.5	30
Glycylglycine	70.6	29
Gly-Gly-Gly	113	37
Gly-$(Gly)_2$-Gly	159	43
Gly-$(Gly)_3$-Gly	215	50
Gly-Ala	71.8	29
Gly-Phe	70.4	30
Leu-Gly-Gly	120.4	38
Lys-Glu	345	62
Glycine betaine	18.2	17
o-Benzbetaine	18.7	18
m-Benzbetaine	48.4	25
p-Benzbetaine	72.4	30

Dielectric dispersion and relaxation times of amino acids and proteins

Dielectric increments of polar molecules must be determined at a frequency low enough to permit the molecules to orient themselves in response to the field. As the frequency increases, the molecules cannot reorient rapidly enough to keep pace with the alternating field, and the value of ϵ declines until, at very high frequencies, it becomes equal to n^2, the square of the optical refractive index. The range of frequencies where this decline occurs gives a measure of the relaxation time of the molecules involved. This time is a function of the molecular dimensions, and of the viscosity of the medium. Proteins, being far larger than the amino acids and peptides listed in Table I, have correspondingly longer relaxation times; and rather different techniques are required to study their dielectric properties at the low frequencies necessary to permit full study of their δ values and relaxation times. Jeffries did an early study of the protein zein from maize [51], dissolved in 70% n-propanol, and did a later study on myoglobin with H. O. Marcy using an improved technique [52]. Such studies have given, from the relaxation times observed, important information concerning the size and shape of the protein, as well as its charge distribution. I mention this work only briefly, since it was J.L. (Larry) Oncley who was the main contributor to this field of work on proteins during the same period, after he joined Cohn's laboratory at Harvard Medical School in 1936, and I must refer the reader to his classical reviews [53,54].

Further implications of the dipolar ion structure: solubility studies, molar volumes, heat capacities, Raman spectra

Edwin Cohn's great interest in the solubility of proteins naturally led to similar studies on amino acids and peptides. Given the dipolar ion structure of the amino acids, it was plain that, in a general way, they should respond to added electrolytes in a manner qualitatively similar to that of proteins; one would expect them to be salted in and salted out. Indeed there were a few data in the literature that gave some evidence of this. The planning of experiments required much thinking and discussion among the group, and we needed a theoretical analysis, essentially an extension of the Debye–Hückel theory of ionic interactions, to cover the interactions of ions with dipoles of high electric moment. Nobody in Cohn's laboratory was capable of this, but fortunately there was a constant interplay of minds with George Scatchard and Jack Kirkwood. Kirkwood in particular, with his combination of mathematical power and physical intuition, developed several models for dipolar ions — spheres or ellipsoids, with suitably placed dipoles in them — which would vary their activity coefficients (and therefore their solubilities) on the addition of ions. The limiting equation for salting in, at low ionic strength (I), was of the form

$$\lim_{I \to 0} \log(S/S_o) = K_R(\epsilon_o/\epsilon)^2 I \tag{4}$$

K_R is a salting-in coefficient, increasing as the dipole moment increases; ϵ_o is the dielectric constant of water, ϵ that of the actual solvent (such as an alcohol-water mixture); S is the solubility of the dipolar ion at ionic strength I, S_o the solubility at $I = 0$. The salting in of dipolar ions by dilute ionic solutions is thus proportional to I, not to $I^{1/2}$, as for ion–ion interactions [55–57].

Experimental studies gave an excellent fit to Kirkwood's equation, and the calculated dipole moment for glycine and other α-amino acids, as determined from the K_R values, was close to 15 D, in good agreement with the conclusions from the dielectric constant studies, and other evidence.

Another long series of studies involved the relative solubility of amino acids and related compounds in water and various organic media; ethanol was the organic solvent most commonly used, but studies were also done in formamide, methanol, n-butanol and acetone. McMeekin played a central role in all this work [58]. The ratio (Q) of solubility in water (x_w) to that in the organic solvent (x_A), where x denotes mol fraction, was the most important variable. The influence of the dipolar ion structure on Q appears on comparing the value for an amino acid with that for an uncharged, though quite polar, isomer: in this case McMeekin and Cohn et al. compared glycine with its isomer glycolamide, and alanine with lactam-

TABLE II

Solubility Ratios (Q) of some amino acids and isomeric substances in water and ethanol at 25°C, and Gibbs energies of transfer (ΔG_t)

Substance	Q[a]	log Q	ΔG_t[b] (kcal·mol^{-1})	$\Delta G_t / \Delta n_{CH_2}$
Glycine	2460	3.39	4.63	-0.74
Alanine	718	2.85	3.89	-0.64
α-Aminobutyric acid	240	2.38	3.25	-0.67
α-Aminocaproic acid	26	1.41	1.92	
Phenylalanine	28	1.45	1.98	
Glycolamide	6.3	0.80	1.09	-0.74
Lactamide	1.8	0.254	0.35	

[a] Q is the solubility ratio: $Q = (x_{water})/(x_{ethanol})$ where x is mol fraction of solute at saturation.
[b] $\Delta G_t = RT \ln Q$. The last column shows the effect of added CH_2 groups on ΔG_t. For the effect of the phenyl ring, compare ΔG_t for phenylalanine with that for alanine: $\Delta (\Delta G_t) = 1.98 - 3.89 = -1.91$ kcal·mol^{-1}. Comparison of amino acids with uncharged isomers:
(ΔG_t) [glycine - glycolamide] = 4.63 -1.09 = 3.54
(ΔG_t) [alanine - lactamide] = 3.89 -0.35 = 3.54

Data from Cohn and Edsall [36] Chapter 7, and McMeekin et al. [58].

ide (Table II)*. The results are almost identical in the two comparisons: Q is roughly 400 times as great for the dipolar ion as for its isomer, and the increase in the Gibbs free energy of transfer (ΔG_t $RT \ln Q$) from water to the organic medium is 3.5 kcal·mol^{-1}. This is a rough calculation, since it neglects corrections for activity coefficients in these highly concentrated saturated solutions, but these corrections are relatively small.

* George Scatchard pointed out in one of our seminars that nitroethane is also an isomer of glycine, with an electronic configuration almost the same as that of glycine. As far as I know, however, no one has measured the Gibbs energy of transfer of nitroethane from water to organic solvents.

McMeekin's measurements also showed strikingly the influence of hydrocarbon side chains on the solubility ratios, also illustrated in Table II. Each added CH_2 group in the side chain diminishes Q by a factor of the order of 3, corresponding to approx. -0.7 kcal·mol^{-1} in ΔG_t. Traube [59], more than 40 years earlier, had noted the same approximate factor of 3 per CH_2 group in his studies of the lowering of surface tension in aqueous solutions, in homologous series of organic compounds, a relation that became known as Traube's rule. Later Langmuir [60] had pointed out the thermodynamic significance of Traube's data in the course of his studies on surface films. The essential phenomenon — the transfer of a compound containing nonpolar groups from water, either into a surface film or into a more hydrophobic solvent — was much the same in either case. McMeekin's studies showed, by comparison of the log Q values for alanine and phenylalanine, that the increment in ΔG_t for the phenyl group (1.9 kcal·mol^{-1}) was a little less than that for three CH_2 groups. All these effects of nonpolar side chains held only if the side chain terminated in a methyl group. A charged or polar group at the end of the side chain, as in lysine, greatly diminished the effect of the intervening CH_2 groups.

All this work served to emphasize the importance of what we would now call hydrophobic interactions. Butler [61] and Hartley [62] in England at the same time made important contributions to this subject. Butler emphasized the striking enthalpy and entropy effects of introducing molecules with nonpolar groups into water. We were well aware of his work, but it was only much later that we came to recognize Hartley's contributions.

In a short note [63], I made another contribution to hydrophobic interactions, comparing studies in other laboratories on the increments, per CH_2 group, of the apparent molar heat capacities in dilute aqueous solutions of homologous series of compounds. These increments (in cal·K^{-1}·mol^{-1}) were consistently 21–25 or more per CH_2 group, whereas the corresponding increments for the same substances as pure organic liquids were only 6 to 8. This remarkable increase in heat capacity, on transferring nonpolar groups from an organic medium to dilute aqueous solution, has been abundantly confirmed in later work (see for instance [64]). I puzzled much over the origin of this effect, but light began to dawn only after the work of Frank [65] and Kauzmann [66] on hydrophobic interactions. The subject is still far from being really understood; Tanford's book [67] is valuable and various investigators are beginning to formulate testable interpretations of what goes on when water molecules are very close to a hydrophobic surface. Hugh McKenzie and I have recently surveyed many of the current developments relating to hydrophobic interactions, and the relations between water and proteins [68].

Another consequence of dipolar ion structure was the electrostriction that results from the orientation and close packing of water molecules around the centers of charge [69]. The phenomenon had been recognized, 40 years before, by Drude and Nernst [70] as an explanation for the volume contraction that accompanies the ionization of such substances as acetic acid, or of water itself. It furnished a valuable criterion for distinguishing dipolar ions from uncharged isomers, by determining their apparent molar volumes (V_m) in dilute solution in water. The effect emerges most clearly when one compares amino acids with their uncharged isomers, the results are shown in Table III.

Even for an α-amino acid, when one compares glycine or alanine with the isomeric hydroxyamides, the electrostriction due to the charged groups amounts to about 13 $cm^3 \cdot mol^{-1}$; when the charges are farther apart the difference becomes significantly greater, rising to 17 $cm^3 \cdot mol^{-1}$ when glycylglycine is compared with methylhydantoic acid. The effect clearly must be due to a volume change in the packing of the water molecules in the neighborhood of the charged groups. We were well aware of the great paper by Bernal and Fowler [71] on the structure of water, demonstrating (among many other things) the very open structure of ice and of liquid water, which would allow the water molecules, in the presence of the intense electric field of an ionic group, to squeeze together more closely and pack into a smaller volume. The same effect was shown by heat capacity measurements, done by Frank T. Gucker at Northwestern University on glycine, α- and β-alanine and their uncharged isomers [72,73]. The charged groups on the amino acids strikingly lowered their apparent molal heat capacities in solution.

TABLE III

Apparent molar volumes (V_m) of dipolar ions and their uncharged isomers at 25·C

All values are in ml/mol. Experimental data from the work of T.L. McMeekin and associates. This table is taken, with slight modifications, from Cohn and Edsall ([36], p. 158).

Substance	*Vm*	E^a
Glycolamide ($CH_2OH \cdot CONH_2$)	56.2	
Glycine ($^+H_3N \cdot CH_2 \cdot COO^-$)	43.5	12.7
Lactamide ($CH_3CHOH \cdot CONH_2$)	73.8	
α-Alanine ($^+H_3N \cdot CH(CH_3)COO^-$)	60.6	13.2
β-Alanine ($^+H_3N \cdot CH_2 \cdot CH_2 \cdot COO^-$)	58.9	14.9
Methylhydantoic acid ($H_2N \cdot CONH \cdot CH(CH_3)COOH$)	94.2	
Glycylglycine ($^+H_3N \cdot CH_2 \cdot CONH \cdot CH_2COO^-$)	77.2	17.0

[a] The electrostriction, E, is given by the observed difference between the V_m values for a dipolar ion and its uncharged isomer, e.g. between glycine and glycolamide.

At 5°C, for instance, the limiting value of $\overline{C}_p$ for β-alanine was 4.0 $cal \cdot K^{-1} \cdot mol^{-1}$; for its isomer lactamide it was 52.4.

I worked several years on Raman spectra of amino acids and related compounds. These revealed, along with many other findings, the characteristic vibrational frequencies of the carboxyl and amino groups, both in the ionized and the un-ionized state [74–76]. The results for isoelectric amino acids were unequivocal; both the amino and the carboxyl groups were in the ionized state at pH values near neutrality. The result was of course not unexpected, but it was at the time most satisfying to be able to see such direct and convincing spectroscopic evidence*.

* All the matters discussed here were later presented in great detail in the first 12 chapters of ref. 36.

Implications for protein structure

All these studies on simpler compounds obviously had far-reaching implications for proteins. They made clear that almost all the ϵ-amino groups in a protein, at neutral pH, would carry positive charges, and the carboxyls would be negatively charged. Some of these groups might be tied up by hydrogen bonding, but in general the region of solvent near the surface of a protein molecule must be subject to local electrostatic forces due to the charged groups, even if the net charge on the molecule as a whole was very small or zero.

In a brilliant pioneer investigation, Linderstrøm-Lang [77] had applied the Debye-Hückel theory of interionic forces to calculate the variation, with ionic strength, of the electrical free energy of a spherical macromolecular ion of variable mean net charge $\bar{Z}$. To simplify the calculations, he had to assume that the charge was spread uniformly over the surface of the ion. R.K. Cannan and his associates, at New York University, applied the Linderstrøm-Lang theory, with great success, to studies of acid-base equilibrium titrations in solutions of several proteins [78,79]. In the meantime, Kirkwood [80] had greatly extended the theoretical treatment, so as to consider the electrical interactions of a spherical macromolecule with any specified distribution of discrete charges located at points on its surface. At that time, and indeed for nearly 30 years more, no such detailed analysis could be applied in practice, until the molecular structures provided by X-ray diffraction became available. In 1957, however, Tanford and Kirkwood [81,82] extended Kirkwood's analysis so as to apply it more effectively to actual protein structures. In recent years Frank Gurd and his associates have applied the theory in further extended detail.

The wartime program of plasma fractionation

This decade of studies on amino acids and peptides, at Harvard Medical School, came virtually to an end in 1940–41. It had been a long excursion from the original aim of the laboratory, which was to study proteins. The increased insight gained during this time had more than justified the excursion, but it seemed time to return intensively to the physical chemistry of proteins. This decision was powerfully accelerated by the outbreak of war in Europe in 1939. The Nazi victories of 1940 alarmed us deeply, and to many of us represented a direct threat to the United States. These concerns led Edwin Cohn, who was in constant consultation with the authorities in Washington, to envisage a large-scale program for the fractionation of blood plasma, in order to make various plasma proteins available for clinical use. Serum albumin for transfusion in cases of shock was the first major product in demand; γ-globulin for temporary immunization against measles and infectious hepatitis followed; and the use of fibrin foam, fibrin film, and thrombin in neurosurgery became a third major field of clinical importance. It was certainly Cohn who envisaged the possibility, and the broad scope, of such a program; and his driving energy and vision that made its realization possible, with immensely generous support from government funds (the laboratory had never received government money before this). All the science and art of protein fractionation, all that we had learned from the earlier studies, provided a background for dealing with these new and overwhelmingly urgent practical problems. It was certainly a period of immense importance in my own life, and in that of dozens of others who were involved. It is far too large a story for further discussion here: Cohn gave a comprehensive picture of the whole enterprise [83], and I discussed the story in terms of its relation to protein chemistry [84].

Jeffries Wyman's first studies on hemoglobin

Jeffries published his first venture into hemoglobin (Hb) chemistry in 1937 but the subject had long been in his mind, and in his teaching. As I have already mentioned, he and I as students, and later, had both been greatly influenced by L.J. Henderson. Henderson's Silliman Lectures on blood [85], and his earlier work that led up to it, had revealed a fundamental picture of a highly adapted and coordinated system in which a change in any one component resulted in changes in all the others. There was a line of great earlier investigators — Christian Bohr in Copenhagen, J.S. Haldane in Oxford, Joseph Barcroft in Cambridge, among others — whose work had shown the cooperative character of oxygen uptake and release, and the coupling between the acid strength of Hb and the binding or release of oxygen. D.D. Van Slyke and his colleagues at the Rockefeller Institute, working in a close cooperative relationship with Henderson at Harvard, had studied the distribution of ions between red cells and plasma, in terms of the Donnan equilibrium. Hastings et al., in Van Slyke's laboratory, had also studied carefully the hydrogen ion equilibrium of oxy- and deoxy-Hb, and had deduced that one acidic group per heme shifted its p*K* value from 8.03 to 6.57 on oxygenation [86,87]. Hastings et al. worked over the pH range 6.8–7.6. They could do electrometric titrations on deoxy-Hb with the hydrogen electrode, but this was obviously impossible for oxy-Hb. In this case they inferred the pH indirectly from the ratio of HCO^{-}_{3} to CO_2 in the solutions, knowing p*K* for carbonic acid.

By 1936, the introduction of the glass electrode had eliminated the need for bubbling hydrogen gas through the solution, and had made it a simple matter to do electrometric titrations on both deoxy- and oxyhemoglobin. Jeffries, with his student Bernard German [88], proceeded to study the hydrogen ion titration curves of horse Hb in both forms, over a much wider pH range than in the earlier work, from about pH 4.3–9.5.

In the alkaline pH range, from about 6–9, they confirmed and extended the earlier work of Hastings et al: oxygenation of Hb resulted in proton release, reaching a maximum of about 2.2 H^+ ions released per Hb tetramer, about pH 7.4, and falling to zero on each side, at pH 6.0–6.1, and at pH 9. On the acid side of pH 6 they found a reversal of the effect of oxygenation; on binding of oxygen Hb-absorbed protons, the maximum absorption being close to 1 H^+ taken up per mol of tetramer oxygenated.

J.S. Haldane, and more particularly L.J. Henderson, had already pointed out the reciprocal relations involved. If binding of oxygen causes release of protons, as happens between pH 6 and 9, then conversely addition of H^+ ions (decrease of pH) must diminish the oxygen affinity. Below pH 6 these relations are reversed; in this range oxygenation decreases the acidity of Hb. From their own data, by a simple integration, German and Wyman calculated the curve for oxygen affinity as a function of pH. Figs. 9 and 10 illustrate the absorption of protons as a function of oxygenation. Two years later Jeffries studied the titration curve of HbO_2 over a temperature range from 6.5 to 37.5° C, permitting the heat of ionization (ΔH) to be calculated as a function of pH [89]. In the pH range from 6–8, ΔH was around 6–6.5 kcal·mol^{-1}, characteristic of imidazole groups. Below pH 5.5, ΔH fell to a small negative value characteristic of carboxyl groups, and above pH 9 it rose to around 11 kcal·mol^{-1}, characteristic of amino groups.

In all this work, and for some years thereafter, the best available data for the oxygen binding of Hb were those of Ferry and Green [24], working in the Physical Chemistry Department of Harvard Medical School. Several years later, Pauling [90] gave a brilliant

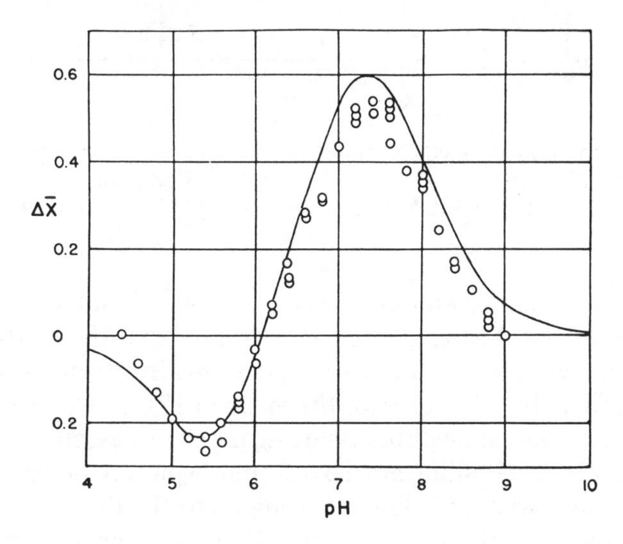

Fig. 9. Differential titration data of German and Wyman. Positive values of $\Delta \bar{X}$ denote H^+ ions released on complete oxygenation; negative values denote absorption of H^+. Smooth curve is calculated from assumed pK values of oxygen-linked acid groups. From [94].

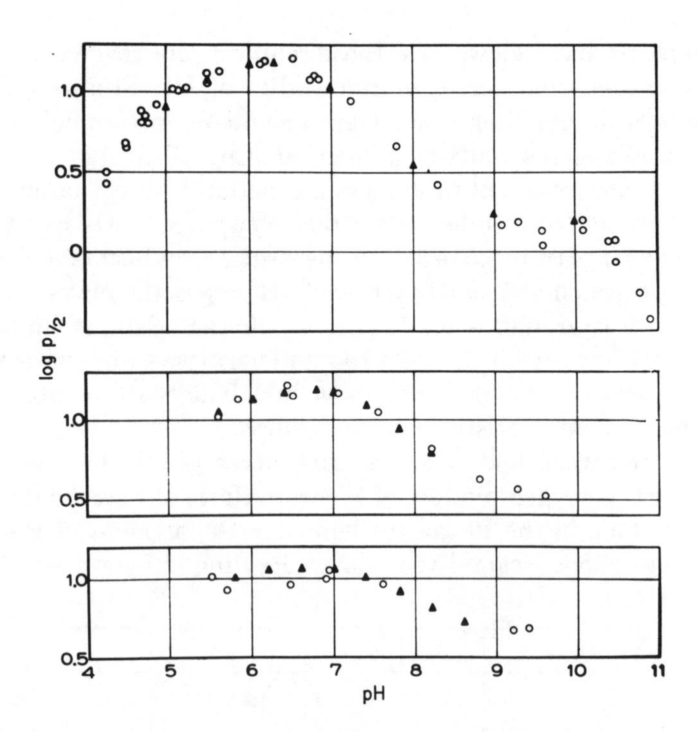

Fig. 10. Comparison of directly observed (0) values of log $p_{1/2}$ with values calculated from differential titrations (▲) such as those of Fig. 9. Upper curve, at ionic strength 0.3; middle curve: 2 M NaCl; lower curve: 5 M NaCl. From [97].

interpretation of their data, using a model involving a square arrangement of the four hemes. He assumed that the binding of O_2 to any one heme would increase the oxygen affinity of its two nearest neighbors by a factor α (of the order of 12); and that a heme in which both nearest neighbors were liganded increased its O_2 affinity by a factor α^2. Pauling also gave a simple picture of the variation of O_2 affinity with pH. The resulting fit to the data of Ferry and Green was remarkably good, and Pauling's conception greatly influenced the thinking of Jeffries Wyman and of others who were studying Hb.

The next step that Jeffries took was to study the heat of oxygenation of Hb, by pH titration of Hb and HbO_2 over a range of temperatures, from 7 to 38°C [91]. Roughton in Cambridge, England had been studying the heat of oxygenation of Hb, both by calorimetry and by measuring the oxygen equilibrium as a function of temperature. Jeffries calculated the heat of oxygenation over the whole pH range from 3–11, using Roughton's value at pH 6.8 as a point of reference, and concluded that the variation with pH could be accounted for entirely in terms of the release (or uptake) of H^+ ions on oxygenation, taking ΔH for the "heme-linked" acid groups as 6.5 kcal·mol^{-1}, a value characteristic of the imidazole groups of histidine. He concluded that the assumption of two such groups per heme could adequately describe the data; the group ionizing in the acid range changing its pK' value from 5.25–5.75 on oxygenation, and that ionizing in the alkaline range from 7.81–6.80 (at 25°C).

In 1941, with E.N. Ingalls [92], he attacked a larger problem: the correlation of the oxygen affinity of Hb, and its variation with pH, with the reversible oxidation of Hb to met-Hb, i.e. for the reaction $Hb \rightleftharpoons Hb^+ + e^-$ (where Hb denotes a single subunit and e^-an electron) or more realistically $Hb_4 \rightleftharpoons Hb_4^{4+} + 4e^-$. Taylor and Hastings [93] had studied this system in detail, and had obtained redox potentials over a range of conditions. To complete the picture of the system, it was necessary to do differential pH titrations of ferri-Hb ($Hb^4{}_4{}^+$) against either deoxy- or oxy-Hb. Jeffries chose the latter.

In this system the variation of the voltage, **E**, multiplied by the factor, F/RT (where F is the Faraday equivalent) corresponds exactly with the term ln pO_2 in an oxygen equilibrium experiment. From his own differential titrations, and the data of Taylor and Hastings, Jeffries could then show that the variation with pH of the oxygenation and the oxidation-reduction experiments were quite concordant within the limits of error; either one could be calculated from the other.

Here I have given only the barest indication of the character of this early work. The rigorous and subtle thermodynamic analysis of what he later termed linked functions was already apparent in this early work; the papers are stamped with his characteristic style. However further detail here would appear superfluous, in view of the far greater range and depth of the later development of his thinking, which will appear later.

Our teaching experiences at Harvard

I have spoken before of the influence of L.J. Henderson's biochemistry course, Chemistry 15, which I took as a senior in College. In 1928, both Jeffries and I served as teaching assistants in the course. Both of us also became involved in tutorial work with individual students, he in the Biology program, I in Biochemical Sciences. We both found this form of teaching valuable and stimulating, with thoughtful and eager students, of whom we had a good many. Henderson had had much to do with launching the Biochemical Sciences tutoral program, about 1925. Some years later I became Chairman of the Board of Tutors, and continued in the job for many years.

Henderson continued to give his course, but his central interest was shifting from biochemistry and physiology to Pareto's general sociology. He gradually relinquished nearly all of the lecturing in his course to a group of the younger biochemists, and Jesse P. Greenstein became the person chiefly responsible for organizing the course and arranging the responsibilities of the lecturers. This he did most effectively. At the same time he was a very busy tutor in Biochemical Sciences and was doing important research in Cohn's laboratory on peptides and on protein denaturation. In 1938 we lost him to the National Institutes of Health, where he was to become the leading biochemical researcher in the National Cancer Institute until his untimely death in 1959.

Harvard Medical School had established a Department of Biological Chemistry in 1908, but apart from Henderson's course there was no biochemistry at the University in Cambridge. By 1935, however, the Department of Biology began to realize that there was an urgent need for the subject; and George Wald, who had come to Harvard from Columbia University, undertook to give a course in general biochemistry. He proved to be an inspiring teacher as well as a brilliant investigator. Jeffries had for years been giving a course in physical biochemistry in the Biology Department. When, about 1940, Henderson finally gave up Chemistry 15, I joined Jeffries in giving the physical biochemistry course. As we worked over the notes that we gave out to the students, year after year, we gradually became convinced that we should develop them into a book. This was the germ of our book *Biophysical Chemistry* [49] which finally saw the light in 1958, six years after Jeffries had left Harvard, as will be told below.

Jeffries was on leave from Harvard, serving with the Navy, in the latter war years (1944–45). He worked in the Pacific area, in the development of smoke screens and other urgent problems of naval warfare, and at the Woods Hole Oceanographic Institute. In these projects he was closely associated with the eminent marine biologist P.F. Scholander. After his return to Harvard, he returned also to his work on Hb. Fortunately I was able to persuade him to write a review, exploring the subject in depth [94]. This turned out in fact to be far more than a review, although it did give an elegant and comprehensive picture of the status of the field. What could be found nowhere else, however, was a deep analysis of the nature of linked functions, which are of such fundamental importance in a highly integrated system such as blood.

The concept of thermodynamic linkage and its uses

The phenomenon of linkage has been an abiding concern in Jeffries' work from the very beginning of his work on Hb. It is notably exemplified by the Bohr effect — the mutual interaction of oxygen (or CO) binding and proton binding. Later research has added other interactions with O_2 binding — direct combination of Hb with CO_2 as carbamate, binding of organic phosphates, and the influence of chloride anions — and all these interactions can be encompassed by the general linkage relations. Linkage is of course a general characteristic of the great majority of thermodynamic systems of two or more components. If the activity of component B is changed by addition of component A, then there must be a reciprocal effect of added B on the activity of A. If there is no interaction, as is at least approximately true in a dilute gas, there is no linkage. In biochemical systems, however, the interaction coefficients are often much larger than in the inorganic world; not surprisingly, since mutations that lead to interactions that serve a use-

organism tend to be preserved by natural selection.

Jeffries has greatly extended and deepened his thinking about linkage, over nearly 50 years. In this brief discussion I will disregard chronology and draw upon some of his later and more powerful approaches whenever it seems convenient.

The idea of thermodynamic linkage arises directly from the work of Willard Gibbs in his memoir [95] of 1876. In a system consisting of t components at constant pressure and temperature a variation $\mathrm{d}G$ in the Gibbs energy depends on variations in the amounts of the components, according to the relation:

$$\begin{aligned} \mathrm{d}G &= \mu_{,}\mathrm{d}n_1 + \mu_2\,\mathrm{d}n_2 + \ldots\ldots \mu_T \mathrm{d}n_T \\ &= \left(\frac{\partial G}{\partial n_1}\right)\mathrm{d}n_1 + \left(\frac{\partial G}{\partial n_2}\right)\mathrm{d}n_2 + \ \ldots\ldots \left(\frac{\partial G}{\partial n_T}\right)\mathrm{d}n_T \end{aligned} \tag{5}$$

n_1, n_2, etc. are the amounts of the components, here expressed in mol (though other units are equally suitable). The chemical potentials μ_1, μ_2, etc. are defined as Equation (5) indicates; i.e.

$$\mu_{\mathrm{i}} = \left(\frac{\partial G}{\partial n_{\mathrm{i}}}\right)_{n_k}$$

the subscript n_{k} denoting that all other n's except n_{i} are held constant. Consider the effect of an increment in another component j, on the potential of i, and the reciprocal effect of i on the potential of j. By differentiation we get (since the order of differentiation is immaterial):

$$\left(\frac{\partial \mu_i}{\partial \mathrm{n}_j}\right)_{n_i,\, n_k} = \frac{\partial^2 G}{\partial n_i \partial n_j} = \left(\frac{\partial \mu_j}{\partial n_i}\right)_{n_j,\, n_k} \tag{6}$$

The variation in the potential of any component is related to the variation in its activity (a_i), by:

$$\begin{aligned} &\mathrm{d}\mu_{\mathrm{i}} = RT\mathrm{dln}a_{\mathrm{i}} \cong RT\mathrm{dln}C_i \\ &\left(\frac{\partial \mathrm{ln}a_i}{\partial n_j}\right)_{n_i,n_k} = \left(\frac{\partial \mathrm{ln}a_j}{\partial n_i}\right)_{n_j,n_k} \end{aligned} \tag{7}$$

For the systems we are considering, we may replace a_i by the molar concentration c_i, to a good approximation. The numerical values of a_i and c_i are defined by reference to a standard state of unit activity or concentration. For gases, which can be assumed to obey Henry's law, we have the relation $c_i = kp_i$, where p_i is the partial pressure of i_j hence $d \ln c_i = d \ln p_i$, and we can use either c_i or p_i, in formulating the potential of the gas. (Since the Henry's law factor k varies with temperature, one must take care in comparing data at different temperatures.) Considering only, in the Bohr effect, the interaction between oxygen binding and proton binding, we may take the H^+ ion potential as:

$$\mu_{H+} = RT \ln [H^+] = -2.303\, RT\, \mathrm{pH}.$$

In this, we disregard the fact that adding H^+ ion also involves adding an anion to the system, or removing a cation; neglect of this factor proves to be usually justifiable as a good approximation (see for instance [96, pp.247–249]).

Then we are dealing with the amounts of two components added to (or removed from) the system: the O_2 bound per mol of Hb, which can be expressed as the mean fractional saturation $\overline{Y}$, and the amount of proton added, also per mol of Hb, denoted here as $\bar{h}$. (The reference point, $\bar{h} = 0$, can be taken at any convenient point on the titration curve). Correspondingly there are two potentials to consider: that of oxygen, expressed as $RT \ln p$ (or $RT \ln c$, for unbound oxygen in the Hb solution) and that of protons, expressed as $RT \ln [H^+] = -2.303\, RT\, \mathrm{pH}$.

$$\left(\frac{\partial \ln p}{\partial \bar{h}}\right)_{\overline{Y}} = \left(\frac{ln[H^+]}{\partial \overline{Y}}\right)_{\bar{h}} = \left(\frac{-2.3\, \partial \mathrm{pH}}{\partial \overline{Y}}\right)_{\bar{h}} \tag{8}$$

Thus one can add acid to the system, keeping $\overline{Y}$ constant, and determine the oxygen pressure required to maintain this constancy. Most commonly this is determined as the pressure required to keep Hb half saturated with oxygen ($\overline{Y} = 0.50$) denoted here as p_m, the mean ligand activity. (A more exact definition of p_m is given later.) On the other hand one can change $\overline{Y}$, by adjusting the oxygen pressure, at constant $\bar{h}$, and observe the resulting pH change. This gives the coefficient on the right-hand side of Eqn. (8), which should be equivalent to the term on the left.

Another linkage equation takes the form (see [96], pp. 226 and 247).

$$\left(\frac{\partial h}{\partial \overline{Y}}\right)_{\mathrm{pH}} = \left(\frac{\partial \ln p}{\partial \ln[\mathrm{H}^+]}\right)_{\overline{Y}} = \left(\frac{\partial \ln p}{2.3 \partial \mathrm{pH}}\right)_{\overline{Y}} \tag{9}$$

The coefficient on the left can be measured by a differential titration, starting with deoxy Hb at a specified pH, and then saturating with oxygen (i.e. going from $\overline{Y} = 0$ to $\overline{Y} = 1$). Then the pH is adjusted to its initial value by titrating with H^+ or OH^- ions, which gives the increment in $\bar{h}$. The coefficient on the right-hand side of Eqn. (9) is given by determining oxygen-binding curves, in buffered solutions, at a series of pH values, and plotting the oxygen pressure at $\overline{Y} = 0.5$ ($\rho_{1/2}$) as a function of pH. This assumes that the form of the curve for $\overline{Y}$ as a function of ln p is independent of pH. Fig. 9 shows the original data of German and Wyman [88] for data obtained by the differential titration method. Fig. 10 shows a plot of data for $p_{1/2}$, directly obtained by the second method, combined with those calculated from data obtained by the first method by Eqn. (9). The excellent agreement shows the validity of the assumptions involved [97].

In his 1948 review [94], Jeffries was led to propose a modification of Pauling's square model [90] for homotropic interactions in oxygen binding by Hb. This model required that the O_2-binding curve should be symmetric about its midpoint, that is, a 180° rotation around the point where $\overline{Y} = 1/2$ should bring the curve into coincidence with itself. The data of Ferry and Green [24] were compatible with such symmetry, but more accurate later data (see [96], Fig. 3, p. 448) showed clear evidence of asymmetry, the curve for $\bar{Y}$ vs. log p rising more steeply above $\overline{Y} = 0.7$ than the reciprocal region between $\overline{Y} = 0$ and $\overline{Y} = 0.3$. From this and other evidence Jeffries proposed a rectangular, rather than a square, model, with the four heme groups coupled in pairs, the members of each pair being involved in strong interactions, while their interactions with members of the other pair were much weaker. At the time of this proposal, the existence of two distinct classes of peptide chains in Hb was still unknown — indeed no complete sequence of any peptide chain from a protein was yet known. In one sense the Wyman rectangular model was therefore an anticipation of the coupling of α and β chains, the two closely coupled pairs in the rectangle might later have been identified with the $\alpha^1\beta^1$ dimer units, and the more loosely coupled with the $\alpha^1\beta^2$ dimer units. However, there is a fundamental difference; the assumption in 1948 was that the dissociated dimer bound ligand in a strongly cooperative fashion, and this assumption remained the favored working hypothesis for the next twenty years. Later work, however, has shown conclusively that the dimer reacts noncooperatively with oxygen and other ligands; the oxygen affinity is high, but the binding at either site is unaffected by the presence or absence of bound ligand at the other.

Conformation of the hemoglobin molecule: its relation to the Bohr effect and cooperative ligand binding

It had already been known, in the 19th century, that Hb of a given species may form crystals belonging to different classes, for the oxy and deoxy forms. This fact was dramatically illustrated by Felix Haurowitz [98],who placed crystals of horse deoxy-Hb under a cover glass; then, on admitting oxygen, the crystals disintegrated and the quite different needle-shaped crystals of the oxy form gradually appeared. Haurowitz inferred that some change in the Hb molecule itself was involved, but at the time it was not possible to be more specific. In the meantime the early X-ray diffraction studies of Max Perutz, long before he was able to deduce a molecular structure, showed plainly a pronounced difference between the oxy and deoxy diffraction patterns. The idea of a molecular conformation change on ligand binding was in the air, though no details were known.

No one, however, had yet seen a connection between these inferred conformational changes and cooperative ligand binding, or the Bohr effect. As Jeffries has told me in a recent letter, the idea of such a connection came to him in a flash one day in 1950. He was far away from the laboratory, on a 6-month assignment for the U.S. State Department, as a scientific visitor and lecturer in Japan. He had just been visiting the famous Zen garden in Kyoto, with its stones and raked sand, when this vision suddenly came to him. It was worked out in detail in a paper with D.W. Allen that appeared a year later [99].

In this paper they pointed out that the interactions leading to cooperative ligand binding must be primarily entropy effects, since the enthalpy of binding changed little during the successive steps of binding. This suggested a conformational change in the molecule as a whole. Moreover, the nearly identical cooperativity for binding of O_2 and CO — as measured by the Hill coefficient n_H — and the very similar value of n_H for oxidation-reduction titrations all fitted the idea of a conformation change. They recalled the well known fact that the oxy and deoxy forms of a given species of Hb often differed widely in solubility; another phenomenon fitting readily the concept of a conformation change. With regard to the Bohr effect, they noted that the magnitude of the effect differs markedly from one species of Hb to another, although the change of bond character on ligand binding was presumably essentially the same in all. Likewise myoglobin, with similar changes on ligand bonding to heme, shows essentially no Bohr effect, and apparently little change in conformation on binding. They accordingly suggested

"...that the reason why certain acid groups are affected by oxygenation is simply the alteration in their position and environment which results from the change of configuration of the hemoglobin molecule as a whole accompanying oxygenation" [99, p.513].

They ended with a suggestion that

"...if we are prepared to accept hemoglobin as an enzyme, its behavior might give us a hint as to the kind of process to be looked for in enzymes more generally" [99 p. 515].

Therefore, they proposed, entropy changes such as those connected with the unusual properties of Hb might prove widely applicable to enzymes in general, or at least to many of them.

This paper, as I remember it, did not immediately arouse wide interest among biochemists; but the proposed concepts percolated gradually into the minds of many investigators, along with evidence along similar lines from other workers. 14 years later these ideas were to form an integral part of a dramatic further development, to which we shall come later.

A different way of life for Jeffries: years in the U.S. Embassy in Paris and as a UNESCO Administrator in Egypt

It was not long after the publication of that seminal article on conformation changes in Hb that Jeffries left the world of scientific research. In 1952 he became Science Attaché at the United States Embassy in Paris, where he served for 3 years. It was a busy and rather hectic time for international scientific relations. In the United States Senator Joseph McCarthy had aroused in many people a state of alarm over the alleged infiltration of communist influence. In the atmosphere of fear that this created within some branches of the U.S. Government, it was often difficult for foreign visitors to get a visa to visit the United States. The atmosphere was not conducive to free international scientific exchange; and the holding of international congresses of science in the United States was seriously impeded. I know that Jeffries did much to keep the lines of communication open, and enabled a good many French scientists to overcome bureaucratic obstacles and get the necessary visas to visit this country. Moreover, he traveled widely within France, visited scientists and their laboratories in many parts of the country, including some institutions that had seldom seen a foreign visitor. Thus he performed a valuable service in international relations. After leaving the post in Paris, he did not return to Harvard except for occasional visits, but soon became Director of UNESCO's Science Cooperation office for the Middle East. His headquarters were in Cairo, but his work covered a vast area, from Morocco in the west to Pakistan in the east. This post, with all its multiple demands, occupied him for nearly 5 years. There were of course intervals when he returned to the United States for visits, and there were vacation trips: the most notable of the latter came when he went into part of West Pakistan and then into Wakan, a region of Afghanistan that was closed to foreigners. After due consideration the principal chieftain of the region decreed that Jeffries must leave, and return to Pakistan. Accompanied by a bodyguard, and treated with all due courtesy, he rode on a yak to the border.

My stay in Cambridge, England as Fulbright Lecturer; return to Harvard and move to Harvard College

It was during those years, however, that Jeffries and I somehow found the time to synthesize a large portion of our teaching notes into a book. My own life had also undergone important changes during those years, though they did not take me out of academic life. In 1952 I spent most of the year in Cambridge, England, as a Fulbright Lecturer. Some of those who had taught and guided me in my student days, notably Keilin and Roughton, were still there, and still provided inspiration and intellectual stimulus. I occupied a small room in the hut adjoining the Institute of Biochemistry, where A.C. Chibnall was in charge of the protein group, having retired from the Biochemistry professorship. The high point of excitement there was the work of Fred Sanger, in the room adjoining mine. He had already worked out the sequence of the B chain of insulin and, with E.O.P. Thompson, was close to completing the sequence of the A chain. One of the great landmarks of biochemistry was nearly complete. Naturally I saw much of Max Perutz and John Kendrew also; the three dimensional structures of myoglobin and hemoglobin were still unknown, but Perutz was on the verge of discovering how to use isomorphous replacements with heavy atoms, to solve the phase problem and determine the structures. Those months were immensely stimulating, and they included memorable travels in England, Scotland, Wales, France and Italy, with my wife Margaret.

On my return to Harvard Medical School, late in 1952, I came to realize that Edwin Cohn was seriously ill, having suffered a series of small strokes. He remained full of activity, and he was much concerned with ideas about protein sequence and structure, which in the end proved unrelated to the great work of Sanger, Kendrew, and Perutz, which was establishing the foundations of the subject. Cohn was also extremely active, and very effective, in establishing the Protein Foundation, an organization set up to continue the work of the laboratory on plasma proteins, in close relation to the work of clinicians, immunologists, and experts in blood typing. That organization, now called the Center for Blood Research, is flourishing and productive in Boston today, under the direction of Dr. Douglas M. Surgenor, who was closely associated with Cohn in its original development.

In October 1953, Cohn died from a stroke at the age of 60. He had suffered from high blood pressure for over 20 years, but had not spared himself. He led a life of intense activity, especially in the war years, when he initiated and directed the great plasma fractionation program, and was involved in all its complicated relations with clinical testing, industrial production of the products initially developed in the laboratory, and the extensive government funding of our work by the Comittee on Medical Research of the Office of Scientific Research and Development. Cohn, with his energy and driving ambition, and his often dominating personality, made many enemies. Some people found him impossible to work with. Nevertheless I found the laboratory that he headed an immensely stimulating place, with new ideas and new experiments to test them constantly emerging, and a lively and congenial group of fellow workers.

I add this account to my previous memoir on Edwin Cohn [21]; an extraordinary man who deeply influenced my life. After his death I had to decide whether to stay in the Department of Biophysical Chemistry (as it had been renamed in 1950) or to move to the Biological Laboratories in Cambridge, as I had previously planned. I finally decided to make the move. This brought me into close relations with a group of fellow chemists, biochemists and biologists. Frank H. Westheimer and Konrad Bloch, about the same time, came to the Harvard Chemistry Department from the University of Chicago. In the Biology Department, George Wald and Kenneth Thimann were active in biochemical research and teaching; and before long Jim Watson arrived from CalTech. For the first time there was a really substantial group of biochemists in Cambridge, and we launched a graduate program in the field. (Such a program had existed at the Medical school since 1908.) Before long, Matthew Meselson and A.M. Pappenheimer Jr. arrived; and Jim Watson's brilliant work in protein biosynthesis drew Walter Gilbert from theoretical physics into Molecular Biology, with results well known to all workers in the field.

Our book on biophysical chemistry

Jeffries and I, though far apart in space for the most part, continued the writing of our book on biophysical chemistry [49]. We added our bit to the load of mail between Boston and Paris, later between Boston and Cairo. Jeffries returned to the United States for a time every year, and I was in Europe for visits nearly every year, so there were often meetings for discussion and criticism. We divided the work between us, he doing the first draft of some chapters, I of others. Finally, late in 1957, we had a manuscript that we were willing to send to the publishers, and we were proud to have it appear the next year. We envisaged it as Vol. I of a two-volume work, and rashly inscribed "Volume I" on the cover and title page. Alas! Vol. II has never appeared, and now we have sadly recognized that it never will. Unexpected events changed our lives. What happened to Jeffries — indeed one of the happiest events of his scientific life — will be told below. I, on the other hand, was persuaded to become Editor-in-Chief of the *Journal of Biological Chemistry*, beginning very early in 1958 for what proved to be a 10-year term, with one year's leave of absence in 1963–64. The burden was at first enormous — I had known nothing like it, except during the war years — but after 6 months or so I learned to live with it, and was able after a while to take some enjoyable and much needed vacations, and also in the laboratory to direct a busy and excellent group of young investigators in research on carbonic anhydrase, which delighted me. I did care deeply about my work as an Editor, and I trust that I served the journal well. Certainly the work brought me closely into touch with a superb group of fellow members of the Editorial Board, and I cherish deeply the lasting friendships that arose from that time of service. In the midst of all this, however, Vol. II fell by the wayside, to the lasting regret of Jeffries and myself.

Actually we did write substantial parts of it, and Jeffries in particular wrote much. We saw it as divided into two parts, the first dealing with the size, shape, structure, and electric charge distribution of macromolecules, and the second with the physical chemistry of blood. I had already written rather extensively on the first of these general subjects, in the book with Edwin Cohn [36] and later in a long chapter in the first edition of *The Proteins* [99a] and I did some work for Vol.II on the optical properties of molecules in relation to their structure. Jeffries, however, wrote at least two major chapters that were nearly in final form. One, on Brownian movement and diffusion, was a really masterly presentation of fundamental concepts and their implications, at a deep level, more advanced in character than most of the book. (There was to have been another chapter, never written, on the experimental study of diffusion). The other was on dielectric dispersion and absorption, the natural complement to the discussion of dielectric constants and dipole moments in Vol. I. It dealt with complex polarizability, as a function of frequency, on Kirkwood's concept of charge fluctuations in macromolecules, the Debye dispersion theory, and the Maxwell-Wagner theory, among other topics. Looking at these two chapters again, I feel an enduring regret that they have never been published.

The chapters dealing with the physical chemistry of blood were to include a discussion of plasma proteins and their role, the structure and contents of the red cell, and the course of its life and eventual breakdown, the physical chemistry of Hb, and the physico-chemical changes in blood during the respiratory cycle. Also we planned to give an updated version of Henderson's nomographic description of the blood. We had discussed and written on these topics, year after year, with our students, when we had been teaching together in our course on physical biochemistry. Here again, in the actual drafting of chapters for Vol. II, Jeffries did substantially more than I.

Naturally we have often been asked when Vol. II was going to appear. Sadly it seems clear that the answer now must be "never", but I must at least note these plans, and the work that we, and Jeffries in particular, did for their partial fullfilment.

Plate 6. Jeffries Wyman (left) and John Kendrew. Caprarola, Italy, June 1976.

Invitation to Jeffries from Eraldo Antonini; return to hemoglobin research; the laboratory in Rome

In late 1959 Jeffries had finished his service with UNESCO and, at John Kendrew's invitation, had temporarily settled in Cambridge, England, at Peterhouse for the autumn term. One day he got a note from Roughton that Quentin Gibson (then in Sheffield) was bringing an "interesting young Italian" with him to lunch with Roughton at Trinity College. Gibson in fact was unable to come, but the young Italian, Eraldo Antonini, was there. He was already known for his work with A. Rossi Fanelli on Hb. As Jeffries has written to me in a recent letter:

> "After lunch, as we strolled about the Backs, I was immediately captivated by Eraldo's engaging personality and irrepressible enthusiasm ... So when Eraldo proposed my paying a visit to the group in Rome I did not hestitate to accept. I well remember the drive down there in early spring and the warm welcome with which I was received by Eraldo and Rossi Fanelli ... When I left I carried back with me a proposal to join the group in Rome as a guest scientist. It was a tempting offer from all points of view and after talking it over with my wife we decided to accept, at least for a trial period."

What began as a "trial period" soon became an absorbing life that has held Jeffries in Rome for the last 24 years. Certainly in modern times it is most unusual, if not unprecedented, for a scientist to leave research for as long as 8 years, becoming deeply involved in other responsibilities, and then return to science, and make his most important research contributions in the years that followed. Yet that is, in fact, what Jeffries achieved.

The laboratory in Rome was an exciting place, full of enthusiasm and high spirits, with plans for new experiments constantly bubbling up in the mind of one researcher or another. That would lead to lively discussion within the group, and generally in a very short time to new experiments. Rossi Fanelli, as head of the Department, provided constant interest, help and encouragement. Eraldo's enthusiasm and warmth were immensely influential in attracting other young people and making them active in the work. Jeffries, with his long experience of Hb, and his searching analytical mind, was constantly involved with the younger members of the group, and the large majority of the papers from the laboratory, during the 1960s and much of the 1970s, involve him. Indeed his close involvement with the group has never ceased. At the same time he was developing his own particular lines of thought. These involved a great extension and deepening of his earlier use of linked functions, leading to the powerful concept of the binding potential; his collaboration with Monod and Changeux on an influential model for allosteric transitions, and his own further development of the concept of allosteric linkage; the use of Legendre transformations to generalize the study of the thermodynamics of multicomponent systems; polysteric linkage in reversibly dissociating systems; linkage in systems involving phase changes, as in sickle-cell Hb; and steady-state phenomena. Many of these studies involved collaboration with other workers, sometimes in Rome, sometimes elsewhere, and his capacity for entering into such collaborative arrangements is notable. Others, like the development of the binding potential and the analysis of allosteric linkage were entirely his own.

In what follows I turn first to the general activities of the group in Rome, and then to the lines of work that Jeffries followed, as mentioned in the last paragraph. Both accounts must necessarily be rather brief, and both deserve far more than I can give them here.

Jeffries and Eraldo worked a great deal on the relation between the oxygenation of Hb and its tendency to dissociate into subunits. It was clear that a connection existed, but they failed repeatedly, especially with deoxy-Hb, to get definitive and reproducible results, as Jeffries has noted in a recent letter. That problem was not to be fully resolved until years later, and then in other laboratories. An unexpected and important finding came in work with Romano Zito, in studying the action of carboxypeptidases A and B (CPA and CPB) in digesting Hb. The action of CPA, which removed the C-terminal His-146 and Tyr-145 from the β chain, was particularly striking. The resulting product (Hb CPA) had lost all cooperative interactions in oxygen binding, the Hill coefficient (n_H) dropping from 2.9 to 1 while the O_2 affinity increased substantially. The Bohr effect also was greatly decreased. On the other hand, removal of the C-terminal arginine from the α chain by CPB produced little change in cooperative oxygen binding, though it increased the oxygen affinity and substantially diminished the Bohr effect [100, 101]. Later Kilmartin and Wootton [102] in Cambridge showed that the removal of Tyr-145 in the β chain was crucial for abolishing cooperative interactions; if only His-146 were removed, the value of n_H was still about 2.5, and the Bohr effect was about half normal. These findings proved to be important later for the structural work of Perutz in interpreting conformation changes on ligand binding.

Work in the Rome laboratory also involved kinetic studies of ligand binding and release. Eraldo Antonini had spent some time with Quentin Gibson, who was then in Sheffield, in learning techniques of stopped-flow measurements, and Gibson also came to visit in Rome. John F. Taylor, who had earlier worked with A.B. Hastings [93] on oxidation-reduction potentials, came over to do further work in Rome in this field. Other visitors who made an active contribution to the thinking and activity of the laboratory included Charles Tanford, W.F. Libby, and Bo Malmström. Maurizio Brunori, who had recently joined the laboratory, also played a major role in the work, and he has been a central figure in the work of the Rome laboratory ever since [103, 104]. Since Eraldo's tragically early death in 1983, he has recently become the head of the Department.

Another American visitor, the immunologist Morris Reichlin, worked with the group on the immunochemical differences between horse oxy- and deoxy-Hb [105] and later on the properties of the isolated α and β chains [106], which had been prepared by a greatly improved method developed by Bucci and Fronticelli [107]. Indeed their work stimulated a whole series of researches, in the Rome laboratory and elsewhere, to define the properties of the separated chains, and this was naturally important for understanding the nature of the interactions between the subunits on assembly into the Hb tetramer. They studied in detail the reassociation of the separated chains into the normal Hb tetramer, observing changes in light absorption, sedimentation, and other properties. (see for instance [108]). The isolated α chains naturally showed no

Plate 7. Jeffries Wyman's (left) 75th birthday. R.W. Noble (centre), E. Antonini (right). Caprarola, Italy, June 1976.

cooperativity in ligand binding, nor did the β tetramer (Hb H) as had been previously shown by Benesch et al. [109], who also detected no Bohr effect.

There were also many studies of oxidation-reduction potentials of Hb's of various species, and of the isolated chains, in which Maurizio Brunori played a leading role. John Taylor was also actively involved, along with Wyman, Antonini, Rossi-Fanelli and others. All these studies, through 1970, are described in the book of Antonini and Brunori [110], an admirable survey of the whole field of Hb and myoglobin studies as they stood at that time.

The numerous international relations of the Rome laboratory led to a series of meetings known as the La Cura conferences. One frequent visitor to the laboratory, Rufus Lumry from the University of Minnesota, through his constant and challenging discussion and questioning, was an important catalyst in arousing Eraldo and the others to invite an international group of Hb researchers, and other related people, to get together. I took part only in the first two conferences, and I have a particularly vivid recollection of the first, in May 1966. La Cura is a small village, north of Rome, and a few kilometers south of Viterbo. It contained a castello that was built a few centuries ago by some local nobleman, and was available to house the participants. Sometimes we met indoors for discussion, but often we preferred to be outside, in the beautiful Italian spring weather, with a blackboard set up to scribble on. Inside or out, the discussions were equally animated. Eraldo Antonini, more than anyone else, was the presiding spirit; his charm and enthusiasm were infectious. Rufus Lumry also bubbled over with ideas. There was much discussion about the dimer-tetramer equilibrium; at that time most of us, including the Rome group, still believed that ligand binding to the dimer was cooperative. Roughton was there, and was able to present evidence that finally settled a dispute over binding of CO_2 as carbamate to oxy- and deoxy-Hb. He and Jeffries had debated this for nearly 20 years. Roughton was convinced, from his earlier work with J.K.W. Ferguson, that deoxy-Hb bound more CO_2 as carbamate than did oxy-Hb, and that this made a significant contribution to CO_2 transport, and CO_2 discharge in the lung. Jeffries had suggested in his 1948 review [94], that this might not be due to an intrinsic change in CO_2 affinity of the amino groups of Hb on oxygenation, but could perhaps be explained simply in terms of pH changes on oxygenation. The question was difficult to resolve, and the disagreement persisted. Finally, at the 1966 La Cura meeting, Roughton was able to present clearcut evidence, from measurements with improved technique by Rossi-Bernardi and himself, that deoxy-Hb bound more CO_2 as carbamate than oxy-Hb, when the pH was carefully held constant [111]. The work was elegant and appeared decisive; Jeffries and the rest of us agreed, and that long-standing dispute was settled.

Lynn Hoard of Cornell, at that meeting, presented his work on the structures of simpler heme compounds that were analogous in ligand binding to Hb, showing that the iron atom could fit into the heme plane, in low-spin forms analogous to oxy-Hb, but was forced out of it in high-spin forms, similar to deoxy-Hb. This major structural change, which later proved so significant for Max Perutz in his studies of Hb structure, was quite new to most of us then.

There was much discussion of other matters, of course, but those points stand out in my recollection. In the afternoons we explored the neighborhood; we were in Etruscan country, and there were fascinating Etruscan remains, including tomb paintings, and Christian churches, as well as the general charm of the Italian countryside in spring. Emilia ("Milina") Chiancone, who was then and still is an important member of the Rome laboratory, was active in these explorations. Eraldo Antonini, with his great zest in life, and his varied intellectual interests, was a central figure in the whole gathering, and Rufus Lumry was a warm and lively companion.

Only the earliest of these conferences were actually held in La Cura. The old castello was in a shaky state, and about 1969 was decreed unsafe for human habitation. In all, however, there were eight La Cura conferences; the last was held in Caprarola, near Rome, in 1980. Eraldo was a leading spirit in all of them, and it is sad to realize that we have now lost that buoyant and eager spirit, when he was only 51 years old [112].

Facilitated diffusion

Jeffries was much involved in all the activities I have described above, but he also had other important involvements. One was with the interpretation of facilitated diffusion — a phenomenon discovered independently (1959–60) by P.F. Scholander and J.B. Wittenberg. Dr. Wittenberg, during a stay in Rome, introduced Jeffries to the results of the experimental studies, in which diffusion of oxygen was measured across a slab of myoglobin (Mb) or Hb solution, supported on a Millipore membrane, with oxygen at a fixed pressure P on one side, and oxygen at nearly zero pressure on the other. At suitable concentrations of Mb or Hb the flux of oxygen across the membrane was greatly enhanced by the presence of the protein. Ferri-Hb (Hb^+) which cannot combine with oxygen, produced no such enhancement; it was evident that reversible combination between the ligand-binding protein and the gas was essential. The diffusion of the protein, carrying oxygen with it, was clearly involved. Other workers had already developed theoretical approaches to the problem; Jeffries pushed the analysis substantially further, although he could not obtain an analytic solution to the fundamental equation that he derived. He considered, but disproved, the possibility that the facilitation might be due to rotary diffusion of Mb or Hb, a process analogous to the motion of a water wheel. It was definitely translational diffusion that was involved. He concluded that the role of Mb in muscle was not merely to act as a storehouse of oxygen; in fact Mb could, by diffusion, take over a large fraction of oxygen transport within the cell, whenever the oxygen pressure dropped to around 10 mm Hg or less [113] (see also Wittenberg's review [114]).

Pondering the mathematical difficulties involved, Jeffries consulted his friend J.D. Murray of Oxford. Murray, using a singular perturbation approach, was able to reduce Wyman's nonlinear second-order differential equation to the form of a quadratic, and to calculate the concentration of free oxygen, and the fractional saturation of the protein, as a function of distance across the membrane. The results were in good agreement with Wittenberg's experimental data, and supported Wyman's conjecture [113] that

Mb or Hb, in these experiments, was everywhere in equilibrium with free O_2 within the diffusing layer [115]. Murray and Wyman [116] then applied Murray's mathematical analysis to the case of carbon monoxide diffusion, for which it was well known [114] that Mb or Hb produced virtually no facilitation of the diffusion of the gas. At first sight this would appear surprising; why should two ligands (CO and O_2), binding at the same sites in the protein, behave so differently? The conclusion of the mathematical analysis, expressed in physical terms, was that the pumping system on the low pressure side of the diffusing layer in the experiments could not keep the CO pressure low enough to cause the protein to release CO to any appreciable extent. This was due to the very high affinity of the protein for CO, determined primarily by the extremely small rate constant for HbCO dissociation. The protein was thus essentially saturated with CO at all points in the diffusing solution. They enunciated a general principle applying to such cases:

"...the carrier molecule can only make a significant contribution (to facilitated diffusion) when its ligand affinity is such in relation to the ambient ligand activity that it is only partially saturated with ligand in some part of the system [116, p.5906]."

Jeffries Wyman and the European Molecular Biology Organization (EMBO)

Although Rome was the center of his scientific life, Jeffries had wider European contacts and responsibilities, notably his major role in the development of EMBO. In the words of John Kendrew in 1976:

"His years of service as the first Secretary General of EMBO, as a member of its Council and, at some time or another, of almost every Committee established by that organization, have given him a key role in the whole community of molecular biology. Of all the activities of EMBO the one that lay nearest to his heart was the foundation of the European Laboratory of Molecular Biology; he played a central part in developing the philosophy underlying the laboratory, in drafting the plans for it to be submitted to Governments and in persuading them to accept these: at the ceremony for signing the Agreement establishing the Laboratory he was described in a speech by one Government delegate as the "grand old man of European molecular biology" — a phrase to the second word of which he took considerable exception but which we may nevertheless think in every other respect particularly appropriate" [117]

(This passage, in a short tribute to Jeffries, on his 75th birthday, was actually written by John Kendrew, who was uniquely qualified, as the first President of EMBO, to know what Jeffries had achieved there).

Allosteric proteins and the model of Monod, Wyman and Changeux

Jeffries first came to know Jacques Monod well during his years as Science Attaché in Paris (1952–55), when he would often join Monod, with André Lwoff, François Jacob, and others, at their informal luncheons; and he went rock climbing with Monod in the Forest of Fontainebleau. A decade later, after he had settled in Rome, he developed a much closer relation with Jacques, as their scientific interests in the role of regulatory proteins in biology — the class of proteins that Monod termed allosteric — converged. Jeffries in 1951 had perceived the role of conformational changes in Hb as a determining factor in cooperative ligand binding and in the Bohr effect [99]. Monod's concerns came from a different direction, from the regulatory aspects of metabolism, especially in bacteria, and the phenomena of feedback inhibition of metabolic pathways by the end products of such pathways, operating so as to inhibit the action of the enzyme that catalyzes the first step in the pathway, thus sparing the production of unneeded end products. Pioneer work in this field had been done in the United States, by A. Novick and L. Szilard, by A.B. Pardee, and by H.E. Umbarger. It was Monod and François Jacob, however, who set forth a systematic conception of allosteric enzymes: proteins that can bind effector molecules at sites quite distinct (and perhaps distant) from the catalytic site, thereby inducing conformational changes that promote or inhibit catalytic activity. In general there need be no chemical similarity at all between the effector and the substrate of the enzyme, since they bind at entirely distinct sites. These concepts found their fullest expression in a masterly review by Monod et al. on *Allosteric Proteins and Cellular Control Systems* [118].

Jeffries has given an account of the events that resulted in the Monod-Wyman-Changeux (MWC) model for allosteric proteins [119]. After completing his review of linkage relations [96], he gave a seminar in Paris, arranged by Changeux in the autumn of 1964, on ligand-linked conformational changes in Hb and other proteins. By that time there was evidence for many proteins that homotropic (cooperative) and heterotropic (regulatory control) interactions could be explained by such conformational changes; and Perutz's structural work on Hb had shown that it existed in two major quaternary forms, one characteristic of oxy-, the other of deoxy-Hb. Perutz's work showed moreover that the four heme groups were not close together; they were separated by distances of the order of 30 Å. Practically everyone had thought in earlier years that, when binding at one site strongly affected the strength of binding at another, the two sites must be close together; their mutual interactions would fall off rapidly with increasing distance between them. Perutz's data on Hb seemed to call for a quite different approach. The heme groups in Hb were so far apart that in either major quaternary structure, they might be expected to bind oxygen quite independently of one another; one should seek to explain cooperativity in terms of a marked difference in ligand affinity between the two conformations.

Jacques Monod and Jean-Pierre Changeux were aroused to activity by Jeffries' seminar. As to the resulting paper, Jeffries writes of

> "Jacques' quickness of response to an idea, which was one of his great qualities. The paper as it stands was written almost wholly by him and was presented to me more or less as a fait accompli for discussion and criticism" [119, p. 223].

It included an extensive discussion of symmetry relations in the assembly of oligomeric proteins, almost entirely due to Jacques. Jeffries, in spite of his long-standing interest in the symmetry (or asymmetry) of ligand binding curves, and their implications, has always remained skeptical concerning these symmetry arguments, which certainly stand on shakier ground than the thermodynamic arguments in the rest of the paper. Before the paper was published, its authors engaged in a vigorous discussion of it with Perutz, Kendrew, Francis Crick and Sydney Brenner

"a rather formidable group of critics"

as Jeffries has remarked. Their fire was particularly directed against the symmetry arguments, but all finally agreed that it should be published as it stood.

The resulting paper [120] must be so well known to most readers of this story that I need only recall its major concepts: the assumption that allosteric proteins are oligomers, composed of several subunits (protomers), the oligomer being capable of existence in (at least) two distinct conformations, denoted as T and R, which differ in their affinity for ligands and effectors. Both T and R contain n ligand-binding sites, and both T and R can exist in $n + 1$ distinct states of ligand binding: T_o, T_1, $T_2 \ldots T_n$ and R_o, R_1, $R_2 \ldots R_n$. The microscopic dissociation constants, K_T, for T, and K_R for R, with the ligand, F, are all identical, and independent of the presence or absence of bound ligand at the other sites. The two fundamental parameters are $L = (T_o)/(R_o)$ and $c = K_R/K_T$. To fit the oxygen-binding data of horse Hb, at pH 7 and 19°C, obtained by R.W.J. Lyster in Roughton's laboratory, Monod et al. had to take $L = 9.054 \times 10^3$, and $c = 0.014$. Thus in deoxy-Hb, the protein must be overwhelmingly in the T form, and in oxy-Hb overwhelmingly in the R form, according to the model. Homotropic interactions, for proteins obeying the postulates of the model, must always be cooperative; heterotropic interactions, which must (in the model) be due exclusively to displacement of the R/T equilibrium by the effector, could involve either activation or inhibition at the catalytic sites (for an enzyme) or at the ligand-binding site (for Hb).

This model has been immensely influential, and has stimulated a great deal of research. It has certainly performed the primary function of a good working hypothesis — to coordinate a large body of experimental data, and to stimulate further research to test the implications of the model.

The best known alternative model is that of Koshland, Nemethy, and Filmer [121] (KNF model). Unlike the concerted conformational change in the MWC model, the KNF model involves sequential changes, as in the original model of Pauling [90], but is elaborated in much greater detail. With suitable choice of constants, the KNF model could fit the equilibrium data for oxygen binding by Hb about as closely as the MWC model. Jeffries has pointed out that both models can be regarded as special cases of a more general, though still quite limited, scheme of ligand binding [122].

Evolution of Wyman's concepts of linkage; the binding potential and other thermodynamic potentials

Allosteric, polysteric and polyphasic linkage

Over nearly 20 years, Jeffries has developed his earlier treatment of linkage phenomena into a more powerful method of analysis, applicable to a wider range of experimental conditions. Drawing on a series of his papers [96, 123–127] and on a paper *Linkage Graphs*, which at the moment is still unpublished*, I will try to indicate the nature of the thinking involved, and indicate something of the uses of the analysis.

The binding potential. Consider a macromolecule, P, with r binding sites for a given ligand X. We use the symbol x for the activity of X, and for this discussion assume the concentration equal to the activity. P can exist in $r + 1$ different states of ligand binding:

$$P_{\text{total}} = (P_o) + (PX) + (PX_2) + \ldots.(PX_r)$$

where (P_o) denotes the concentration of P molecules with no ligand bound; (PX) is the total concentration of all the r microscopic forms of P with one ligand bound, etc. The equilibrium between P_o and the molecules of class (PX_i) is given by:

* This paper has now appeared in Quart. Rev. Biophys, 17 (1985) 453–488.

$P_o + iX \rightleftharpoons PX_i;\ (PX_i) = K_i(P_o)x^i$

Then the ratio $(P_{total})/(P_o)$ is given by:

$$(P_{total})/(P_o) = \sum_{i=0}^{r} K_i x^i = 1 + K_1 x + K_2 x^2 + \qquad (10)$$
$$K x^n = Q(x)$$

The series $Q_r(x)$ represents a partition function, the term in x^i denoting the relative amount of P_{total} that is in the form PX_i, (P_o) being normalized to unity. Note that $K_o = 1$. Wyman has termed this series a binding polynomial.

Szabo and Karplus [128], and other authors since, have made powerful use of generating functions, closely related and often identical to the binding polynomials of Wyman. Their use often greatly simplifies the analysis of complicated binding equilibria [129].

The mean number of X molecules bound, $\bar{n}_x$, per molecule of P, is given by differentiation (here we call the partition function simply Q for brevity):

$$\bar{n}_x = \frac{\mathrm{d} \ln Q}{\mathrm{d} \ln x} = \frac{x\,\mathrm{d}Q}{Q\,\mathrm{d}x} = \sum_{i=0}^{r} i\,K_i x^i \Big/ \sum_{i=0}^{r} K_i x^i \qquad (11)$$

For hemoglobin ($r = 4$), taking x as O_2, this gives the Adair equation:

$$\bar{n}_{O_2} = \frac{K_1 x + 2\,K_2 x^2 + 3\,K_3 x^3 + 4\,K_4\,x^4}{1 + K_1\,x + K_2 x^2 + K_3 x^3 + K_4 x^4} = 4\,\bar{X} \qquad (12)$$

In such cases as this, when we can describe the process of ligand binding by a binding polynomial such as $Q_r(x)$, the Wyman binding potential, Λ, is simply equal to $RT \ln Q_r(x)$, and $\mu_x = \mathrm{RT} \ln x$ is the chemical potential of x. (Strictly speaking $\mu_x = RT \ln (x/x_o)$ where x_o is the activity of x in a standard state, chosen for convenience. If we set $x_o = 1$, in the appropriate units, the x_o term vanishes; in any case it vanishes on differentiating μ_x.) Then on differentiating Λ with respect to μ_x, we obtain the amount of bound ligand $\bar{n}_x$, per mol of macromolecule, as follows obviously from Eqn. (11):

$$\left(\frac{\partial \Lambda}{\partial \mu_x}\right)_{p,\, T,\, \mu_j \neq x} = \frac{R\, T\, \partial \ln Q}{R\, T\, \partial \ln x} = \bar{n}_x \tag{13}$$

The differential in Eqn. (13) is written as a partial derivative, since Λ is a function, not only of the potential of x, but of all the other potentials of molecules in the system that can be regarded as ligands of the macromolecule. Thus the total derivative of Λ at constant p and T is given (for t ligands)
by:

$$\mathrm{d}\,\Lambda = \sum_{i=1}^{t} \left(\frac{\partial \Lambda}{\partial \mu_i}\right) \mathrm{d}\,\mu_i = \sum_{i=1}^{t} \bar{n}_i \mathrm{d}\mu_i \tag{14}$$

Here the summation extends over all the components of the system, except the reference component (the macromolecule) which we designate as component zero. Eqn. (14) is simply related to the well known Gibbs–Duhem equation, which applies to all the $n + 1$ components of the system, at constant pressure and temperature:

$$\sum_{i=0}^{t} n_i \mathrm{d}\mu_i = 0; \text{ or } - n_o \mathrm{d}\mu_o = \sum_{i=1}^{t} n_i \mathrm{d}\mu_i \tag{15}$$

If we divide every term in Eqn. (15) by n_o, we have:

$$- \mathrm{d}\mu_o = \sum_{i=1}^{t} (n_i/n_o) \mathrm{d}\mu_i = \sum_{i=1}^{t} \bar{n}_i \mathrm{d}\mu_i \equiv \mathrm{d}\Lambda \tag{16}$$

Thus the binding potential Λ is equal, with change of sign, to the chemical potential of the reference component [126].

The phenomenon of heterotropic linkage arises from the interaction of two (or more) ligands binding to the same macromolecule. In addition to ligand X for which P contains r binding sites, there may be another ligand Y, for which P carries s binding sites, which are distinct from the binding sites for X. Thus the P molecules can exist in $(r + 1)$ $(s + 1)$ distinct states of binding,

$$\left[P\right]_{\text{total}} = \sum_{i=0}^{r} \sum_{j=0}^{s} P X_i\, Y_j$$

The binding polynomial then becomes

$$Q(x,y) = \sum_{i=0}^{r} \sum_{j=0}^{s} Kijx^i y^j$$

By the same reasoning as before, this gives

$$\bar{n}_x = \left[\frac{\partial \ln Q(x,y)}{\partial \ln x}\right]_y \text{ and } \bar{n}_y = \left[\frac{\partial \ln Q\,(x,y)}{\partial \ln y}\right]_x \qquad (17)$$

Since d ln $Q_{(x,y)}$ is a perfect differential, this gives immediately the fundamental linkage relation:

$$\left(\frac{\partial \bar{n}_x}{\partial \ln y}\right)_x = \left(\frac{\partial n_y}{\partial \ln x}\right)_y = \frac{\partial^2 Q(x,y)}{\partial \ln x . \partial \ln y} \qquad (18)$$

It may be of course that the binding of the ligands is unlinked. In that case the coefficients in Eqn. (18) are equal to zero. If linkage does occur, however, the reciprocal relations must hold; if we know the effect of x on the binding of y, we can immediately calculate the effect of y on the binding of x, and vice versa.

Relations of the same form are readily extended to three or more ligands. Jeffries treated them in detail [96] before developing the concept of the binding potential; but the use of binding potentials and binding polynomials increases the power of the method, especially in more complicated cases. Some examples of the treatment are given by Hess and Szabo [129], including a derivation of the equations for the MWC model by a different formalism from that of the original authors [120].

Eqn. (14), in which all the n ligands of the macromolecule are defined in terms of their potentials, was used in the original definition of the binding potential [123]. This is, however, only one of a whole class of such potentials, as Jeffries soon recognized. In planning an experiment in a system of several components we may choose to specify some of them by fixing their potentials — for a gas, for example, by equilibrating the solution with a gas phase at specified partial pressure, or for a solute by equilibrating the solution with the pure solute in a crystalline phase. We may choose to define other components in terms of their amounts (per unit amount of reference component). Thus the amount of water in the system is commonly fixed by the way we make it up; we control the quantity of it, rather than the potential (which we could determine, if we wished, by measuring its vapor pressure; but for most purposes an attempt to measure and regulate its vapor pressure would be cumbrous and unnecessary).

In a later analysis [126], Jeffries broadened his original concept of the binding potential to include a large class of such potentials, in which some components of the system are defined in terms of their individual potentials (the μ's) and others in terms of their quantities (the n's).

To avoid confusion, I note that the term "potential" is used in thermodynamics in two different but related senses. In Eqn. (5) and later equations it denotes the increment in Gibbs energy per unit quantity of a component added to the system, at constant p and T. This is the most familiar use. The term is also used to denote several quantities characteristic of a system as a whole, such as the internal energy (E), the enthalpy (H), the Helmholtz energy (A), and the Gibbs energy (G). All these quantities were originally defined by Gibbs [95, pp.85–89], who, however, used a different notation.

To understand the basis of the analysis of the various binding potentials [126], it is best to begin with the internal energy E. Following Gibbs, we note that E is a function of the entropy and volume of the system, and of the relative amounts, $n_o, n_1, n_2 \ldots n_t$, of the components in a $t+1$ component system. Then a differential increment of E is given by:

$$\mathrm{d}E = T\mathrm{d}S - p\mathrm{d}V + \sum_{i=0}^{t} \mu_i \, \mathrm{d}n_i \tag{19}$$

Here $T\mathrm{d}S$ is the heat absorbed by the system, $-p\mathrm{d}V$ is the pressure volume work done by the system on its surroundings (and therefore carries a negative sign) and μ_i is the increment in E per unit amount of component i added, at constant entropy and volume:

$$\mu_i = (\partial E/\partial n_i)_{S,V\ n_j \neq n_i}$$

Note that every term on the right of Eqn. (19) is of the form $X\mathrm{d}\bar{Y}_l$ where the X's are all intensive quantities; that is, they are independent of the size of the system and depend only on its temperature, pressure, and composition. The Y's, on the other hand, are extensive properties; the values of S and V increase in direct proportion with the size of the system, and so does the sum of all the n's (at constant composition). Now, by a simple transformation, one can obtain new functions in which one or more of the $X\mathrm{d}Y$ terms is replaced by $-Y\mathrm{d}X$. Gibbs in this way derived the potentials H, A, and G from E.

The procedure is a standard one, and is given in almost any text on thermodynamics. Consider for instance the Helmholtz energy $A = E - TS$. Its derivative is:

$$\mathrm{d}A = \mathrm{d}E - T\mathrm{d}S - S\mathrm{d}T \tag{20}$$

Inserting the value of $\mathrm{d}E$ from Eqn. (13) this gives at once:

$$\mathrm{d}A = -S\mathrm{d}T - p\mathrm{d}V + \sum_{i=1}^{t} \mu_i \mathrm{d}n_i \tag{21}$$

Thus A is a function of the intensive variable T, and the extensive variable V, and of the n's. It is particularly suitable for study of systems at constant volume and temperature. Comparing Eqn. (15) with Eqn. (13), the only difference between $\mathrm{d}E$ and $\mathrm{d}A$ is that $T\mathrm{d}S$ has been replaced by $-S\mathrm{d}T$. We obtain the Gibbs energy $G = E + pV - TS$ by a further step, employing the same kind of transformation. This converts $-p\mathrm{d}V$, in the expression for $\mathrm{d}E$, into $+V\mathrm{d}p$ in the expression for $\mathrm{d}G$.

$$\mathrm{d}G = -S\mathrm{d}T + V\mathrm{d}p + \sum_{i=1}^{t} \mu_i \mathrm{d}n_i \tag{22}$$

Thus G is a function of the two intensive variables, pressure and temperature, and in practical work is generally the most convenient function to employ.

The transformations described above are known to mathematicians as Legendre transformations, though Gibbs, in making use of them, did not call them by that name. Now, as Jeffries pointed out in 1975, one can immediately extend the use of these transformations to any or all of the terms in the μ's and n's, involving the potentials and the quantities of the components of the system, thereby obtaining an expression in which a given $\mu_i\,\mathrm{d}n_i$ in Eqn. (22)

is converted into $- n_i\, \mathrm{d}\mu_i$. We can do this for any or all of these terms, in any combination. For practical purposes we consider just the class of potentials in which pressure and temperature are the controlling variables, as with the Gibbs energy G.

I follow Jeffries, for systems containing a macromolecule, such as Hb by taking it as the reference component, and I denote it by subscript zero. The amounts of the other t components of the system, which we denote as ligands of the macromolecule, are then expressed as molar ratios $\bar{n}$, relative to n_o. Thus $\bar{n}_i = n_i/n_o$.

An increment in the Gibbs free energy, G, as usually defined, is given by Eqn. (22). For simplicity consider the case of a three-component system composed of the macromolecule and two ligands, denoted as components 1 and 2. Then, using Legendre transformations, we can derive four different functions for increments of Gibbs free energy; these all involve p and T as independent variables, and each involves one of the four possible combinations of the two $\bar{n}$'s and the two μ's.

$$\begin{aligned}
-\mathrm{d}\bar{G}^{\mathrm{I}}(p,T,\mu_1,\mu_2) &= \bar{S}\mathrm{d}T - \bar{V}\mathrm{d}p + \bar{n}_1\mathrm{d}\mu_1 + \bar{n}_2\mathrm{d}\mu_2 \equiv \mathrm{d}\Lambda^{\mathrm{I}} \\
-\mathrm{d}\bar{G}^{\mathrm{II}}(p,T,\mu_1,\bar{n}_2) &= \bar{S}\mathrm{d}T - \bar{V}\mathrm{d}p + \bar{n}_1\mathrm{d}\mu_1 - \mu_2\mathrm{d}\bar{n}_2 \equiv \mathrm{d}\Lambda^{\mathrm{II}} \\
-\mathrm{d}\bar{G}^{\mathrm{III}}(p,T,\bar{n}_1,\mu_2) &= \bar{S}\mathrm{d}T - \bar{V}\mathrm{d}p - \mu_1\mathrm{d}\bar{n}_1 + \bar{n}_2\mathrm{d}\mu_2 \equiv \mathrm{d}\Lambda^{\mathrm{III}} \\
-\mathrm{d}\bar{G}^{\mathrm{IV}}(p,T,\bar{n}_1,\bar{n}_2) &= \bar{S}\mathrm{d}T - \bar{V}\mathrm{d}p - \mu_1\mathrm{d}\bar{n}_2 - \mu_2\mathrm{d}\bar{n}_2 \equiv \mathrm{d}\Lambda^{\mathrm{IV}}
\end{aligned} \tag{23}$$

Here the bar superscripts denote quantities per mol of component zero (the macromolecule). The quantity denoted G^{IV} is the Gibbs energy increment as usually defined. That denoted as G^{I}, with negative sign, is the binding potential Λ, as Jeffries originally defined it [123]. However, each of the four potentials, as defined in Eqn. (23) provides an equally legitimate basis for calculating relations derived from experimental work. If there were three ligands, instead of two, there would be 8 $(=2^3)$ possible combinations of terms involving the μ's and $\bar{n}$'s, instead of the 4 in Eqn. (17); for t ligands, there would be 2^t such combinations.

Experimentally we may choose to fix the potentials of some of the components at certain values, as by specifying the pH and the partial pressure of oxygen (see the discussion in connection with Eqns. (6)-(9) inclusive). Other components (e.g., water) might be fixed by specifying the amount present ($\bar{n}_i = n_i n_o$) per mol of macromolecule.

This formulation greatly extends the range of analysis for the thermodynamics of multicomponent systems; it is a natural extension of the original treatment of Gibbs. In retrospect it may appear an obvious extension, but to my knowledge no one before Jeffries Wyman ever pointed it out. The basic Eqn. (13) for the binding potential still holds for any component which is defined in one of the equations in terms of its potential μ_i:

$$\left(\frac{\partial \Lambda}{\partial \mu_i}\right)_{p,\, T,\, \text{other variables}} = \bar{n}_i \tag{24}$$

However, the other variables that must be held constant in evaluating $\bar{n}_i$ for any particular component depend upon the way we have set up the experiment. For Λ^{I} in Eqn. (23) above, Eqn. (24) applies to either component 1 or 2, the potential of the other one being held constant, since both are expressed in terms of their potentials. For Λ^{II}, Eqn. (24) applies to component 1, on condition that the quantity (not the potential) of component 2 is held constant; it does not apply to component 2, which is expressed in terms of $\bar{n}_2$, not μ_2. For Λ^{III} the relations are reversed: Eqn. (24) holds for component 2, holding the quantity of 1 constant. For Λ^{IV}, Eqn. (24) does not apply at all. In this case the system is defined in terms of the quantities of the components and the potential of a component is given by the usual equation in terms of Gibbs energy

$$\mu_i = (\partial G/\partial \bar{n}_i)_{p,T,nK} \equiv -\left(\frac{\partial \Lambda^{IV}}{\partial n_i}\right)_{p,T,nK}$$

One must note that not all interactions with ligands can be well expressed in terms of a binding polynomial, as we have done here in Eqns. (10) and (17). Such polynomials are particularly suitable for ligands, such as oxygen in Hb solution, for which the amount of "free" ligand is small compared to that chemically bound. The general relations for the binding potentials, such as Eqn. (24), remain valid, no matter whether the "binding" involves chemical combination or whether it is a much looser form of association. Such considerations apply notably to water, which is indeed the most important of all ligands for biopolymers in general. Some water molecules, in a protein solution, may be quite tightly bound; others, some distance from the protein molecule, may behave much as if they were in pure water, but all are undergoing constant diffusion and interchange. The general thermodynamic relations still hold. Charles Tanford, who spent several months with the hemoglobin group in Rome, has given a thermodynamic analysis of hydration in protein solutions in terms of the Wyman linkage functions [130].

Some other aspects of linkage

Hill coefficient, median ligand activity, negative cooperativity, allosteric linkage, polysteric and polyphasic linkage

The general linkage relations just described are fundamental; on other important work that Jeffries has done in this field I must touch more briefly. His use of the Hill coefficient, n_H — named after A.V. Hill, who presented in 1910 an "aggregation theory" of oxygen binding by Hb (for background see [131]) — has proved a very powerful method for characterizing cooperativity in homotropic interactions. For the binding of a ligand X by a macromolecule, n_H is defined by the relation:

$$n_H = \frac{d \ln [\bar{X}/(1-\bar{X})]}{d \ln x} = \frac{1}{\bar{X}(1-\bar{X})} \frac{d\bar{X}}{d \ln x} \tag{25}$$

For a set of equivalent and independent binding sites, $n_H = 1$ in all cases. For equivalent sites with cooperative binding it is greater than 1 for most of the binding curve, usually reaching a maximum near half saturation of the ligand-binding sites ($X = 0.5$), but this maximum value cannot be greater than the total number of binding sites. As $\bar{X}$ approaches either zero or unity, n_H approaches unity at either end. The form of a typical Hill plot for the binding of oxygen by hemoglobin is shown in Fig. 11.

Jeffries developed this type of plot, in which log $[(\bar{X}/(1-\bar{X})]$ is plotted against log $\bar{X}$ (or log p, for a gaseous ligand) and two parallel diagonal lines are drawn, along the limiting slopes of the curve, at $\bar{X} = 0$ and $\bar{X} = 1$. The perpendicular distance between these parallel lines, multiplied by $RT \sqrt{2}$, gives the Gibbs free energy of

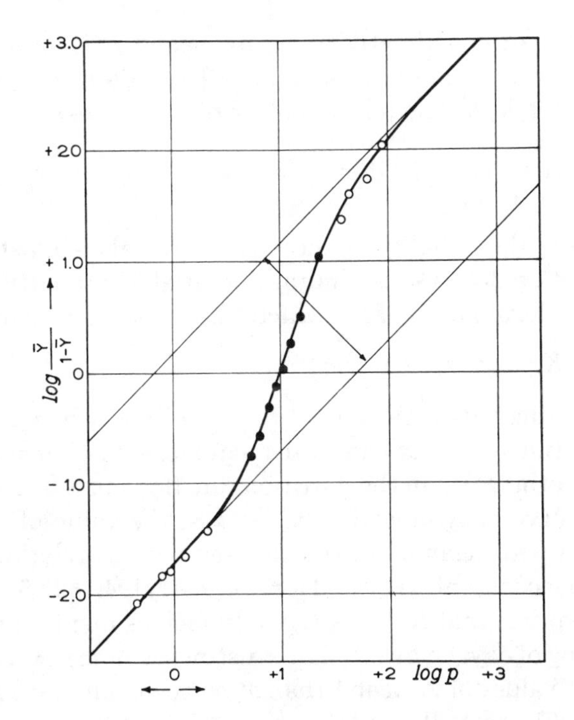

Fig. 11. Hill plot of oxygen equilibrium of horse hemoglobin in 0.6 M phosphate buffer, pH 7.0 and 19°C. Points are divided into three groups, according to techniques employed for different ranges of $\bar{Y}$. Total ΔG of interaction 2.6 kcal·mol^{-1}; $n_H = 2.95 \pm 0.05$ at $\bar{Y} = 0.5$. Data from unpublished measurements of R.L.J. Lyster in Roughton's laboratory; from [96].

interaction between the binding sites, ΔG_I. The procedure is described in detail in [96, pp.234–238]. In general, for a macromolecule containing r binding sites $\Delta G_I = RT \ln (k_r/k_1)$, where k_1 and k_r are the first and last intrinsic binding constants for the ligand, corrected for statistical effects; $k_1 = K_1/r$ and $k_n = rK_r$, where K_1 and K_r are the macroscopic binding constants, as given for instance in Eqn. (10). For human Hb A, n_H at $\overline{X} = 0.5$ is ordinarily in the range of 2.7–3.0, and ΔG_I of the order of 3 kcal/mol. (ΔG_I is taken as positive when the interactions are stabilizing, i.e. $k_n > k_1$). For a non-cooperative Hb, such as Hb H, $n_H = 1$ at all $\overline{X}$ values and $\Delta G_1 = 0$.

The median ligand activity, x_m for ligand x, is another fundamental concept, first formulated in [96]. It is defined, for the binding curve for X, by the relation (for r binding sites):

$$\int_{x=0}^{x_m} \bar{n}_x \mathrm{d}\ln x = \int_{x=x_m}^{\infty} (r - \bar{n}_x)\,\mathrm{d}\ln x \tag{26}$$

Graphically the equality of these two integrals is illustrated in a diagram [96, p.238]. Using binding potentials [124], Jeffries has given a simple evaluation of x_m, which leads to the relation:

$$x^{-r}_m = K_r;\ \text{or}\ K_r x^r_m = 1 \tag{27}$$

On comparing this with Eqn. (10) we see that, when $x = x_m$, the concentrations of the unliganded macromolecule (Po) and the fully liganded form (PX_r) in the partition function must be equal. If the binding curve is symmetrical $x_m = x_{0.5}$, the value of x for 50% saturation with ligand. Jeffries pointed out the relation between the binding constants required for symmetry [94,132]; for 4 binding sites it requires that $K_1K_4 = K_2K_3$. In fact, as mentioned already, the binding of oxygen by Hb shows distinct asymmetry; the curve is steeper at values of $\bar{X}$ near 1 than at corresponding values near 0. Weber [133] and Peller [134] have considered further the significance of this asymmetry. Weber employs an asymmetry index, defined as ln (K_1K_4/K_2K_3) which is also equal to $\ln(k_1k_4/k_2k_3)$. (Weber uses dissociation constants, whereas here I am using association constants; hence the ratio is written as the reverse of his.) For oxygen binding by Hb the asymmetry index is always positive, which is physiologically advantageous, since it means that the usual unloading curve for oxygen release in the tissues, from $\bar{X} = 1$ to about $\bar{X} = 0.7$, is particularly steep. The asymmetry is markedly increased by the binding of organic phosphates to Hb. Both Weber and Peller have discussed the significance of the asymmetry and its relation to the coupling of similar subunits (α or β) in the Hb tetramer.

Negative cooperativity. Values of $n_H < 1$ are often found in studies of rate constants in enzymes with several binding sites for substrate. (Kinetic data of course stand on a different footing from true equilibrium data, but they can still be utilized with reasonable assumptions, as in the MWC paper [120], and many others.) In true negative cooperativity, the binding at one site tends to inhibit binding at others, in some cases by an allosteric mechanism. However, as Jeffries has been careful to point out, when the various binding sites have different intrinsic affinities for ligand, the binding curve becomes depressed below that for equivalent and independent sites. If the successive binding constants are markedly different, the curve will separate into several distinct steps; but if they are closely similar the curve may mimic one for equivalent sites with negative cooperativity, even though the sites are actually independent. Indeed there may be concealed cooperative interactions between the sites, masked by the fact that their affinities are different. It therefore may be unwise to conclude that cooperative interactions are absent from a system, simply because it gives Hill coefficients below unity. Even in a system like Hb, with n_H substantially greater than unity, the value of n_H will be somewhat depressed if the α and β chains differ significantly in ligand affinity — as, under some circumstances, they certainly do.

Allosteric linkage. Jeffries has treated this form of linkage in great detail [124] and has presented the major features of the analysis as part of a more general picture [127]. If a macromolecule can exist in r different conformations, all in equilibrium with one another, we can describe the binding potential of the system, Λ, in terms of the binding potentials of the individual forms, by:

$$\Lambda = RT \ln \sum_{i=1}^{r} \nu_{io}\, e^{\Lambda i/RT} \tag{28}$$

Here ν_{io} is the fraction of the macromolecule in the i'th conformation, in the absence of ligand. Thus $\Sigma\nu_{io} = 1$. (Note that, if there were only one conformation, Eqn. (28) would simply reduce to $\Lambda = \Lambda_i$.) In contrast, if the different conformations were frozen, and could not undergo reversible transitions, the binding potential would be simply the sum of the Λ's for the individual forms:

$$\Lambda = \sum_{i=1}^{r} \nu_i \Lambda_i \tag{29}$$

For the MWC model, with 2 quaternary conformations and 4 identical and independent binding sites in each, one can show that Eqn. (28) becomes, for a ligand x:

$$\Lambda = RT \ln [\nu_{10} (1 + k_{11}x)^4 + \nu_{20} (1 + k_{12}x)^4]$$

$$\Lambda = RT \ln [1 + k_{11}x)^4 + L_o (1 + k_{12}x)^4] \tag{30}$$

Here L_o is the ratio T_o/R_o, as previously defined in the discussion of the MWC theory; and k_{11} and k_{12} are the ligand **association** constants for the T and R forms, respectively (the earlier discussion employed dissociation constants). The binding curve corresponding to Eqn. (30) is steeper at every point than a simple titration curve. The total Gibbs energy of binding is always reduced by the effect of the allosteric transition, due to what Jeffries has called

"...the progressive yielding of the macromolecule to the ligand" [127, p.139].

If the two conformations were frozen as in Eqn. (29), the binding curve would be biphasic, with two separate steps, the high affinity form reacting first.

Polysteric linkage. Macromolecules may undergo ligand-linked dissociation and association, as with the tetramer-dimer equilibrium in Hb, which is shifted toward dissociation by binding of ligand, with the converse effect that the dimer has a higher ligand affinity than the tetramer. Colosimo et al. [135] have analyzed such situations in detail, with respect to the homotropic and heterotropic effects that are observed. Qualitatively the binding patterns for ligands are similar to those in allosteric systems with no dissociation, but the form of the curves is naturally dependent on the total macromolecule concentration. The Hill coefficient, n_H, is generally smaller than for an otherwise equivalent allosteric system. When the number of binding sites is large, as for some of the hemocyanins, the cooperativity of homotropic interactions can become very large, even when the total Gibbs energy of interaction is quite small; and heterotropic control of the binding of one ligand by another can produce sharp transitions. By a somewhat different approach, the work of Gary Ackers and his colleagues has provided probably the most extensive and accurate evidence on the linkage of oxygen binding and tetramer-dimer dissociation in Hb [136].

Woolfson and Bardsley [137] have provided an analogue of the binding polynomial for the general case of ligands binding to an aggregating macromolecule.

Polyphasic linkage. Linkage relations can also be of importance in a system composed of two or more phases. This is strikingly illustrated by sickle-cell Hb (Hb S) which Jeffries has studied in a long-standing collaboration with Stanley J. Gill, working together both in Rome and at the University of Colorado. Hb S, above certain concentrations and at sufficiently low oxygen pressures, precipitates out of solution, to form what appear to be microtubules. In one of their several papers together, Wyman and Gill [138] have given a detailed analysis of polyphasic linkage, both for this particular system and in more general terms. If a solution of Hb S, at a sufficiently high concentration, is progressively deoxygenated, deoxy Hb S begins to precipitate at a critical oxygen pressure, the value of which is a function of the concentration. Thus the liquid phase becomes progressively more dilute as deoxygenation proceeds, and the vapor pressure (activity) of water must therefore increase, as more Hb S passes from the liquid into the solid phase. This transfer is of course oxygen-linked, but the data show that there is no link between the activity (potential) of water and that of oxygen in the system. Wyman and Gill obtain a phase diagram by plotting μ_{H_2O} against μ_{O_2}. The chemical potential of Hb S can be obtained from activity coefficient measurements, by osmotic pressure or sedimentation. The authors conclude with a more general treatment of polyphasic systems, showing how one can apply the concepts of linkage and binding potentials to such systems, as a natural extension of the treatment of allosteric and polysteric linkage.

From equilibrium to steady states

Equilibrium theory, powerful as it is, cannot be applicable to the dynamics of living cells. Much of the time, however, they may be regarded as being in a steady state, requiring energy consumption for its maintenance, but with an essentially constant concentration of many intermediates in vital processes persisting over considerable time intervals. Here, even in simple cases, the analysis becomes decidedly more difficult than in a true equilibrium. Jeffries has become increasingly concerned with such problems in recent years. His first theoretical analysis [139] dealt with a very simple model system, designated as "The Turning Wheel". This relates to an enzyme that can exist in two conformations, M_1 and M_2, both of which can combine with a substrate L at a single site. In the working cycle the enzyme-substrate complex dissociates irreversibly, giving rise to an end product in which only one of the conformations functions as an intermediate. The diagram below illustrates the model.

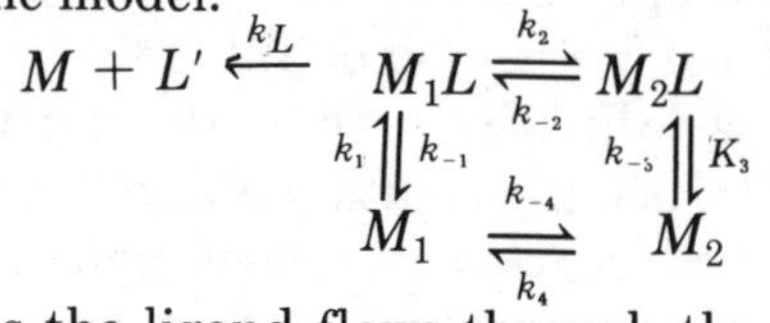

$$\begin{array}{ccccc} M + L' & \xleftarrow{k_L} & M_1L & \underset{k_{-2}}{\overset{k_2}{\rightleftharpoons}} & M_2L \\ & & k_1 \upharpoonleft\downharpoonright k_{-1} & & k_{-3} \upharpoonleft\downharpoonright K_3 \\ & & M_1 & \underset{k_4}{\overset{k_{-4}}{\rightleftharpoons}} & M_2 \end{array}$$

As the ligand flows through the system, the net velocity of each indicated transition must settle down to a value independent of time, both in magnitude and sense. In the case illustrated above, where the irreversible process upper left involves the form M_1L, the cyclic process will operate counterclockwise (if it had involved M_2L the flow would have been clockwise). The amounts of the enzyme-substrate complex relative to the free enzyme, in both forms, prove to be given by a simple equation, which holds in the steady state:

$$\frac{M_1L + M_2L}{M_1 + M_2} = \frac{A_1L + B_1L^2}{A_2 + B_2L} \qquad (31)$$

The four A and B coefficients are third-degree functions of the velocity constants in the diagram; these are given explicitly by the analysis. The equation above is of the same general form as one for true equilibrium of a macromolecule containing not **one**, but **two**, binding sites for its ligand.

In a slightly more complex form of the model, the free energy released by an irreversible process, at one point in the cycle, can be used to drive backward another irreversible (endergonic) process, provided that the coupled overall process involves a negative ΔG value (i.e. is exergonic).

Jeffries later extended this type of analysis further, to a model steady state system involving a macromolecule MN, in which M is a dissociable prosthetic group of the protein N, and X is a ligand of free concentration x. Simple as this may sound, the mathematical analysis is formidable; it was carried out by two Italian mathematicians, Gaetano Fichera and Maria Sneider [140,141]. The outcome, for this restricted case, was clear: there is always a unique and asymptotically stable critical point:

"the solution of the differential system will always approach this critical point by some kind of relaxation process however complex" [140, p.2].

Very recently, in collaboration with the biophysicist Giorgio Careri, Jeffries has considered how the energy input in such a system as the "turning wheel" may be stored [142]. They propose that it may be in the form of a soliton: a solitary wave that propagates as a localized, mobile, dynamically self-sufficient energy packet. A soliton is a stable entity; it disappears only when triggered by some extra energy source, then decaying into thermal phonons. A.S. Davydov has suggested that the energy packet of a soliton may be associated with the amide portion of an α-helix (essentially $C{=}O$ stretching) coupled to longitudinal sound waves. Recent work in Careri's laboratory [143] on acetanilide crystals, which contain an organized system of amide groups analogous to that in an α-helix, has shown the presence of solitons there. Several lines of evidence suggest that the situation in protein systems may be similar, especially for enzymes with bound coenzyme subject to release at some stage in the cycle. Direct evidence for this suggestion is still lacking, but the hypothesis certainly deserves further exploration.

Fish hemoglobins and the expedition to the Amazon

The study of Hb of fish, and later of amphibia, has been a major interest of the Rome laboratory in recent years; Maurizio Brunori has played a particularly important part in this [144,145]. Much of the work centered on trout Hb's; of the four distinct forms of Hb in trout blood, the two major ones, trout I and IV, are of particular interest. Trout I shows no Bohr effect, the oxygen affinity is independent of pH. The heat of reaction with oxygen, which for most Hb's is exothermic throughout the entire course of ligand binding, is for trout I endothermic at low levels of ligand binding, exothermic at high levels; presumably this change is associated with a conformation change [146].

Trout IV is very different; its oxygen affinity is very sensitive to pH, falling sharply as the pH decreases; in slightly acid solutions even pure oxygen cannot saturate the binding sites. This behavior, commonly known as the Root effect, a rather extreme form of the Bohr effect, has an important biological function; input of acid into the blood, as it circulates through the swimbladder, causes release of oxygen into the bladder, serving to maintain the specific gravity of the fish in the water [147,148]. The combination of these two major trout proteins, I and IV, with their very different properties, plays an important biological role. The respiratory system of the fish would have difficulty in functioning with either one alone; each provides part of the total balanced function. Jeffries has provided a subtle thermodynamic analysis of the functional roles of both.

A major scientific expedition to study the fishes of the Amazon — a sequel to several earlier expeditions — occurred in November and December of 1976. More than 2000 species of fish and other aquatic animals have been described in the Amazon, and the remarkable seasonal changes in water level (up to 10 meters) create unusual living conditions in hypoxic waters that have required special adaptations on the part of these organisms. The scientific interest of studies on them is therefore great.

The expedition was headed by Austen Riggs, of the University of Texas, and included 22 scientists from six countries. From the Rome group Wyman and Brunori were involved; also R.W. Noble and J. Bonaventura, who had earlier done much work with them in Rome. The expedition's work was done in the splendidly equipped research vessel, the *Alpha-Helix*, managed by the Scripps Institution of Oceanography in San Diego. Thus conditions for research were excellent, and the workers were enthusiastic. The findings were far too numerous to discuss here; they were presented in a series of papers in *Comparative Biochemistry and Physiology* (1979). Jeffries wrote an introduction to the whole report [149] and Austen Riggs [150] provided a survey of the results obtained in the whole series of papers. These studies provide material of great interest for both molecular and evolutionary biology.

Recent activities of Jeffries and myself

In recent years Jeffries and I have followed quite different paths in our major interests. He is pursuing the deep problems of equilibrium thermodynamics, of linkage and of steady-state phenomena, that I have sketched briefly here. Each year he comes for a month from Rome to the University of Colorado, where he works in collaboration with Stanley Gill and Paul Phillipson. With Stanley Gill he is writing a book, to bring together these fundamental lines of thought and present them in unified fashion. He no longer makes expeditions into the wilder places of the world, as he has done in the past among Alaskan eskimoes, and in Afghanistan, Sudan, the Atlas Mountains, the Amazon basin, and elsewhere. Instead he spends the summer in France, in a section of Burgundy to which he is deeply attached. In Rome he remains close to the group in the laboratory, now headed by Maurizio Brunori, but works largely on his own.

I gave up the running of a laboratory in 1972. I had already retired in 1968 from editing the *Journal of Biological Chemistry*, after ten strenuous and rewarding years, and later I wrote a short history [151] of the journal for its 75th birthday in 1980. It was about 1960 that I first became involved with the study of carbonic anhydrase and that was the center of my research interests for more than a decade. In spite of my own heavy involvement in edi-

Plate 8. Jeffries Wyman and John Edsall, at Anne Wyman's house, May 1984.

torial work, the research made excellent progress, thanks to the gifted group of young investigators whom I was fortunate to have in the laboratory. I name them here in admiration and affectionate regard for their work: Egon E. Rickli, Barbara H. Gibbons, Lynn M. Riddiford, S.A.S. Ghazanfar, Dirck V. Myers, J.Mc.D. Armstrong, Jacob A. Verpoorte, Carole Lindblow, Louis E. Henderson, Dennis L. Drescher, Shelby L. Bradbury (who now publishes as Shelby L. Berger), Julia F. Clark, Anna J. Furth, Philip L. Whitney, Allan J. Tobin, Friedrich Dorner, Pierre Henkart, and Raja G. Khalifah, who has remained deeply involved with carbonic anhydrase ever since. Some of this group, such as Dennis Drescher and Julia Clark, never published a paper from the laboratory, but all of them made significant contributions.

Because of my long-standing interest in the history of science, I had conducted a small seminar in the history of biochemistry at Harvard for several years before I became Emeritus in 1973. During a stay at the National Institutes of Health as a Fogarty Scholar in 1971 I studied the early history of blood and Hb in a paper that appeared a year later [131]. I have pursued the subject ever since [152], though with many interruptions for other activities, and the completion of a monograph on the subject is now the chief aim of my historical research.

In another related activity, I became Chairman of a Committee working to save the unpublished correspondence and other archival papers of important workers in biochemistry, molecular biology, and biomedical science generally. The work was centered at the Library of the American Philosophical Society in Philadelphia, with David Bearman as the principal worker, ably assisted by Margaret Miller and Matthew Konopka. The principal result was a monograph, listing and describing some 600 collections [153].

Much more in this field remains to be done; our monograph represents only a beginning. The physicists are far ahead of the biochemists and molecular biologists in recognizing the importance of preserving the correspondence, notebooks, and other unpublished records of those who have played a significant part in the advancement of their science, as may be seen in the collections of the Niels Bohr Library of the American Institute of Physics in New York. In physiology the notebooks of Claude Bernard, for instance, have proved extraordinarily revealing of the development of his research, in studies by historians such as M.D. Grmek and F.L. Holmes.

My interest in the remarkable properties of water and their biological significance has led to a long-distance collaboration, lasting for over 6 years, with my Australian friend and colleague, Hugh McKenzie of Canberra. The resulting review, in two parts, on *Water and Proteins* [68] required much time, thought and effort by both of us. The plan had germinated in 1970, when I was a Visiting Lecturer at the Australian National University for a few months. I did not get to Australia again, but Hugh paid several visits to the United States during our joint work. As I think of this, and of my work on carbonic anhydrase, I can say that much of my thought in recent years has centered on two triatomic molecules, water and carbon dioxide. My sense of their fundamental importance goes back to my teacher, L.J. Henderson, whose pathbreaking book of 70 years ago [154] devoted a special chapter to each of them.

I served several years on the Interunion Commission on Biothermodynamics, for which Ingemar Wadsö of Lund was an admirable Chairman. One offshoot of its work was a small monograph aimed primarily at biochemists, by Herbert ("Freddie") Gutfreund and myself [154a]. The Commission gave us encouragement to undertake the work, but only the authors are responsible for the result.

Always, and especially since the era of nuclear weapons began in 1945, I have been concerned about the state of the world in general. I had grown up in a household where political events were constantly discussed, so my heightened concern after 1945 came naturally. Even apart from nuclear weapons the enormous and rapidly increasing influence of science and technology on political affairs and world ecology was becoming apparent. All this has led me to be, for many years past, an active member of the Federation of American Scientists; I have served on one occasion as its Vice Chairman and for several years as its Secretary. Also, in the American Association for the Advancement of Science, I served for over 5 years as a member, and for 2 years as Chairman, of its Committee on Scientific Freedom and Responsibility [155,156], whose active concerns include the rights and responsibilities of "whistle blowers" who publicly call attention (sometimes at considerable risk to themselves) to technological problems that may endanger the public. Also the Committee has been urgently concerned with violations of human rights of fellow scientists in other countries.

Sometimes I find myself torn between my deep interest in the history of modern science, and my concern over the forces that appear to be pushing the world toward disaster. In a more peaceful and stable era I could have devoted myself happily to scholarly pursuits; but the insanity of the present arms race alarms and appals me. I am keenly aware of the recent evidence that prolonged darkness and cold from smoke and dust arising from a nuclear war, might precipitate a 'nuclear winter' that could threaten human survival and life support systems in general. This simply reinforces the overwhelming evidence that nuclear weapons are not to be regarded as weapons of war at all, since their use would probably be suicidal for both the contending sides, and neither could achieve any political aim. Yet I feel no confidence that the present political and military leaders, either in the United States or in the Soviet Union, understand these fundamental consequences of the revolutionary developments of our time.

Jeffries does not take part in these political involvements. He remains, on the one hand, preoccupied with the deep intellectual problems that I have attempted to sketch here. His life, however, is certainly not centered on such abstractions; he has a keen appreciation of the fascinations of the visible world; its sights and sounds and smells. Indeed during several periods in his life he has been a painter of obvious talent, although he has made no use of this particular gift for some years past. His range of interests in history, literature and art is wide, as are his personal friendships. All this Jeffries and I share in common. We have lived in a century full of upheavals and of horrors, but also one of marvelous creativity, especially in science. Both of us can say that, in spite of troubles and difficulties, we have lived exceptionally fortunate lives.

Acknowledgments

I am indebted to NSF Grant SES8308892 for support in my historical research.

I have discussed everything in this article with Jeffries Wyman, whose comments and suggestions have naturally been essential; but the responsibility for any errors is mine alone.

REFERENCES

1 J.T. Edsall, Annu. Rev. Biochem., 40 (1971) 1-28.

2 J.T. Edsall, in M. Kageyama et al. (Eds.), Science and Scientists, Essays by Biochemists, Biologists and Chemists, Japan Scientific Societies Press, Tokyo, 1981, pp. 335-341.

2a J.C. Aub and R. Hapgood, Pioneer in American Medicine: David Linn Edsall of Harvard. Harvard Medical Alumni Association, Harvard University Press, Cambridge, MA, 1970.

3 A.S. Packard, Biog. Mem. Natl. Acad. Sci. USA, 2, (1886) 75-126.

4 G.W. Allen, William James: a Biography, Viking Press, New York, 1969, p. 95.

5 Sir F.G. Hopkins, Autobiography, in Hopkins and Biochemistry, J. Needham and E. Baldwin (Eds.), Heffer, Cambridge, U.K., 1949, pp. 3-25.

6 R.E. Kohler, Isis, 69, (1978) pp. 331-355.

7 J. Needham, Persp. Biol. Med., 6, (1962) 2-46.

7a N.W. Pirie, Sir Frederick Gowland Hopkins, in G. Semenza (Ed.), Selected Topics in the History of Biochemistry, Vol. 35, Elsevier, Amsterdam, 1983 pp. 103-128.

8 R.W. Clark, The Life and Work of J.B.S. Haldane, Hodder and Stoughton, London, 1968 (paperback reprint Oxford University Press, London, 1984).

9 N.W. Pirie, J.B.S. Haldane (1892-1964), Biog. Mem. Fellows Roy. Soc. London, 12, 1966, pp. 219-249.

10 G.S. Adair, Proc. Roy. Soc. London, A 109, (1925) 292-300.

11 G.S. Adair, Proc. Roy. Soc. London, A 120, (1928) 573-603.

12 P. Johnson and M. Perutz, Gilbert Smithson Adair (1896-1979), Biog. Mem. Fellows Roy. Soc., 27, (1981) 1-27.

13 Q.H. Gibson, Francis John Worsley Roughton (1899-1972), Biog. Mem. Fellows Roy. Soc., 19, (1973) 563-582.

14 B.S. McEwen, Alfred Ezra Mirsky (1900-1974), Am. Phil. Soc. Yearbook, (1976) pp. 100-103.

15 J.T. Edsall, Adv. Protein Chem., 24 (1970) vii-x.

16 D. Keilin, History of Cell Respiration and Cytochrome, Cambridge University Press, Cambridge, 1966.

17 T. Mann, David Keilin (1887-1963), Biog. Mem. Fellows Roy. Soc. 10 (1964) 183-197.

17a E.F.Hartree, Keilin, cytochrome, and the concept of a respiratory chain. in G. Semenza (Ed.), Of Oxygen, Fuels, and Living Matter, Part 1, Wiley, Chichester, 1981, pp. 161-227.

18 B. Katz, Archibald Vivian Hill (1886-1977), Biog. Mem. Fellows Roy. Soc. 24 (1978) 71-149.

19 J. Wyman, J. Physiol. 61 (1926) 337-352.

20 A.K. Smith and C. Weiner, (Eds.) Robert Oppenheimer: Letters and Recollections, Harvard University Press, Cambridge, MA, 1980, pp. 85-98.

21 J.T. Edsall, Biog. Mem. Natl. Acad. Sci. USA., 35 (1961) 47-84.
22 F. Hofmeister, Arch. Exp. Pathol. Pharmakol., 24 (1888) 247-260.
23 A.A. Green, J. Biol. Chem., 93 (1931) pp. 495-516, pp. 517-542; 95 (1932) 47-66.
24 R.M. Ferry and A.A. Green, J. Biol. Chem., 81, (1929) 175-203.
25 S.P. Colowick, Science, 128, (1958) 519-521.
26 J.T. Edsall and W.H. Stockmayer, Biog. Mem. Natl. Acad. Sci. USA, 52, (1981) 335-377.
27 G. Scatchard, Equilibrium in Solutions and Surface and Colloid Chemistry. See introduction by I.H. Scheinberg, pp. vii-xviii and Scatchard's autobiographical note, pp. xix-xxxiv. Harvard University Press, Cambridge, MA, 1976.
28 P. Debye and E. Hückel, Phys. Z., 24, (1923) 185-206.
29 E.Q. Adams, J. Am. Chem. Soc., 38, (1916) 1503-1510.
30 N. Bjerrum, Z. Physik. Chem. 104 (1923) pp. 147-173; English translation in Bjerrum, N., Selected papers, Munksgaard, Copenhagen, 1949, pp. 175-197.
31 J.T. Edsall and M.H. Blanchard, J. Am. Chem. Soc., 55, (1933) 2337-2353.
32 J.T. Edsall, Nature, 285, (1981) p. 58.
33 J.T. Edsall and A. Meister, Jesse P. Greenstein (1902-1959) in J.T. Edsall (Ed.), Amino Acids, Proteins and Cancer Biochemistry, Academic Press, New York, 1960, pp. 1-8.
34 J.T. Edsall, Trends in Biochem. Sci., 7, (1982) 414-416.
35 G. Scatchard, J. Chem. Phys., 33, (1959) 1279-1281.
36 E.J. Cohn and J.T. Edsall, Proteins, Amino Acids and Peptides as Ions and Dipolar Ions. Reinhold, New York, 1943 700 pp.
37 A. von Muralt and J.T. Edsall, J. Biol. Chem., 89, (1930) 315-386.
38 J.T. Edsall and A. von Muralt, Trends in Biochem. Sci., 5, (1980) 228-230.
39 P. Debye, Polar Molecules, Chemical Catalog Co., New York, 1929, 172.
40 J. Wyman, Phy. Rev., 35, (1930) 623-634.
41 J.T. Edsall and A. Meister, Jesse P. Greenstein (1902-1959) in J.T. Edsall, (Ed.), Amino Acids, Proteins and Cancer Biochemistry, Academic Press, New York, 1960 1-8.
42 J. Wyman and T.L. McMeekin, J. Am. Chem. Soc., 55, (1933) 908-914.
43 J. Wyman, Chem. Rev., 19, (1936) 213-239.
44 J.P. Greenstein, J. Wyman and E.J. Cohn, J. Am. Chem. Soc., 57, (1935) 637-642.
45 J.T. Edsall and J. Wyman, J. Am. Chem. Soc., 57, (1936) 1964-1975.
46 J. Wyman, J. Am. Chem. Soc., 58, (1936) 1482-1486.
47 L. Onsager, J. Am. Chem. Soc., 58, (1936) 1486-1493.
48 J.G. Kirkwood, J. Chem. Phys. 7, (1939) 911-919.
49 J.T. Edsall and J. Wyman, Biophysical Chemistry. Academic Press, New York, 1958.
50 J.G. Kirkwood, in [36] Chapter 12, p. 296.
51 J. Wyman, J. Biol. Chem., 90, (1931) 443-476.

52 H.O. Marcy and J. Wyman, J. Am. Chem. Soc., 64, (1942) 638-643.
53 J.L. Oncley, Chem. Rev., 30, (1942) 433-450.
54 J.L. Oncley, in [36], Chapter 22.
55 J.G. Kirkwood, J. Chem. Phys., 2, (1934) 351-361.
56 J.G. Kirkwood, in [36], Chapter 12.
57 J.T. Edsall, Trends in Biochem. Sci., 7, (1982) 414-416.
58 T.L. McMeekin, E.J. Cohn and J.T. Weare, J. Am. Chem. Soc., 57, (1935) 626-633.
59 J. Traube, Ann. Chem. Pharm., 265, (1891) 27-55.
60 I. Langmuir, J. Am. Chem. Soc., 39, (1917) 1848-1906, (see 1889-1895).
61 J.A.V. Butler, Trans Faraday Soc., 33, (1937) 229-236.
62 G.S. Hartley, Aqueous Solutions of Paraffin Chain Salts: a Study of Micelle Formation. Hermann, Paris, 1936.
63 J.T. Edsall, J. Am. Chem. Soc., 57, (1935) 1506-1507.
64 N. Nichols, R. Sköld, O. Spink, J. Suurkusk and I. Wadsö, J. Chem. Thermodyn. 8, (1976) 1081-1093.
65 H.S. Frank and M.W. Evans, J. Chem. Phys., 13, (1945) 507-532.
66 W. Kauzmann, Adv. Protein Chem., 14, (1959) 1-63.
67 C. Tanford, The Hydrophobic Effect, 2nd ed., Wiley, New York, 1980.
68 J.T. Edsall and H.A. McKenzie, Adv. Biophys., 10, (1978) 137-208; 16, (1983) 53-183.
69 E.J. Cohn, T.L. McMeekin, J.T. Edsall and H. Blanchard, J. Am. Chem. Soc., 56, (1934) 784-794.
70 P. Drude and W. Nernst, Z. Phys. Chem., 15, (1894) 79-85.
71 J.D. Bernal and R.H. Fowler, J. Chem. Phys., 1, (1933) 533-548.
72 F.T. Gucker, W.L. Ford and C.E. Moser, J. Phys. Chem., 43, (1939) 153-168.
73 F.T. Gucker and T.W. Allen, J. Am. Chem. Soc., 64, (1942) 191-199.
74 J.T. Edsall, J. Chem. Phys., 4, (1936) 1-8.
75 J.T. Edsall, J. Chem. Phys. 5, (1937) 225-237.
76 J.T. Edsall and H. Scheinberg, J. Chem. Phys., 8, (1940) 520-525.
77 K. Linderstrøm-Lang, C.R. Trav. Lab. Carlsberg., 15, No. 7 (1924).
78 R.K. Cannan, Chem. Rev., 30, (1942) 395-412.
79 E.J. Cohn and J.T. Edsall, in [36], Chapter 20.
80 J.G. Kirkwood, in [36], Chapter 12.
81 C. Tanford and J.G. Kirkwood, J. Am. Chem. Soc., 79, (1957) 5333-5339.
82 C. Tanford, Adv. Protein Chem., 17, (1962) 69-165.
83 E.J. Cohn, in E.C. Andrus et al., (Eds.), Advances in Military Medicine, Vol. I, Little Brown, Boston, 1948 pp. 364-443.
84 J.T. Edsall, Adv. Protein Chem., 3, (1947) 383-479.
85 L.J. Henderson, Blood: a Study in General Physiology. Yale University Press, New Haven, CT, 1928.
86 A.B. Hastings, D.D. Van Slyke, J.M. Neill, M. Heidelberger and C.R. Harington, J. Biol. Chem., 60, (1924) 89-153.

87 A.B. Hastings, J. Sendroy Jr., D.C. Murray and M. Heidelberger, J. Biol. Chem., 61, (1924) 317-335.
88 B. German and J. Wyman, J. Biol. Chem., 117, (1937) 533-550.
89 J. Wyman, J. Biol. Chem., 127, (1939) 1-13.
90 L. Pauling, Proc. Natl. Acad. Sci. USA, 21, (1935) 186-191.
91 J. Wyman, J. Biol. Chem., 127, (1939) 581-599.
92 J. Wyman and E.N. Ingalls, J. Biol. Chem., 139, (1941) 877-895.
93 J.H. Taylor and A.B. Hastings, J. Biol. Chem., 131, (1939) 649-662.
94 J. Wyman, Adv. Protein Chem., 4, (1948) 407-531.
95 J.W. Gibbs, On the Equilibrium of Heterogeneous Substances. The Collected Works of J. Willard Gibbs. Vol. I. Thermodynamics, London, Toronto: Longmans Green, New York, 1931 pp. 55-353.
96 J. Wyman, Adv. Protein Chem., 19, (1964) 223-286.
97 E. Antonini, J. Wyman, M. Brunori, E. Bucci, C. Fronticelli and A. Rossi-Fanelli, J. Biol. Chem., 238, (1963) 2950-2957.
98 F. Haurowitz, Z. Physiol. Chem., 254, (1938) 266-274.
99 J. Wyman and D.W. Allen, J. Polymer Sci., 7, (1951) 499-518.
99a J.T. Edsall, in H. Neurath and K. Bailey (Eds.), The Proteins, Vol. IB, Academic Press, New York, 1953, 549-726.
100 E. Antonini, J. Wyman, R. Zito, A. Rossi-Fanelli and A. Caputo, J. Biol. Chem., 236, (1961) PC 60-63.
101 E. Antonini and M. Brunori, Hemoglobin and Myoglobin in their Reactions with Ligands, North Holland, Amsterdam, 1971, 296-297.
102 J.V. Kilmartin and J.F. Wootton, Nature, 228, (1970) 766-767.
103 E. Antonini, J. Wyman, M. Brunori, J.F. Taylor, A. Rossi-Fanelli and A. Caputo, J. Biol. Chem., 239, (1964) 907-912.
104 M. Brunori, E. Antonini, J. Wyman, R. Zito, J.F. Taylor and A. Rossi-Fanelli, J. Biol. Chem., 239, (1964) 2340-2344.
105 M. Reichlin, E. Bucci, E. Antonini, J. Wyman, J. Mol. Biol., 9, (1964) 785-788.
106 M. Reichlin, E. Bucchi, C. Fronticelli, J. Wyman, E. Antonini and A. Rossi-Fanelli, J. Mol. Biol. 12, (1965) 774-779.
107 E. Bucci and C. Fronticelli, J. Biol. Chem., 240, (1965) PC 551-552.
108 E. Antonini, E. Bucci, C. Fronticelli, E. Chiancone, J. Wyman and A. Rossi-Fanelli, J. Mol. Biol., 17, (1966) 29-46.
109 R.E. Benesch, H.M. Ranney, R. Benesch and G.M. Smith, J. Biol. Chem., 236, (1961) 2926-2929.
110 E. Antonini and M. Brunori, Hemoglobin and Myoglobin in their Reactions with Ligands, North Holland, Amsterdam, 1971, 436.
111 L. Rossi-Bernardi and F.J.W. Roughton, J. Physiol., 189, (1967) 1-29.
112 M. Brunori, C. Chiancone and J. Wyman, Trends Biochem. Sci., 9, (1984) 12-13.
113 J. Wyman, J. Biol. Chem., 241, (1966) 115-121.
114 J.B. Wittenberg, Physiol. Rev., 50, (1970) 560.

115 J.D. Murray, Proc. Roy. Soc. Lond. B, 178, (1971) 95–110.
116 J.D. Murray and J. Wyman, J. Biol. Chem., 246, (1971) 5903–5906.
117 J.T. Edsall, J. Mol. Biol., 108, (1976) 269–270.
118 J. Monod, J.-P. Changeux and F. Jacob, J. Mol. Biol., 6, (1963) 306–329.
119 J. Wyman in A. Lwoff and A. Ullmann, (Eds.), Recollections of Jacques Monod, in Origins of Molecular Biology: a Tribute to Jacques Monod, Academic Press, New York, 1979, pp. 221–224.
120 J. Monod, J. Wyman and J.-P. Changeux, J. Mol. Biol. 12, (1965) 88–118.
121 D.E. Koshland, G. Némethy and D. Filmer, Biochemistry, 5, (1966) 365–385.
122 J. Wyman, Curr. Topics Cell Reg., 6, (1972) 209–226.
123 J. Wyman, J. Mol. Biol., 11, (1965) 631–644.
124 J. Wyman, J. Am. Chem. Soc., 89, (1967) 2202–2218.
125 J. Wyman, Quart. Rev. Biophys., 1, (1968) 35–80.
126 J. Wyman, Proc. Natl. Acad. Sci., USA, 72, (1975) 1464–1468.
127 J. Wyman, Biophys. Chem., 14, (1981) 135–146.
128 A. Szabo and M. Karplus, J. Mol. Biol., 72, (1972) 163–197.
129 V.L. Hess and A. Szabo, J. Chem. Educ., 56, (1979) 289–293.
130 C. Tanford, J. Mol. Biol., 39, (1969) 539–544.
131 J.T. Edsall, J. Hist. Biol., 5, (1972) 205–257.
132 D.W. Allen, K.F. Guthe and J. Wyman, J. Biol. Chem., 187, (1950) 393–410.
133 G. Weber, Nature, 300, (1982) 603–607.
134 L. Peller, Nature, 300, (1982) 661–662.
135 A. Colosimo, M. Brunori and J. Wyman, J. Mol. Biol., 100, (1976) 47–57.
136 G.K. Ackers, Biophys. J., 32, (1980) 331–346.
137 R. Woolfson and W.G. Bardsley, J. Mol. Biol., 136, (1980) 451–454.
138 J. Wyman and S.J. Gill, Proc. Natl. Acad. Sci. USA, 77, (1980) 5239–5242.
139 J. Wyman, Proc. Natl. Acad. Sci. USA, 72, (1975) 3983–3987.
140 G. Fichera, M.A. Sneider and J. Wyman, Atti Acad. Natl. Lincei, Serie VIII, Vol. XIV, (1977) 26.
141 G. Fichera, M.A. Sneider and J. Wyman, Proc. Natl. Acad. Sci. USA., 74, (1977) 4182–4186.
142 G. Careri and J. Wyman, Proc. Natl. Acad. Sci. USA, 81 (1984) 4386–4388..
143 G. Careri, U. Buontempo, F. Carta, E. Gratton and A.C. Scott, Phys. Rev. Lett., 51 (1983) 304–307.
144 M. Brunori, Curr. Top. Cell Reg., 9, (1975) 1–39.
145 M. Perutz and M. Brunori, Nature, 299, (1982) 421–426.
146 J. Wyman, S.J. Gill, L. Noll, B. Giardina, A. Colosimo and M. Brunori, J. Mol. Biol., 109, (1977) 195–205.
147 M. Brunori, M. Coletta, B. Giardina and J. Wyman, Proc. Natl. Acad. Sci. USA, 75, (1978) 4310–4314.
148 J. Wyman, S.J. Gill, H.T. Gaud, A. Colosimo, B. Giardina, H.A. Kuiper and M. Brunori, J. Mol. Biol. 124, (1978) 161–175.
149 J. Wyman, Comp. Biochem. Physiol., 62A, (1979) 9–12.

150 A. Riggs, Comp. Biochem. Physiol., 62A, (1979) 257-271.
151 J.T. Edsall, J. Biol. Chem., 255, (1980) 8939-8951.
152 J.T. Edsall, Fed. Proc., 39, (1980) 226-235.
153 D. Bearman and J.T. Edsall (Eds.) Archival Sources for the History of Biochemistry and Molecular Biology. American Philosophical Society, Philadelphia, 1980.
154 L.J. Henderson, The Fitness of the Environment, Macmillan, New York, 1913, (republished 1958 by Beacon Press, Boston).
154a J.T. Edsall and H. Gutfreund, Biothermodynamics, The Study of Biochemical Processes at Equilibrium, Wiley, Chichester and New York, 1983.
155 J.T. Edsall, Scientific Freedom and Responsibility. American Association for the Advancement of Science, Washington, DC, 1975.
156 J.T. Edsall, Science, 212, (1981) 11-14.

APPENDIX

A letter of Jeffries Wyman to John T. Edsall on the occasion of the latter's 80th birthday

Dear John,

I cannot tell you how badly I feel not to be at your 80th birthday celebrations, as you and Margaret were at mine in Caprarola over 6 years ago when I was 75. I had thought that nothing could prevent me coming but, as things turned out, it has become next to impossible.

Occasions like this are a time for old friends to revisit the distant world of their youth, which takes on an increasingly platonic reality of its own as years pass. Ours goes back to the days when we were both undergraduates at Harvard. You were studying chemistry with an eye to the Medical School; I was concentrating in Philosophy under the guidance of Raphael Demos but with my mind turning more and more to biology.

Like all of your set, we were having all sorts of love affairs in a new world of ideas so different from school. I remember your enthusiasm at Henderson's course on the History of Science — your interests were already setting in that direction.

Do you recall the little discussing group, which used to meet, somewhat irregularly, once or twice a month in one or the other of our rooms? Alfred Mirsky and Tim Anson, later of the Rockefeller Institute, were members. So also was Otto Koenig and sometimes Raphael Demos would come too.

For each meeting we would invite some glamorous member of the faculty to come and talk to us about his latest interests, and I cannot recall ever being refused. On the evening when Henderson came, he began his talk by remarking, without explanation, that he had been worrying all day about a mathematical problem. That wetted our curiosity. I now know it was the construction of his nomogram to describe the properties of the circulating blood. But the evenings were not limited to science.

After graduation in 1923 we both spent another year at Harvard, you at the Medical School where your father was Dean; I in the graduate school, making up deficiencies in physics and chemistry.

It was then that Bob Oppenheimer entered the picture. He and I took several courses together and I can remember studying with him for finals. He was amazingly quick and a most stimulating companion, but not always without affectation. On one extremely hot June afternoon he stopped at my room with the remark "What weather! I have been unable to do anything all day but read Jeans' Dynamical Theory of Gases".

In June 1924 we set out together for Europe. We were to study biochemistry under Hopkins in Cambridge. But, before going there, we had plans to spend the summer in Graz, learning German. We set sail from East Boston on a small ship called the Winnifredian, which also carried cattle. The voyage lasted nearly 10 days — a pleasant oasis in life which brought together a small group of people never to meet again. How different from modern travel!

We landed in Liverpool, whence we made our way to Vienna, via the Hook of Holland, with several stops on the way. I remember the one in Heidelberg where we watched the sun go down behind the Palatinate from the terrace above the Castle.

Armed with letters from your father, we were well received in Graz by Otto Loewi and Professor Pregl, both Nobel Laureates, which impressed us greatly. Through Loewi we arranged to be paying guests in two families living close together on opposite sides of the Rosenbergstrasse. Mine was that of a doctor named Uranitch. Do you remember our amusement at my having to sign a Hausordnung in which I agreed "keine Freundin in der Nacht zu bekommen". Those were the days when a year's rent was the price of a postage stamp in Austria.

Loewi was extremely kind to us. Among other things, he took us on a trip to the Oetztal where the last stage of the journey was in a horse-drawn omnibus.

During our stay there (in Obergurgl) we arranged the Ramolkogl with guides. That meant spending the night in the refuge under the peak. Next afternoon, when we got back to the inn, somewhat tired from the climb, we both tumbled into bed without showing ourselves to anyone. Hardly had we closed our eyes when Loewi burst into the room. When he saw us safe and sound, he threw up his hands, exclaiming "Gott sei Dank". I fear we were more amused than penitent. Years later, after the War, I met Loewi, then an old man, in Woods Hole.

At the end of the summer in Graz we spent a week or so in Vienna, where Mrs. Washburn, wife of the American ambassador, arranged for us to live in the ground floor of the old palace which was then the residence. The stay in Vienna gave us our first look at the Breugels. While in Vienna we made a trip by river steamer down the Danube to Budapest where we had a letter to Jeremiah Smith, financial administrator for Hungary under the terms of the truce.

When we got back to England, before proceeding to Cambridge, we spent a day or two with the Days in Rochester. Mr. Day was an ecclessiastical lawyer and they lived in the Close next to the Cathedral. Mrs. Day was a Bostonian, a niece of old President Eliot of Harvard, and she looked for all the world like him. I remember the round tin hat tub and the accompanying pitcher of hot water, which was brought into our room in the morning by a servant. It was all straight out of Trollope.

In Cambridge we were members of St. Johns. You had rooms in the second court, I in the third, overlooking the river — I could have fished out of my window. I still remember the smoky coal fire and the chilly morning trip over the Bridge of Sighs to the college showers.

Every day we went on our bicycles to the laboratory in Tennis Court Road, where we were enrolled in the Part II course of Biochemistry. The day would open with a lecture by Hopkins, sometimes replaced by Jack Haldane, the reader. Hopkins was not a very good lecturer, but a very impressive figure. I remember him saying one day, prophetically, "...and then the Nucleic Acids — there is a great future there."

Jack Roughton was responsible for the laboratory. One of the exercises was on the flow method he and Hartridge had developed for measuring the reaction of oxygen and hemoglobin. Do you remember how one of the connections broke and the 5-gallon carboy of a solution of blood squirted out all over the room? Adair was at that time working in the physiology laboratory and we saw less of him.

Every afternoon there was laboratory tea where we used to listen to the banter between Jack Haldane and Majorie Stephenson, the microbiologist. I remember Stephenson remarking one day that you reminded her of Pierre in "War and Peace". I have always felt that she had a point, for you both combine a slight awkwardness of manner with an impression of unbounded benevolence.

For that first Christmas vacation we were invited to stay with a family of distant connections, the Thomassons, in London. They had a large house in Mayfair, just behind Park Lane, and two charming daughters, Margery and Barbara. That visit gave me my first view of the extreme luxury of English upper class life which survived into the first quarter of this century — two Rolls Royces, with a chauffeur for each, and endless house servants.

The following summer, while you were wandering on the Continent, I spent a month with the Thomassons at their place "Halsteads" on Ullswater, where Wordsworth is said to have written his poem "Daffodils". It was there that I made my acquaintance with crag fox hunting after the fashion of John Peale.

Before the end of my first term in Cambridge I had decided that I wanted to work with A.V. Hill, then Fullerton Professor of the Royal Society at University College.

Soon after the Christmas vacation I made the change, moving to London where I arranged to live as a paying guest with a charming motherly woman named Mrs. Bratten who had a comfortable house in St. Johns Wood. I am sure you remember her, for my move by no means separated us. You used often to come up to London on weekends to stay with me, while I on the other hand would visit you in Cambridge, always staying with the statistician Udney Yule who had ample quarters in New Court. On those occasions, he would invariably arrange a breakfast party on Sunday morning, inviting some new acquaintance such as Gregory Bateson (who made a great impression on both of us). Yule was a delightful man, full of whimsy. He came, I believe, from an Anglo-Indian family and used to describe pig-sticking in the Raj.

Bob Oppenheimer was at this time in Cambridge, working on an experimental problem with J.J. Thomson. It was before he went to Max Born in Göttingen. Things were not going well and he was deeply unhappy. During one of the long Spring vacations we three made that famous walking trip in Corsica from which he returned abruptly because of the poisoned apple (he said he'd left a poisoned apple on Blackett's desk), while you and I went on to Sardinia and thence to Italy.

See, I am well down on my fourth page without mentioning a fraction of the things that flood my mind. I fear my visit to the platonic world of our student days has been too long. Indeed I feel like Tristram Shandy who spent seven years describing the first seven days of his life. At any rate, there is no time to go on to the next chapter, when we were both back in America, you with Edwin Cohn in the Medical School, I in Cambridge, MA, where the Whiteheads and the circle of friends that formed around them — the Greens, Raphael Demos, Philip Johnson and others — bulk so large.

So it remains only to congratulate you on being 80 years old and to assure you of my affectionate best wishes for the next decade.

Jeffries Wyman
Paris, November 1982.

G. Semenza (Ed.) Selected Topics in the History of Biochemistry: Personal Recollections (Comprehensive Biochemistry Vol. 36) ©1985 Elsevier Science Publishers

Chapter 4

*The BAL-labile Factor in the Respiratory Chain**

E.C. SLATER

Laboratory of Biochemistry, B.C.P. Jansen Institute, University of Amsterdam, Plantage Muidergracht 12, 1018 TV Amsterdam (The Netherlands)

First acquaintance with BAL (1942)

In 1942, during the darkest days of the War so far as Australia was concerned, I was in charge of a small team of chemists at the Chemical Defence Division of the Munitions Supply Laboratories at Maribyrnong, near Melbourne. Our main responsibility was to adapt to conditions of the Pacific War knowledge available to the Allied armies, particularly the British, concerning the detection of war gases and of methods of treating contamination with such gases. One of the assignments given to my team was to prepare for testing a quantity of a new compound which was then known under the code name DTH in the secret reports that were available to us. This compound, 2,3-dimercaptopropanol, had been developed by brilliant work of R.A. Peters, R.H.S. Thompson and L.A. Stocken [1] in Oxford, England, as an extremely effective antidote to the war gas, lewisite, preventing vesication of the skin when applied up to 30 min after contamination with lewisite. Peters and colleagues ascribed the antidote working to the reaction:

---

*This factor is also known as "Slater factor" (Ed.).

Plate 9. E.C. Slater.

$$\text{Cl-CH} = \text{CH-As}\begin{matrix}\text{Cl}\\ \text{Cl}\end{matrix} + \begin{matrix}\text{HS-CH}_2\\ \text{HS-CH}\\ \text{OH-CH}_2\end{matrix}$$

Lewisite DTH

$$\downarrow$$

$$\text{Cl-CH} = \text{CH-As}\begin{matrix}\text{S-CH}_2\\ \text{S-CH}\\ \text{HO-CH}_2\end{matrix} + 2\,\text{HCl}$$

In this way, lewisite was prevented from exerting its toxic effect to act on a dithiol group involved in the oxidation of pyruvic acid.

Although trained as an organic chemist, I had been appointed in 1939 Biochemist at the Australian Institute of Anatomy, Canberra, from which position I was seconded in 1942 to the Munitions Supply Laboratory. I remember being fascinated by the biochemical basis on which DTH had been developed, but my job in 1942 was as an organic chemist to make the compound and, in true military fashion, I passed on my assignment to one of my team, Jim Lincoln, who duly synthesized it, following Len Stocken's directions, and we handed it over to the biologists for testing. The latter included the biochemists Jack Legge and Hugh Ennor and later Robert Thompson from England (but by that time I had switched from chemical defence to analysing the nutritional value of Assault Rations).

The American services found it necessary for some reason or other to give their own code name for DTH. They generously chose BAL (British Anti-Lewisite) and that name has stuck.

Research project at Molteno Institute (1946)

Shortly after the War, I had the great good fortune to be accepted, as a British Council Scholar, by David Keilin to work at the Molteno Institute, University of Cambridge. Except for a break of 8 months at Severo Ochoa's laboratory in New York, I stayed at the Molteno Institute until 1955, when I was appointed to succeed

B.C.P. Jansen at the University of Amsterdam. The 9 years that I spent at Cambridge (between the ages of 29 and 38) were the most exciting and the happiest of my life and laid the foundation for my career in Amsterdam. David Keilin was the greatest scientist I have ever known, an inspiring teacher and the kindest of men.

When I had been appointed Biochemist at the Australian Institute of Anatomy at the ripe age of 22, I had never studied biochemistry. W.J. Young (of the Harden and Young ester), then Professor of Biochemistry at the University of Melbourne, gave me a crash biochemical practical course (the one followed by medical students) and otherwise I had to teach myself biochemistry. By the time I left Canberra in 1946, I was heading a group of 3 or 4 biochemists and had published about a dozen papers mostly in biochemical or medical journals. When I reached Cambridge, however, I soon realized that I had either been a poor teacher or poor student or both, because I had little knowledge and less real understanding of modern biochemistry. It was David Keilin and a number of distinguished biochemists in the Biochemistry Department (then headed by A.C. Chibnall), whose lectures I followed, who enabled me to come to grips with the subject for the first time.

In view of my inexperience, Keilin gave me a small project as a starter. In 1938, Frederick Gowland Hopkins and coworkers [2] had published a well-known paper in which they demonstrated that succinate oxidation is inhibited by incubation with oxidized glutathione and the activity is restored by incubation with reduced glutathione. They concluded that succinate dehydrogenase contains an essential thiol group that is oxidized (to a disulphide) by oxidized glutathione and that glutathione reduces the inactive disulphide to the active enzyme. As a matter of fact, I was aware of this work, since as part of my M.Sc. study in organic chemistry at the University of Melbourne in 1939, I was assigned "glutathione" as my colloquium project. Keilin was, however, not completely satisfied with Hopkins' conclusion. I remember vividly his words:

"You know, Hoppy used to use lumps of meat as his enzyme. It seems to me possible that a large molecule like oxidized glutathione just covers up the lumps of meat and prevents access of substrate. Try it on our heart-muscle preparation."

The latter was what became known for a long time as the Keilin and Hartree heart-muscle preparation [3]. Now we would call it heart sub-mitochondrial particles or even inverted vesicles.

First, I had to isolate glutathione from yeast, which was a messy preparation [4] involving precipitation as the mercury salt and subsequent removal of the mercury as the insoluble sulphide with the use of a Kipps apparatus on the roof of the Molteno Institute. I oxidized the glutathione to the disulphide form by adding a trace of $FeSO_4$ and bubbling air through the solution for 3 h. Sure enough, I found that after 4 h incubation at 37°C of the heart-muscle preparation with a high concentration of oxidized glutathione (30 mM) extensive inactivation of the succinate oxidase activity had occurred. Now, of course, it was time to try if the activity could be restored by incubation with glutathione.

Discovery of BAL-labile factor

My first experiment, dated November 15, 1946, was only partially successful (Table I), mainly due to an unexpected inhibition by glutathione itself. I next (November 19) confirmed this latter finding in a direct experiment in which I noted that glutathione took up oxygen before I added the substrate succinate. The only conclusion that I drew (I quote from my note-book) was "The activity of suc-

TABLE I

First experiment (November 15, 1946) on effect of GSH on GSSG-treated heart-muscle preparation

Treatment	Succinate oxidase activity (μl O_2/h)
None	152
GSSG (31 mM)	42
GSSG (31 mM) followed by GSH[a] (12 mM)	66
GSH (12 mM)	107

[a] After removing GSSG

cinic dehydrogenase in the presence of GSH cannot be measured in this way".

The next few weeks I spent improving my technique of making the heart-muscle preparation and measuring the succinate oxidase activity and testing conditions for inhibiting it with GSSG. These experiments consumed a lot of glutathione and since I could not face making another batch, I decided after Christmas to use commercially available oxidants and -SH-combining reagents and to reserve my precious GSH for the reversal experiments.

On what I now see was my thirtieth birthday (January 16, 1947), I obtained a nice inhibition-concentration curve with iodosobenzoate, which was known to oxidize -SH groups in a number of enzymes. On January 20, I attempted to reverse this inhibition by a subsequent treatment with GSH (after separating excess iodosobenzoate and reaction products by acid precipitation of the enzyme). The results of this experiment are shown in Table II. It was clear that I had completely failed to reverse with GSH the iodosobenzoate-induced inhibition. On the contrary, the inhibition increased after GSH treatment. Moreover, GSH, especially at the higher concentration, inhibited to an even greater extent than in the earlier experiment. I noted:

"The effect of GSH added to the control ... must be due to GSSG formed on oxidation, but is a surprisingly large inhibition in view of previous experience with

TABLE II

Effect of GSH on iodosobenzoate-treated heart-muscle preparation (January 20, 1947)

Treatment	Succinate oxidase activity (μl O_2/10 min)
None	67
Iodosobenzoate (3.3 mM)	19
Iodosobenzoate (3.3.mM) + GSH (1.8 mM)	10
Iodosobenzoate (3.3 mM) + GSH (10 mM)	6
GSH (1.8 mM)	54
GSH (10 mM)	30

GSSG as an inhibitor" and "it is obvious that the GSH must be added under anaerobic conditions and must be removed".

As I have described elsewhere [5], not much experimental work could be carried out in February, 1947, since owing to an exceptionally severe winter and a fuel shortage, all heating of buildings was stopped and aqueous buffers froze on the laboratory bench. When I returned to experimental work, I started a systematic investigation of the inhibition of succinate oxidation by three types of -SH-reacting inhibitors: iodosobenzoate, *p*-aminophenylarsenoxide and *p*-chloromercuribenzoate and of the relative effectiveness of glutathione and BAL (kindly provided by Len Stocken and Robert Thompson) in reversing the inhibition. BAL also was found to be inhibitory under aerobic conditions, but it was possible to overcome this by incubating concentrated arsenical-inhibited particles with BAL under anaerobic conditions and then diluting about 50-fold before measuring the succinate oxidase activity manometrically. This formed the basis of the fifth of a series of papers on succinate oxidase published in the Biochemical Journal in 1949 [6].

By the end of March, it is clear from my note-books that I was becoming more interested in the inhibition by glutathione or BAL, and since my stocks of glutathione were nearly exhausted and in any case BAL, which does not require neutralization, was much more convenient than glutathione, all future experiments were carried out with this compound.

In fact, the effect of BAL on succinate oxidation had previously been studied by two groups. Edwin Webb and Ruth van Heyningen [7] had found no effect of BAL when succinate oxidation was followed by oxygen uptake with methylene blue as carrier. Barron and coworkers [8], however, had found that BAL does inhibit when succinate oxidation is followed without the addition of methylene blue. I was soon able to confirm both results: there was negligible inactivation with methylene blue as acceptor, except after prolonged incubation (see Fig. 12 – experiments of May 1, 1947 – my wife's birthday) whereas the succinate oxidase activity was rapidly, but not instantaneously, inactivated.

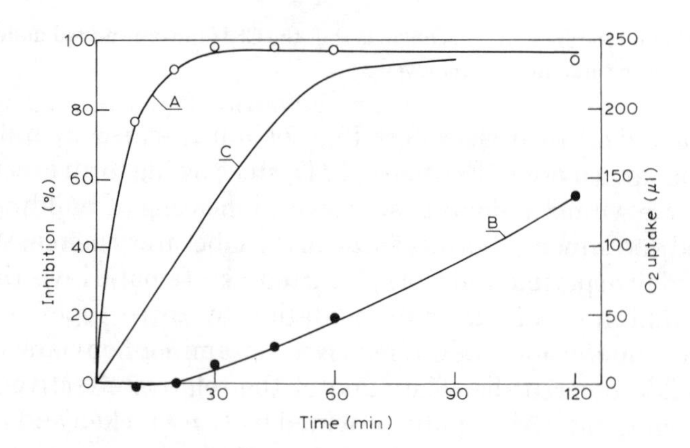

Fig. 12. Effect of treatment with BAL on succinate oxidase system and succinate dehydrogenase in heart-muscle preparation. Preparation shaken in air with 9.4 mM BAL at 37°C for various periods of time. (A) Inhibition of succinate oxidase system; (B) inhibition of succinate dehydrogenase (methylene blue and cyanide added); (C) oxygen uptake by BAL. From [21].

From this observation, it was clear that succinate dehydrogenase, which catalyses the oxidation of succinate by methylene blue, is not the target of the BAL treatment but that this must lie in that part of the respiratory chain lying between succinate dehydrogenase and oxygen. I was soon able to eliminate both cytochrome *c* and cytochrome oxidase, so I concluded that the site of action of BAL must lie between succinate dehydrogenase and cytochrome *c*. Moreover, since BAL had no action when incubated under anaerobic conditions and BAL previously oxidized by molecular oxygen (using particles depleted of succinate dehydrogenase by alkali extraction in order to oxidize the BAL) had a much smaller effect, I concluded that the inactivation is brought about by the coupled effect of BAL and oxygen.

The investigation now proceeded in two directions: (1) the identification of the inhibitory species arising during the oxidation of BAL; (2) the identification of the site of action of BAL within the respiratory chain.

Since some H_2O_2 was detected (with catalase) when BAL was oxidized by oxygen in absence of heart-muscle particles, it seemed possible that this was the inhibitory species. However, the enzyme was not protected by catalase, either in presence or absence of ethanol, or by pyruvate which would be expected to remove the hydrogen peroxide, nor did other H_2O_2-delivering systems (D-amino acid oxidase or glucose oxidase in the presence of their substrates) have the same effect as BAL and oxygen. It was concluded that the inactivation was due to the coupled oxidation by oxygen of BAL and a component of the succinate oxidase system. Inactivation of this component could be prevented by adding cytochrome *c*, which functioned as electron carrier between BAL and oxygen via cytochrome *c* oxidase and thereby greatly increased the rate of oxidation of BAL. It was not, therefore, the oxidation of BAL itself that caused the inactivation but the oxidation of BAL via some component preceding cytochrome *c* in the electron-transfer chain from succinate.

The first thing that one did in the Molteno Institute with a new inhibitor of the succinate oxidase system was to look at its effect on the absorption bands of the cytochromes, making use of Keilin's microspectroscope. This instrument has been fully described in Keilin's book [9].

My notebook for May 5 reads:

"Spectroscopic observations of heart muscle preparation after treatment with BAL.

5-5-47

A - 1 ml heart muscle preparation + 0.2 ml H_2O

B - .... + 0.2 ml 0.1 M BAL

Heated with shaking in Barcroft for 30 min.

No bands were visible in either A or B. On adding succinate, A gave the normal picture. The bands appeared rather slowly in B, but appeared to be normal. On shaking B with air, band *b* remained reduced while the other bands disappeared and reappeared again with extreme slowness (about 10 min).

Hydrosulphite* added to A and B gave the normal picture".

I did not note that in the control it is not possible to see the bands disappear because by the time the tube is placed under the microspectroscope all the O_2 that one has managed to shake into the suspension has been exhausted by the succinate.

---

*Now called dithionite ($S_2O_4^{2-}$).

I was extremely excited by being able to see with my own eyes that BAL cuts the chain somewhere between cytochromes *b* and *c*. Band *b* (due to reduced cytochrome *b*) remained visible whereas bands *a* and *c* were no longer visible since the corresponding cytochromes are oxidized. However, cytochrome *b* could be oxidized by dichloroindophenol. At the first opportunity, I showed the result to Keilin and no doubt to everyone else I could persuade to look down the microspectroscope in Keilin's laboratory on the first floor of the Molteno Institute.

Specificity of effect of BAL

The conclusion that incubation with BAL in the presence of oxygen causes the coupled oxidation of a factor required for electron transfer between cytochromes *b* and *c* could already be drawn, but a lot of controls had to be done before a paper could be ready for publication. In particular, the specificity of the effect had rigorously to be examined. This was not the first time that impaired electron transfer between cytochromes *b* and *c* had been observed. Already in his first paper on cytochrome in 1925, Keilin [10] had observed that addition of narcotics (such as phenylurethane) had the same effect. However, comparing the two effects in the microspectroscope, that of BAL was quantitatively so much more dramatic that I felt quite confident that I had really destroyed a hitherto unknown component of the respiratory chain.

On the other hand, Keilin and Hartree in their classical paper on the succinate oxidase system in 1940 [11] had emphasized that since all the components of the respiratory chain are firmly bound to small particles (it was not known then that they consist of vesicles formed by fragments of the mitochondrial inner membrane), inhibition of electron transfer is to be expected not only when one of the components of the chain is specifically inhibited but also when the "colloidal" properties of the particle are altered in such a way that the mutual accessibility of the components of the chain is affected. They explained in this way the effects of a number of inhibitors of the succinate oxidase system examined both in their

1940 paper and in extensive subsequent experiments which were not published until 1949 [12].

It is perhaps useful for the present-day reader to put this "accessibility" concept of Keilin and Hartree into a modern context. We now know largely as a result of the work of David Green and his associates [13] that the so-called respiratory chain consists of a number of multi-subunit large proteins (called complexes by Green) and two smaller components, the lipid-soluble ubiquinone and the relatively small protein cytochrome *c*. The succinate oxidase system of Keilin and Hartree can now be described by the reaction chain

$$\text{succinate} \rightarrow \boxed{\text{II}} \rightarrow \text{Q} \rightarrow \boxed{\text{III}} \rightarrow c \rightarrow \boxed{\text{IV}} \rightarrow O_2$$

II = succinate : Q oxidoreductase
Q = ubiquinone
III = QH_2 : cytochrome *c* oxidoreductase
c = cytochrome *c*
IV = cytochrome *c* oxidase

It is the QH_2 : cytochrome *c* oxidoreductase that contains the cytochrome *b* (and, as we shall see, much more).

The picture that is now emerging is that the large proteins (II, III and IV) can move freely in the phospholipid bilayer of the inner mitochondrial membrane and that cytochrome *c* shuttles along the outside of the bilayer, receiving electrons when it binds to reduced III and delivering electrons (from the same binding site) when it binds to oxidized IV. In sub-mitochondrial particles, which are "inside-out", the cytochrome *c* presumably moves along the inside face of the vesicle.

Translated into terms of today's picture of the enzyme system, Keilin and Hartree's concept means that any (physical) treatment that affects the orientation of the electron-transferring components within the large molecules to one another or affects the specific interactions of the components shown in the above scheme will inhibit electron transfer. As an example of the latter effect might be mentioned increasing the rigidity of the bilayer by lowering the temperature or adding cholesterol.

In any case, I got the idea that if the effect of an inhibitor or treatment is purely to affect the physical interaction between cytochrome *b* (now say enzyme III) with cytochrome *c*, this could be overcome by adding a large excess of purified cytochrome *c*, the only purified component of the respiratory chain then available. If, however, the treatment specifically destroys a component of the system (in present-day terms an electron-transferring subunit in enzyme III), the degree of inhibition would not be affected by flooding the system with cytochrome *c*. As a model for a non-specific inhibitor I used a surface-active compound, namely the haemolytic substance discovered by Hans Laser [14] in the Molteno Institute and later identified as *cis*-vaccenic acid. I found that inhibition by this compound was greater in the absence of cytochrome *c* than in its presence and moreover that, as would be expected from a non-specific effect on the particles, succinate dehydrogenase (measured with methylene blue) and cytochrome *c* oxidase were also inhibited by somewhat higher concentrations. On the other hand, BAL treatment had only a very slight effect on succinate dehydrogenase or cytochrome *c* oxidase, and cytochrome *c* (added after the BAL treatment) had no effect whatsoever on the degree of inhibition by BAL.

These experiments convinced me that the effect of BAL is specific. They also apparently impressed Keilin, since in a paper [12] submitted on June 25, 1948, Keilin and Hartree write in the summary:

"The various claims to have demonstrated the existence of new 'links' in the succinic system (comprising dehydrogenase, cytochromes *b*, *c*, *a* and a_3) have been examined in detail. All are shown to be based on insufficient evidence except that of Slater (1948), who has produced reasoned arguments in favour of an oxido-reduction catalyst linking *b* and *c*".

Re-reading my paper [15] with the hind-sight of nearly 40 years, I am now not quite as impressed as I was then with my criterion for distinguishing between specific and non-specific inhibition of the respiratory chain. The argument is, I still believe, essentially valid. Of course, now we can see that a treatment affecting the quaternary structure of enzyme III in such a way as to impair electron flow between subunits of the enzyme would not be revealed by determining the effect of added cytochrome *c*, but at that time the only known component between succinate dehydrogenase and cytochrome *c* was cytochrome *b*. Thus, the conclusion that there is an additional electron carrier between *b* and *c* drawn at that time remains a logical conclusion from the experiments. However, the effect of cytochrome *c* on inhibition by *cis*-vaccenic acid is quite small - e.g., 60% inhibition with no added cytochrome *c*, 44% with the addition. More impressive was the effect of lowering the ionic strength below the optimal - 73% inhibition in the absence of cytochrome *c* and 42% in its presence. The problem is that I still do not know why lowering the ionic strength has such an effect (addition of EDTA abolishes it [16]), nor why cytochrome *c* has such a large effect (3-fold stimulation at low ionic strength) on sub-mitochondrial particles, in which according to present-day concepts the cytochrome *c*-binding site is inaccessible within the vesicle.

Nature of BAL-sensitive factor

In any case, having convinced myself and my surroundings that I had a new component of the respiratory chain to look for, it was necessary to identify it. I was at that time sharing a room on the top floor of the Molteno Institute with an old school and university friend from Geelong and Melbourne, respectively. Jack Legge had worked with Rudolph Lemberg on the in vitro decomposition of haemoglobin into bile pigments by the coupled oxidation with ascorbate [17]. I was impressed by the possibility that the BAL-labile factor is a haemoprotein, of the type of haemo(myo)globin which, unlike the cytochromes, do not show strong absorption bands in the visible spectrum and would not, therefore, be detected in the microscope. Indeed, Keilin had already found (but not published) that the amount of protohaem, measured as pyridine haemochrome, in the heart-muscle particles far exceeds that of cytochromes *b* and *c* combined. Sure enough, I found in exhaustive (and exhausting) experiments that the total protohaem content, measured visually as pyridine haemochrome in the microspectroscope, declined by 15–20% after BAL treatment. In these measurements, which I repeated many times, Jack Legge helped by giving me at random (and blind) control and BAL-treated samples to measure, since it is easy to deceive yourself in visual colorimetry, especially when you desire a particular result.

First publications

I was now ready to publish. The problem was what to call my precious factor. So I consulted Keilin. I can still vividly see him before me with the faintest of smiles at the corners of his mouth pondering how to reply without hurting my feelings. In any case, I got the idea that he thought it better to wait until I found out what it was before naming it, so in my publication in *Nature*, submitted as a letter, but published to my satisfaction on March 13, 1948, as a paper, the title is *A factor in heart-muscle required for the reduction of cytochrome* c *by cytochrome* b [18]. It never entered my head that not naming a factor was the best way of getting your name attached to it. So far as I know, it was Van Potter [19] who first referred to the factor as the "Slater factor" and this took on and although the name has long disappeared from the literature and is unknown to the younger biochemist, I am still known by not-so-old colleagues in various parts of the world, who probably learned their biochemistry from the older text-books, as the Slater of the "Slater-factor".

On re-reading the *Nature* paper, there is nothing that I would like to change except that it would have been wise to have written "possibly" instead of "probably" in the final conclusion.

"These observations suggest the existence of a BAL-labile factor, probably a haematin compound, which transfers electrons from cytochrome *b* to cytochrome *c* and is destroyed by coupled oxidation with reducing agents (cf. the action of ascorbic acid on haemoglobin, Lemberg et al.) when heart-muscle preparation is treated with reducing agents in the presence of air".

The next publication appeared in French [20] under the title "Un catalyseur respiratoire exigé pour la réduction du cytochrome *c* par le cytochrome *b*". In fact, this paper was presented in English between October 6th and 8th (I do not remember the exact day) to the 8e Congrès de Chimie Biologique in Paris and was translated for later publication in the *Bulletin de la Société de chimie biologique*. I must have been very anxious to present this work, since, as Professor Jean Courtois is fond of recalling, this was the first of the many Congresses that he organized and mine was the first extract that he received. The title of my paper was *A respiratory catalyst required for the reduction of cytochrome* c *by cytochrome* b, and I think that I would have used the word "nécessaire" rather than "exigé" in the title, but my French-speaking friends would find it hilarious if I were to consider myself an expert in their language.

The full paper with the same title as that read to the Paris meeting was submitted soon afterwards (October 22) to the *Biochemical Journal*, as the third [21] of a series to be published together. The first was *A comparative study of the succinic dehydrogenase-cytochrome system in heart muscle and in kidney* [22], and the second the paper already referred to on *The action of inhibitors on the system of enzymes which catalyse the aerobic oxidation of succinate* [15]. In my attempts to characterize the factor, I had carried out detailed studies on the mechanism of oxidation of BAL in the presence of the heart-muscle particles and in model systems containing possible candidates, such as copper (then known to be present in the particles, but not yet identified as a component of cytochrome oxidase). I included these experiments, as well as many other control experiments, in the paper, which as a consequence, became very long. In a letter dated January 6, 1948 (but this was a typographical error for 1949), the Chairman of the Editorial Board of the *Biochemical Journal* stated that the paper was acceptable but that certain alterations were necessary. In particular, the editors and referees considered:

> "that a number of points are included, which though interesting in themselves, are not relevant to the title of the paper. For these reasons they recommended that the paper be returned to the author for re-writing in a more concise form".

How often have I as Managing Editor of *Biochimica et Biophysi-*

ca Acta written such a letter, and how often has it been received with the same dismay as I felt? I returned the manuscript on January 10 (sic), after removing 2500 words and 3 tables. To my consternation I received a letter (January 21) from Robert Thompson, who was the editor concerned, with the passage:

"The work contained in this paper is of very great interest, and while I agree that the amendments which you have already made improve the clarity of the paper, I still feel that the importance of your results tends to be obscured by the manner of presentation which includes some data and arguments which I do not think are really relevant or necessary. On the other hand you, who have done the work, may not agree with the various points that I have to raise, and so I feel that, in order to expedite matters, it would really be best if we could sit down and discuss rather than engage in the slow process of writing letters to each other".

So, I visited Robert Thompson at Guy's Hospital, London, and agreed that the part dealing with ferricyanide as acceptor (partial inhibition by BAL) could also be omitted. Even so, it remained a long paper and, in fact, the three papers together occupied the first 30 pages of a 120-page issue of the *Biochemical Journal* when they finally appeared. The trouble that the referee (who revealed himself to me nearly 30 years later) and Robert Thompson took with a young and difficult author to improve his paper (successfully I think) was illustrative of the consideration that I always received from the *Biochemical Journal* at that time, and was the model that I tried to live up to when I became Managing Editor of *Biochimica et Biophysica Acta*.

I still rather like reading this paper and would not now wish to alter it, except of course the conclusion that the factor is a haem compound, and perhaps the section on the inhibition of the oxidation of *p*-phenylenediamine. This compound (and related compounds, such as the later more often used tetramethyl-*p*-phenylenediamine) differs from other non-enzymic reductants of cytochrome *c* in that it is able rapidly to reduce cytochrome *c* present in the particles (the so-called endogenous cytochrome *c*). Thus, compounds related to *p*-phenylenediamine are rapidly oxidized by tissue preparations, with the formation of brightly coloured oxidation products, a fact known to Paul Ehrlich in 1885. However, the rate of oxidation is further increased by added cytochrome *c* [23]. When the measured rates are extrapolated to infinite cytochrome *c* concentration, the same maximum rates are obtained with both types of substrate [24].

I found that BAL treatment had no effect on the cytochrome *c* oxidase activity, measured with either ascorbate or *p*-phenylenediamine as reductant for cytochrome *c*, when the data were extrapolated to infinite cytochrome *c* concentration. However, the oxidation of *p*-phenylenediamine in the absence of added cytochrome *c* was affected somewhat and, furthermore, more added cytochrome *c* was necessary to saturate the system after BAL treatment.

These effects are second-order effects compared with the effect of BAL on the succinate oxidase system, and I would have done better to have taken the comments of the editor and referee more to heart and to have considerably cut this section. As it is, I think that we can now explain the inhibition of *p*-phenylenediamine oxidation by postulating two entries of electrons from this substrate – one direct to cytochrome *c* and one to one of the intermediates of the QH_2: cytochrome *c* reductase reaction, possibly a semiquinone. In my paper, I rejected the possibility that the factor is necessary (i.e., obligatory) for the reduction of cytochrome *c* by *p*-phenylenediamine, but did not consider the possibility of a subsidiary pathway.

The only other still unexplained point in this paper is why catalase stimulates inactivation of the succinate oxidase system by H_2O_2 produced by glucose and glucose oxidase (one of the control experiments).

Relation to other "factors"

As already mentioned, a number of claims to have demonstrated additional components of the respiratory chain had been dismissed by Keilin and Hartree [12]. One of these: Straub's "SC factor" is worthy of further discussion.

During the war, Bruno Straub, then at Szeged, Hungary, published a paper [25] in which he described the fractionation of heart-muscle preparation with ammonium sulphate after dispersal in cholate. The fraction precipitating at intermediate ammonium sulphate concentrations was found to contain an active succinate dehydrogenase and cytochrome *c* oxidase, but was unable to oxidize succinate aerobically. Succinate oxidase activity could be restored by adding a preparation made by heating the particle preparation at pH 9.0 to 55°C for 15 min. Straub concluded that the latter preparation contained a factor (SC factor), which links succinic dehydrogenase to the dehydrogenase system and which is split off the enzyme by the action of bile salts.

I repeated these experiments, and found that the SC factor reactivated not only the succinate oxidase activity, but also the succinate dehydrogenase. I concluded that the cholate acted as a non-specific inhibitor and that the SC factor reversed this inhibition. Keilin and Hartree [12] showed, in fact, that the SC factor was not a specific reactivator since calcium phosphate was a much more effective reactivator. It seems likely, then, that the heated heart-muscle particles ("SC factor" preparation), like calcium phosphate, acted simply by removing the inhibitory cholate.

In any case, Bruno Straub's experiments provided no evidence for an additional factor. Nevertheless, they were important because they were the first to attempt ammonium sulphate fractionation of detergent-dispersed particles and they led others following up his work (Straub as the discoverer of actin had moved into the muscle field), to the discovery of cytochrome c_1 (see below) and to the fractionation by Lucile Smith, Wainio and David Green of the enzymes of the respiratory chain.

During the course of my experiments on BAL, two papers appeared on the effects of inhibitors which were described as affecting a link between succinic dehydrogenase and cytochrome oxidase. The most important of these was the paper by Eric Ball, Chris Anfinsen (later to obtain the Nobel Prize for his work on protein structure) and O. Cooper [26], published in April 1947. They found that certain antimalarials with the structure 2-hydroxy-3-alkyl-1:4-naphthoquinone inhibit succinate oxidase without any effect on the cytochrome oxidase, whereas succinate dehydrogenase was partially inhibited. The last paragraph of this summary reads:

"The conclusion is reached from manometric and spectrophotometric observations that these naphthoquinones act below cytochrome *c* and above cytochrome *b* in the main chain of respiratory enzymes. The possibility that their inhibitory action is the result of their combination with an unknown enzyme which mediates the reaction between cytochrome *c* and cytochrome *b* is discussed."

I did not know quite what to do about this paper. I thought that there was reason to believe that the naphthoquinones act non-specifically, since only hydroxynaphthoquinones with long aliphatic side chains were effective inhibitors. As I put it in my paper, the length of an aliphatic side chain could not appreciably affect the chemical properties of a naphthoquinone, but would profoundly affect its physical properties. The fact that succinate dehydrogenase was inhibited to a certain extent was also in agreement with this interpretation.

The other inhibitor - described by Case and Frank Dickens at a meeting of the Biochemical Society [27] in London on the day before my first paper appeared in *Nature* - is 4 : 4′-dihydroxystilbene. In my full paper, I pointed out that this compound is not very reactive chemically, but is strongly adsorbed on protein films.

I concluded that further work is required to elucidate the mechanism of action of these interesting new inhibitors, but that the findings reported to date, like those of previous workers discussed by Keilin and Hartree, do not, in themselves, provide any evidence for the existence of a factor between the dehydrogenase and oxidase. However, in the table, I left the matter open, in the sense that I put a question mark in the column "probable mechanism of action".

In the light of our present knowledge, my comments on the "physical" nature of these inhibitors and their lack of chemical reactivity are completely unjustified. However, it should be remembered that, at that time, we had no idea that the proteins of the respiratory chain are embedded in a phospholipid bilayer. I knew that the particles contain a lot of lipid, but I had no idea of its function. Lipid solubility was known to be an important parameter for the effect of pharmacologically active compounds, but this was thought to be at the level of the cell membrane. That the respiratory chain is located in mitochondria in other tissues was just becoming known from the work of Albert Claude (whom I heard lecture in Stockholm in June 1947), but the origin of the Keilin and Hartree heart-muscle particles was not clear until 1953 [28]. The discovery of ubiquinone in David Green's laboratory was not made until 1957 [29].

Now we know that Eric Ball's quinones, like a number studied by Karl Folkers, quite specifically affect the binding of ubiquinone to its binding sites in the ubiquinol : cytochrome *c* oxidoreductase. The inhibition of succinate dehydrogenase can probably be explained by a similar effect on succinate : Q oxidoreductase.

I am not aware of any further work on dihydroxystilbene.

Cytochrome c_1

I have described the story of my association with cytochrome c_1 in a paper published in a book on the occasion of Professor Okunuki's retirement in 1969 [5]. Since it has recently been somewhat misrepresented by my friend Tsoo King [30], I shall repeat the story here.

Although I had, partly because of the power restrictions in England in the winter of 1947 referred to above, spent many hours in the library and was confident that I was familiar with the entire literature in the field, I had never come across the paper by Yakushiji and Okunuki entitled *On a new cytochrome component and its function*, that had been published in the Proceedings of the Imperial Academy in Tokyo in 1940 [31]. In the index to my notebooks entitled *Index to work carried out in Molteno Institute 1.10.1946 to 6.8.1948*, cytochrome c_1 does not appear as an entry.

It must have been later in the summer of 1948 that Keilin showed me a reprint of Yakushiji and Okunuki's paper. Why he had not done so earlier I do not know. How long he had the reprint I also do not know. If he had read it, he must have forgotten it, since at that time I was in almost daily contact with him and it is inconceivable to me that he purposely did not refer to it. In any case, King's statement that I "was not able to associate his (i.e. mine, ECS) work with that being conducted by Keilin and Hartree, just one flight down from him" is a misrepresentation of the very close relations that I had with Keilin.

In their paper, Yakushiji and Okunuki proposed that a new cytochrome (c_1) is necessary for electron transfer between cytochromes b and c (as well as between cytochromes b and a in the absence of cytochrome c).

In the paper entitled *Activity of the succinate dehydrogenase-cytochrome* c *system in different preparations* [12] that was submitted on June 25 (probably while I was on vacation in Scotland, just after having my Ph.D. degree conferred), Keilin and Hartree pointed out the possibility that my BAL-labile factor might be closely related to cytochrome c_1. Probably Keilin came across Yakushiji and Okunuki's reprint while writing this paper.

That I was unaware of Yakashiji and Okunuki's paper is I think excusable, since the journal in which it was published was not widely known outside Japan and, so far as I am aware, no reference was made of it in the literature until 1949, when both Keilin and Hartree [12] and I [32] referred to it. In any case, sometime in the late summer or early autumn of 1948 (I have a full report in my notebook, but it is undated), I set out to repeat Yakushiji and Okunuki's preparation and to test the effect of BAL. I used as starting material both the heart-muscle preparation used by the Japanese workers (described by Ogston and Green [33]) and the Keilin and Hartree preparation. I confirmed their observations that the weak c-band present in their preparation of cytochrome c_1 was at a position about 2 nm further to the red than the strong c-band in the Keilin and Hartree preparation. Since I found, however, that CO moved the band back to 550 nm, I ascribed the shift to the presence of denatured protein haemochromogens whose spectrum is weakened on addition of CO. More important, from my point of view, was that I found absolutely no difference between cytochrome c_1 preparations prepared from normal and BAL-treated preparations. Thus, there was no doubt in my mind that cytochrome c_1 is not the BAL-labile factor.

I should have left it at that but I felt duty-bound to publish my conclusion that the evidence for the existence of cytochrome c_1 was unsatisfactory. In my heart, I did not believe that it was possible to draw such an important conclusion from the observation with a microspectroscope of a difference of only 2 nm in the position of an absorption band, and I gave too much credence to the shift that I found with CO (which was also only 2 nm). On January 9, 1949, I sent a note to *Nature* entitled *Cytochrome* c_1 *of Yakushiji and Okanuki* in which, to my everlasting embarrassment, I concluded that "Yakushiji and Okunuki's evidence for the existence of cytochrome c_1 is, therefore, unsatisfactory". This note was published in April, 1949, and it at least brought cytochrome c_1 to the attention of the scientific world.

It was not until 1955 that Keilin and Hartree [34], after a thorough analysis of the c-band in heart-muscle preparation, concluded that it represents the fused bands of two components - cytochrome *c* and what they had previously (August 1949), on the basis of low-temperature spectroscopy, described as cytochrome *e* [35]. They realized that cytochrome *e* is identical with the cytochrome c_1 of Yakushiji and Okuniki and showed, in disagreement with my conclusion, that it is not possible to obtain a fused band at 552 nm by mixing cytochrome *c* (550 nm) with haemochromogen (556–558 nm). They therefore dropped the name of cytochrome *e* in favour of cytochrome c_1.

By that time I had moved to Amsterdam. Ted Hartree gave me a copy of their paper during the Third International Congress of Biochemistry in Brussels. I had the consolation that Keilin and Hartree had confirmed that cytochrome c_1 is not the site of action of BAL.

NADH oxidase system

After writing 5 papers for the *Biochemical Journal* on cytochrome oxidase and the succinate oxidase system, I now turned my attention to the NADH oxidase system (we called it then dihydrocozymase, since the compound had first been described under this name as a coenzyme of fermentation). A Beckman spectrophotometer had recently arrived in the Molteno Institute, and I soon found that, although at that time it was unheard of to use cloudy suspensions in a photoelectric spectrophotometer, the activity of the NADH oxidase was so great and the sensitivity of the assay system so high that it was possible with a very dilute suspension of heart-muscle preparation, added to both the reference and test cuvettes, to measure directly at 340 nm the rate of oxidation of NADH. Since NAD was not at that time available commercially, a lot of time was spent isolating it from yeast and reducing it enzymically (for this purpose a crude preparation of glutamate dehydrogenase was used).

Since this was the first direct demonstration of the NADH oxidase system using NADH, I enjoyed myself studying the effect of the classical inhibitors, including the light-sensitive inhibition by CO. More relevant for the present chapter is the effect of BAL. Of course, if cytochrome *b* were involved in the oxidation of NADH, as it is for succinate, BAL would be inhibitory. However, there was no agreement in the literature concerning the rôle of cytochrome *b* in the NADH oxidase system, and I found that cytochrome *b* was incompletely reduced when NADH in concentrations far above those necessary to saturate the NADH oxidase system were added to heart-muscle preparation in the absence of oxygen.

I concluded then that cytochrome *b* is on a side-path for NADH oxidation. Later, it was shown that cytochrome *b* is readily reduced when much higher concentrations of NADH are used. Indeed, according to present concepts, the QH_2 : cytochrome *c* oxidoreductase, which contains the cytochrome *b*, should not be able to distinguish between QH_2 delivered by succinate : Q oxidoreductase and by NADH : Q oxidoreductase. We know now also that cytochrome *b* contains two protohaem groups with different potentials (referred to as *b*-562 and *b*-566) and that it is difficult to reduce *b*-566 by succinate unless cytochrome *c* is kept completely oxidized. This is explainable on the basis of the Q cycle which will be described later. It is still not clear, however, why NADH with a mid-point potential more than 300 mV lower than that of *b*-566 is unable to reduce this haem.

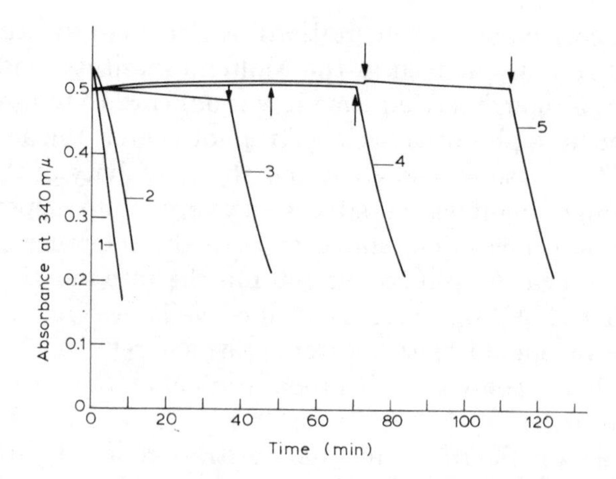

Fig. 13. Effect of treatment with BAL on NADH oxidase and diaphorase activity of heart-muscle preparation (H.M.P.). H.M.P. 1, untreated. H.M.P. 2, shaken in air at 39°C for 15 min, then diluted. H.M.P. 3, shaken with 20 mM BAL in air at 39°C for 15 min.

Curve	Added at		
	Zero time	First arrow	Second arrow
1	NADH, H.M.P. 1		
2	NADH, H.M.P. 2		
3	NADH, H.M.P. 3	Methylene blue	
4	NADH, KCN (10 mM)	H.M.P. 3	Methylene blue
5	NADH, KCN	H.M.P. 2	Methylene blue

From [40].

Since the NADH oxidase system is rather more unstable than the succinate oxidase, I had to change the conditions of BAL treatment somewhat. As shown in Fig. 13, BAL treatment almost completely inhibits NADH oxidation and the inhibition, like that found for succinate oxidation, is completely restored by methylene blue. From these experiments, I concluded that the BAL-labile factor, but not cytochrome *b*, is required for NADH oxidation and, in fact, that the factor is the link between the two oxidase systems. In confirmation of this, I found that, although NADH can be oxidized by fumarate in the presence of the heart-muscle particles, the velocity of this reaction is only about 1–2% of that of NADH oxidation by oxygen and that it is partly sensitive to BAL. Accordingly, in a paper presented to the Biochemical Society at Oxford on May 7, 1949 [36] (and also to the First International Congress of Biochemistry in Cambridge, August 19–25, 1949), I presented the following scheme

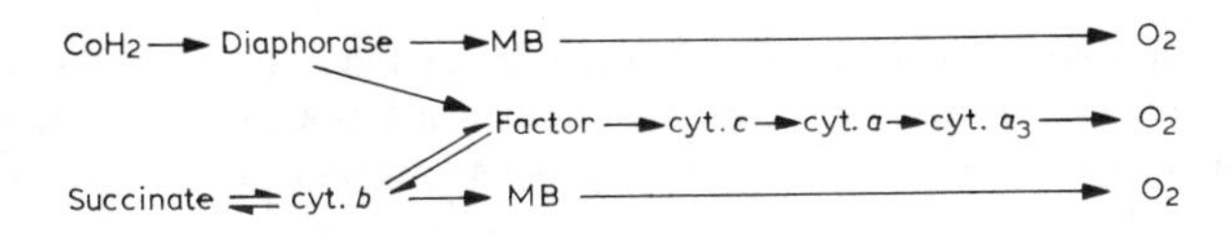

I could understand the slow oxidation of CoH_2 by fumarate if the factor had a mid-point redox potential about half-way between that of cytochrome *b* (then thought to be about -40 mV) and cytochrome *c* (260 mV).

Diaphorase was the name given to the enzyme catalysing the oxidation of NADH (CoH_2) by methylene blue, which was assumed to be the dehydrogenase part of the NADH oxidase system. It had been isolated as a pure protein by Bruno Straub working in the Molteno Institute just before the war [37]. In 1958, Vince Massey showed that the physiological function of diaphorase is, in fact, to catalyse the reduction of NAD by lipoic acid derivatives and not the oxidation of NADH by oxygen [38]. Its high reactivity with methylene blue is an artefact resulting from a copper-induced conformation change in the protein, as was shown by my student Cees Veeger with Vince Massey [39]. As Tsoo King and Tom Singer showed even later, NADH dehydrogenase is best studied with ferricyanide as acceptor.

However, I could not know all this in 1949, before even lipoic acid had been discovered. With hind-sight, it is now clear that, by using methylene blue as acceptor, I had not properly tested whether NADH dehydrogenase is affected by the BAL treatment. We now know that NADH dehydrogenase is, in fact, somewhat susceptible to BAL treatment and this is the reason for the partial inhibition of the oxidation of NADH by fumarate. Why fumarate and succinate:

Q oxidoreductase oxidize QH_2 produced by NADH : Q oxidoreductase so slowly is still somewhat of a mystery.

In the discussion of my full paper [40] on the NADH oxidase system, submitted on November 8, 1949, the day before I left to work as a Rockefeller Fellow in Severo Ochoa's laboratory in New York, I considered in some detail arguments for the participation of diaphorase in the NADH oxidase system, in the light of Lockhart and Potter's [41] opinion published in 1941 that

"the participation of diaphorase in the cytochrome reduction remains to be shown, since its mere presence in the complete system is not enough to warrant the conclusion that it is an essential link in the transport mechanism between $CoIH_2$ and cytochrome *c*".

I pointed out that

"since diaphorase mediates the oxidation of $CoIH_2$ by methylene blue, it is presumably reduced by $CoIH_2$. The objection of Lockhart and Potter (1941) implies then that reduced diaphorase might be oxidized in vivo by a mechanism not involving the cytochrome system. Since the aerobic oxidation of $CoIH_2$ is completely sensitive to cyanide or to treatment with BAL, this hypothetical mechanism cannot be involved in the aerobic oxidation, and the function of diaphorase must be in some anaerobic oxido-reduction, which should be sufficiently rapid to account for the very high diaphorase activity of various tissues. Since no such reaction has been demonstrated, it seems more likely that the function of diaphorase is in the aerobic oxidation of $CoIH_2$."

In support of this conclusion, I brought forward the fact that the rate of reduction of diaphorase by NADH (calculated from the rate of oxidation at infinite methylene blue concentration) was greater than that of the oxidation of NADH by oxygen, and the results of an experiment in which the addition of partly purified diaphorase to particles suspended in sub-optimal phosphate substantially increased the rate of oxidation of NADH. Since Walter Bonner later found that EDTA also stimulates under these conditions [16], it is probable that my preparation of Straub's diaphorase was acting as a metal binder. Indeed, in my paper I kept open the possibility that diaphorase was acting as an indifferent protein rather than as a specific electron carrier.

In any case, Lockhart and Potter were justified in their doubts, and I was correct in pointing out their implications, but I did not forsee that a new function of diaphorase was to be discovered and that methylene blue is not a suitable electron acceptor for NADH dehydrogenase.

However, I was correct in assigning the BAL-sensitive factor a role in the NADH oxidase system. This was now becoming controversial since a number of soluble cytochrome *c* reductases with NADH or NADPH as donor were appearing in the literature. Before dealing with this, however, I would like to quote another passage from the 1949 paper [40].

> "The factor is the BAL-sensitive component of the system and is probably the same as the factor previously found necessary for the oxidation of succinate (Slater, 1948; 1949a). This proposed scheme does not, of course, exclude the possibility that further components, not yet identified, may also be required (cf. Slater, 1949a). In particular, the possible role of the substance responsible for the e band (Keilin and Hartree, 1949b) of cytochrome remains to be investigated."

So much for King's [30] statement that I was not able to associate my work with that being conducted by Keilin and Hartree, just one flight down from me.

Relation of BAL-labile factor to cytochrome *c* reductases

Already in 1940, Haas et al. [42] had published the isolation from yeast of a soluble enzyme that catalyses the oxidation of NADPH by cytochrome *c*. In 1949, Leon Heppel - then working with Bernie Horecker - reported to the Federation Meeting the isolation of a similar enzyme for NADH [43]. When I went to work in the U.S.A. in 1949, I found that enzymologists there were only interested in soluble, not structurally bound, enzymes. The establishment view was that NADH is oxidized by cytochrome *c* reductase (a flavoprotein), cytochrome *c* and cytochrome oxidase. The latter was a bit of a problem, but in any case Wainio [44] and Lucile Smith [45] were showing that it could be solubilized and even purified with bile salts.

In a companion paper to that on the NADH oxidase system, I published a study on the properties of the NADH-cytochrome *c* reductase in the heart-muscle particles, using cyanide to inhibit the oxidation of ferrocytochrome *c* [46]. As was to be expected, the reductase activity was inhibited by BAL. I calculated that the rate of reduction of added cytochrome *c* was 1200 times less than that of the endogenous cytochrome *c* in the particles. However, with sufficient cytochrome *c* (about 50 μM) I was able to oxidize NADH with about the same rate as with oxygen.

My calculation of the relative ineffectiveness of added compared with endogenous cytochrome *c* made quite an impression when I reported it to the Federation Meeting in 1950, but in fact to quote the text of my talk.

> "There is nothing very surprising about this. The cytochrome *c* is firmly bound to the particles. Another way of looking at the figure 1200 is to consider that the rate constants are the same, but that the effective concentration of cytochrome *c* on the granules is 1200 times that using the total volume of the suspension as the basis of the calculation."

Now that we know that the cytochrome *c*-binding site on QH_2 : cytochrome *c* oxidoreductase is on the inside of the membrane of the submitochondrial particles, it is perhaps surprising that the relative activity is not more than 1200.

I gave another lecture during my stay in New York, to the Enzyme Club, founded by David Green in 1942. In a recently published history of this Club, Phil Siekevitz [47] writes:

> "E.C. Slater's presentation of cytochrome *c* reduction probably focussed on his discovery of the 'BAL-sensitive' factor involved in the linkage of the dehydrogenases to cytochrome *c* of the electron transport chain, the first indication of the existence of cytochrome *b*!"

Of course, cytochrome *b* was well and truly discovered by David Keilin in 1925 at the same time as he discovered cytochromes *c* and *a*, and it was not necessary for me to prove its existence. I think that what Siekevitz has in mind was that my discovery of a factor operating between cytochrome *b* and cytochrome *c* also focussed atten-

ating between cytochrome *b* and cytochrome *c* also focussed attention on cytochrome *b*, which it was not receiving at that time.

My experiments on the NADH oxidase system were carried out in Cambridge between February 9 and November 4, 1949. A few days after the last experiment, my wife and I sailed in the Île de France for New York, where I joined Severo Ochoa's laboratory to be introduced into the mysteries of oxidative phosphorylation. After my return to the Molteno Institute in the summer of 1950, I continued on this line of research and did not again work on the factor until after moving to Amsterdam in May, 1955.

In the intervening period, further progress was being made in the isolation of the so-called cytochrome reductases. I reviewed these in the chapter on *Biological Oxidations* in *Annual Review of Biochemistry* in 1953 [48]. Two groups in David Green's laboratory (one containing Osamu Hayashi, the other Henry Mahler) had isolated from pigeon-breast muscle and bovine heart muscle, respectively, BAL-insensitive soluble enzymes that catalysed the oxidation of NADH by cytochrome *c*. I suggested that tissues might possess two systems for the oxidation of NADH:

> "...first that of Green's group, which is a single enzyme, reacting directly with cytochrome *c* and second, the system found in the Keilin and Hartree heart-muscle preparation which comprises two components, Straub's flavoprotein and the BAL-sensitive and antimycin-sensitive factor. This possibility is given support by Potter and Reif's finding of both antimycin-sensitive and antimycin-insensitive pathways for the oxidation of DPNH in heart muscle, liver and kidney homogenates."

The nature of the antimycin-sensitive factor will be discussed in the next section. That there are antimycin- and BAL-insensitive pathways for the oxidation of NADH is now established. In most tissues, however, they are present in the endoplasmic reticulum and not the mitochondria* and are involved in hydroxylations or detoxifications rather than energy conservation. However, we now also know that the preparations of cytochrome reductase isolated

---

*There is also a slow antimycin-insensitive pathway in mitochondria whereby O_2 is reduced to O^-_2 [49]. This pathway is sensitive to BAL [50].

from Green's laboratory in 1952 were derived from the mitochondria, but that their ability to reduce cytochrome *c* was an artefactual property introduced during the isolation, a possibility which in fact Henry Mahler had kept open. Later, the same laboratory was responsible for establishing the function of ubiquinone in intracellular respiration and for isolating the two enzymes involved in the reduction of cytochrome *c* by NADH, namely NADH : Q oxidoreductase (also called Complex I) and QH_2 : cytochrome *c* oxidoreductase (also called Complex III). We know now that the latter enzyme contains cytochrome *b*, the BAL-sensitive factor and the antimycin-sensitive factor.

In 1950, Bernie Horecker extended his previous work on the isolation of the NADPH : cytochrome *c* oxidoreductase to liver [51]. This enzyme was later extensively studied by Henry Kamin. It is known now that it functions in hydroxylations and that the natural acceptor is not cytochrome *c* but cytochrome P-450.

Relation of BAL-sensitive factor to antimycin-sensitive factor

Antimycin A was introduced as an electron-transfer inhibitor by Van Potter and Reif in 1952 [19]. To quote from the Introduction to this paper:

"Antimycin A appears to be a fairly selective inhibitor for an electron-transport component that has been studied by a great many investigators in terms of the same general phenomenon, namely, inhibition of overall electron transport systems without inhibition of the components at either end. This middle component, which appears to be involved in cytochrome *c* reduction, has been extensively studied by Slater and is currently referred to as the 'Slater factor'. However, since the complete identity of the factors cannot be assumed, we shall use the term 'antimycin A-blocked factor' to denote the hypothetical component which is here involved."

Perhaps not surprisingly, I had less hesitation in accepting the reality of the "antimycin A-blocked factor" and in my review in the *Annual Review of Biochemistry* [48] assumed that the antimycin- and BAL-sensitive factors are identical.

However, Potter and Reif's caution in identifying the two factors with one another was justified. The first indication that this might be the case came from Paul Greengard's observation (reported in [52]) that antimycin, in contrast to BAL treatment, has no effect on the oxidation of NADH by fumarate. Now we can see that this was, in fact, not conclusive since the inhibition by BAL may be due to an effect on NADH dehydrogenase. More convincing was Thorn's finding [53] that BAL-treated preparations bind antimycin normally. The final proof was Thorn and Deul's [54] finding in 1962 that antimycin prevents the reduction of cytochrome *b* in BAL-treated preparations (see later).

Early attempts to identify the BAL-labile factor

1. Investigation of possibility that it is a haem compound

In my first papers on the factor, I had tentatively identified it with what we now call a high-spin haemoprotein, whose spectrum would not be visible with the microspectroscope but which would contribute to the total protohaematin content measured as pyridine haemochrome. In support of this conclusion, I brought forward the following considerations.

(a) The heart-muscle particles contain unidentified haematin compounds (not haemochromes), and it is not unreasonable to expect that these might be concerned in electron transfer.
(b) Non-haemochrome haemoproteins such as catalase, peroxidase and methaemoglobin are destroyed by treatment with -SH compounds, including BAL, in the presence of air.
(c) The total protohaematin content of heart-muscle particles was decreased by BAL treatment, but not by a source of H_2O_2 (glucose + glucose oxidase).

By the time that I submitted, on August 15, 1957, a long review on *The constitution of the respiratory chain in animal tissues*, I could only write:

"... this possibility has been further investigated without producing additional evidence in its favour; it still remains open."

During the 9 years that I was in the Molteno Institute, I was either working alone or, in the latter part of this period, at different times with a research student, Francis Holton, and the "post-doc's" Ken Cleland and Paul Greengard. During the last few years, I had the assistance of Anne Searle (later Kandrach). My resources were too limited to follow two lines simultaneously, and I concentrated on studies on mitochondria and oxidative phosphorylation rather than on the electron-transfer system itself.

My appointment to the chair in Amsterdam meant a substantial increase in my physical resources, although it is questionable whether these compensated the loss in intellectual stimulation that the move away from Cambridge entailed. However, at the age of 38, it was time once again to strike out on my own.

In Amsterdam, I was able to pursue both lines of my research, what may be termed the Keilin and Ochoa lines. The first line concentrated on the identification of the BAL-sensitive factor.

Since 1950, three papers had appeared on the effect of BAL. In a careful study, Thorn [53] confirmed that succinate dehydrogenase is not affected, and Stoppani [56] showed that, as was to be expected, competitive inhibitors of succinate dehydrogenase, do not protect against BAL. Elmer Stotz and co-workers [57] had rather alarmingly found that in a succinate : cytochrome *c* reductase preparation made by treating BAL-inactivated heart-muscle particles with cholate, followed by ammonium sulphate fractionation, the cytochrome *b* band was shifted from its normal position at 563 nm to 559 nm and that cytochrome *b* was no longer reduced by succinate. Since my observations on the lack of effect of BAL treatment on the reducibility of cytochrome *b* had been made with the microspectroscope, I felt obliged to record the complete spectrum of BAL-treated and untreated particles in a photoelectric spectrophotometer. These spectra, which I presented in a symposium to the Biochemical Society in Leeds on July 13, 1956 [52], confirmed my original conclusion.

My student, Dirk Deul did find that BAL treatment of heart-muscle particles solubilized by addition of bile salts is no longer specific. It seems likely that the shift of the spectrum of cytochrome *b* observed by Stotz and his co-workers was due to residual BAL still present after solubilization.

Dirk Deul, who was the second of my graduate students in Amsterdam, graduated on July 8, 1959, with a thesis entitled *De BAL-gevoelige factor in de intracellulaire ademhaling* (The BAL-sensitive factor in intracellular respiration), in which he examined possible candidates for the factor, particularly my idea that it might be a haem compound. He found that, as predicted, the model compound methaemoglobin is converted to choleglobin and sulphhaemoglobin by BAL treatment [58]. He and Thorn [54] could also confirm that the total protohaem content is decreased by about 20% by BAL treatment, but the amount correlated with the amount of myoglobin that had been identified by two other graduate students (Jo Colpa-Boonstra and Koen Minnaert) to be present in the preparation [59]. This made the identification of the BAL-labile factor with a non-haemochrome haemoprotein rather unlikely.

However, Deul found effects on the reducibility of cytochrome *b* that were at that time difficult to interpret. It had been shown by Britton Chance in 1958 [60] that the amount of cytochrome *b* reducible by succinate is increased by addition of antimycin. Moreover, the absorption peak is shifted to the red. Deul found that BAL treatment prevented the antimycin-induced reduction of what at that time was variously called "modified cytochrome *b*" [Chance] or "cytochrome *b'*" [Slater] and suggested that cytochrome *b'* is destroyed by BAL treatment and that this might be the site of action of BAL.

I reported this finding at a symposium on haematin enzymes held in Canberra, Australia, on September 4, 1959 [61]. This led Britton Chance to ask a question which appears in the published Discussion of this symposium under the heading *On the redox potential of cytochrome* b, *the kinetics of reduction of cytochrome* b, *and the existence of Slater's factor* [62]. The last part of his question reads:

"Lastly, I would be pleased to hear more details of the destruction of cytochrome *b'* by BAL. How is this related to the tentative identification of the BAL-sensitive factor as a haematin? Does not this experiment remove all data in favour of a haematin 'factor'."

My answer was:

"The haem which, as measured by the intensity of the pyridine haemochrome band, disappears on treatment of a heart-muscle preparation with BAL in the presence of air is probably largely accounted for by the destruction of the myoglobin and cytochrome *b*′ which are present in the preparation. This destruction cannot be related to the complete inactivation of the respiratory chain which also results from the BAL treatment, unless cytochrome *b*′ is in some way involved in the chain. This appears to us rather unlikely, although not impossible, in view of Chance's finding that in the presence of antimycin, cytochrome *b*′ is rapidly reduced by succinate and NADH."

There was, in fact, an important misinterpretation in Deul's experiments which was later corrected by Deul and Thorn [54]. We can now see that cytochrome *b*′, presently identified with cytochrome *b*-566 (or *b*-558), the low-potential haem of the two-haem cytochrome *b*, is not reduced by succinate unless cytochrome *c* is kept highly oxidized (the so-called oxidant-induced reduction of cytochrome *b*), and, for reasons which will later become apparent, experimentally it is not easy to observe this unless antimycin is added. What Deul took to be a destruction of cytochrome *b*′ was, in reality, an inhibition by BAL of its reduction by succinate, although they were at that time not able to decide between the two possibilities, since cytochrome *b*′ was defined as being reducible only in the presence of antimycin. Moreover, Deul and Thorn showed that the reduction of not only cytochrome *b*′ but also that of cytochrome *b* itself (now identified as the high-potential haem of cytochrome *b*) was inhibited by the addition of antimycin to BAL-treated preparations. This removed the last piece of evidence that the BAL-labile factor is a haem compound, but not of course that the factor exists.

Deul and Thorn's experiment was of great significance. From my old work, it was known that BAL treatment prevents the *oxidation* of cytochrome *b*. Potter [18] and also Chance [63] had found that, in this respect, antimycin behaves in the same manner. When, however, antimycin is added to BAL-inactivated particles – the so-called "double kill" experiment – it is the reduction of cytochrome *b* that is inhibited.

Deul and Thorn concluded that

"... these observations are difficult to explain in terms of conventional representations of the respiratory chain."

We shall return to this later.

2. *Investigation of the possibility that the factor is non-haem iron*

I already knew in 1948 that the heart-muscle preparation contains large amounts of copper (some of which, but not all, was later established to be a component of cytochrome *c* oxidase), but it was a great surprise when in 1955 a number of reports appeared from Green's laboratory showing that a similar preparation contains large amounts of non-haem iron [64]. At a meeting of the Society of Experimental Biology in Oxford in September, 1955, I confirmed this for the Keilin and Hartree heart-muscle preparation [65]. This raised the possibility that BAL, in the presence of oxygen, removes the iron. In the review in *Advances in Enzymology* [55], I stated that experiments of Tsou and Wu [66] had completely disproved this possibility. They had shown: first that after inactivation with BAL, the NADH oxidase was not reactivated by the addition of large amounts of the soluble NADH-cytochrome *c* reductase isolated by Mahler from heart-muscle particles and shown by him to contain iron and, second, that, after this treatment, Mahler's enzyme could be isolated with normal properties and still containing its iron.

Since we know now that Mahler's enzyme is only a part of the NADH : Q oxidoreductase and that its reaction with cytochrome *c* is artefactual, this evidence is now seen to be far from conclusive. It was not then known that the iron is liganded to inorganic and cysteine sulphur in a number of different Fe–S clusters in the electron-transfer chain, so the possibility that it was the ligand rather than the iron of one of these clusters that is specifically attacked by BAL could not be imagined.

3. Investigation of the possibility of a lipid-soluble component as the factor

As already mentioned, when I first started working with the succinate oxidase system, nothing was known about the structure of the particles or of the role of the lipid. In 1953, Ken Cleland and I established that the particles are derived from the fragmentation of the membrane of heart-muscle mitochondria (then known as sarcosomes) [28]. In view of the known importance of lipids in membranes, it was now clear that it was no longer possible to ignore the possible role of lipids in the succinate oxidase system. Indeed, I became concerned that inactivation by BAL was caused by peroxidation of unsaturated fatty acids, which had specifically been suggested as required for electron transfer, especially since Meyerhof [67] had found a catalytic effect of thiols and iron on the peroxidation of lipid and, as already mentioned, it was now known from the work of David Green that heart-muscle particles contain large amounts of non-haem iron. I became worried that the whole BAL effect might rest on a change of lipid structure consequent on lipid peroxidation – which would not be very interesting, at least to a non-lipid biochemist.

Therefore, I asked Leslie Wheeldon, a post-doc from Melbourne, to investigate this possibility. He found indeed that BAL does catalyse the peroxidation of unsaturated fatty acids (measured with thiobarbituric acid) but that there was no correlation between the magnitude of this reaction and the inactivation of the succinate oxidase [68]. Addition of iron, cytochrome *c* or haemoglobin substantially increased the degree of lipid peroxidation, but protected the succinate oxidase. Serotonin and quercitin, known inhibitors of lipid peroxidation, did not protect the enzyme. EDTA and cyanide protected the enzyme with little effect on the lipid peroxidation. He concluded, then, that the peroxidation of unsaturated fatty acids is a side reaction and is not an important cause of the inactivation of succinate oxidase. The protection by EDTA and cyanide reported by Wheeldon should be followed up.

In 1953 Martius [69], on the basis of inhibition by dicoumarol and other substances related to vitamin K, suggested that vitamin

K (phylloquinone) is a component of the respiratory chain, functioning in the same place as envisaged for the BAL-sensitive factor. Cormier and Totter [70] and Wosilait and Nason [71] found that BAL reacts with 2-methyl-1,4-naphthoquinone (menadione) so as to prevent its reduction by a flavoprotein menadione reductase. As I pointed out in my review in *Advances in Enzymology* [55], this reaction is clearly a direct one between the -SH groups of BAL and the quinone, which would be expected to take place equally well under anaerobic as under aerobic conditions, whereas inactivation by BAL of succinate or NADH oxidation requires oxygen. Furthermore, I referred to a paper by Bouman and myself [72], in which we reported that analyses carried out in Dam's laboratory in Copenhagen had shown that the heart-muscle particles did not have any vitamin K activity.

Martius [73] later modified his scheme to place vitamin K between NADH and cytochrome *b* in the phosphorylating chain with a non-phosphorylating by-pass involving a flavoprotein between NADH and the factor (as in my scheme). In agreement with this scheme my student Jo Colpa-Boonstra [74] found that reduced menadione can act as a substrate for the electron-transfer chain, and that this oxidation is antimycin sensitive, but that the oxidation of NADH by menadione is much slower. She concluded, then, that reduced menadione enters the respiratory chain in the region of flavoprotein or cytochrome *b*. She left open the possibility that reduced menadione is a model for a naturally occurring vitamin or acts like many other artificial electron donors. Nowadays, duroquinol is often used as an artificial donor to the ubiquinol : cytochrome *c* oxidoreductase, reacting with both endogenous ubiquinone and directly with cytochrome *b*.

The paper by Bouman and myself [72] referred to above in connection with vitamin K dealt mainly with our finding that the particles do contain another fat-soluble vitamin - the chromanol α-tocopherol (vitamin E). Moreover, we found that the concentration of α-tocopherol was nearly trebled by reduction with ascorbic acid - HCl, a procedure known to reduce and cyclize tocopherolquinone to tocopherol. We concluded, then, that the particles contain tocopherolquinone and that the redox pair tocopherol/tocopherolquinone might be involved in electron transfer. However, this is not a simple redox reaction, since it involves cyclisation of the quinol, and I was more interested in the possibility that the reaction is specifically involved in the coupling between electron transfer and phosphorylation, particularly since some of the symptoms in muscular dystrophy resemble those to be expected from an uncoupling of oxidative phosphorylation. The fact that a large part of the supposed total α-tocopherol (oxidized and reduced) was found in the reduced form in a preparation in which all known components of the electron-transfer chain are oxidized precluded a simple redox rôle. The idea of an involvement of a quinone–chromanol interconversion in oxidative phosphorylation was later taken up by Brodie [75].

Alvin Nason was also interested in the idea of α-tocopherol being involved in electron transfer, indeed he suggested that it is the antimycin-sensitive factor [76]. However, we could not accept his experimental evidence, namely the reactivation by α-tocopherol of NADH-cytochrome *c* reductase activity in iso-octane-extracted particles. We were able to show that neither the inactivation of NADH-cytochrome *c* reductase by iso-octane extraction nor reactivation by α-tocopherol is specific for the enzyme or reactivating agent, respectively [77]. We suggested that the inhibition is caused by adsorption of iso-octane on the surface of the enzyme and that α-tocopherol and other lipid-soluble substances reverse the inhibition simply by dissolving and desorbing the iso-octane. We were unable to confirm experimentally the competition between antimycin and α-tocopherol reported by Nason and co-workers.

However, in my review in *Advances in Enzymology* [55], I concluded that although these experiments did not provide good evidence that added tocopherol acts in a catalytic fashion, they did not disprove this possibility and that there were other reasons for keeping it open.

In retrospect, we can now see that little remains intact of our efforts and those of Martius and Nason to find a role for fat-soluble vitamins in intracellular respiration. However, they did help us to see the importance of a new quinone found by Fred Crane and co-workers in David Green's laboratory in 1957 [29]. I ended the section on Vitamin E with the passage [55]:

"Crane et al. have recently announced in a preliminary note the isolation from beef-heart mitochondria of a quinone of unknown structure. It appears to be identical with a compound previously isolated by Morton and recently named ubiquinone. The spectra of this quinone and especially of its hydroquinone are similar to those of the corresponding α-tocopherol compounds, but some other properties are different. The possible relationship between this quinone and α-tocopherolquinone is being investigated".

The results of this investigation, a fine piece of work by Harry Rudney, then a post-doc in my group, and Jan Links, are unfortunately only published in the printed report of a symposium held during the Fourth International Congress of Biochemistry in Vienna on September 6, 1958 [78]. This showed that what we had previously reported as an increase in α-tocopherol, resulting from reduction and cyclisation of α-tocopherolquinone by treatment with ascorbic acid–HCl, was in reality the chemically similar ubichromanol resulting from the cyclization of ubiquinone. No tocopherolquinone is present. Thus the view held by Martius and ourselves that a quinone is in some way involved in electron transfer was correct, but we had the wrong quinone and the wrong redox reaction. I never doubted the importance of ubiquinone, but it took a long time to establish its precise rôle. The pool function of the quinone/quinol redox system between flavoproteins and what is now known as ubiquinol : cytochrome *c* oxidoreductase was first clearly formulated by Martin Klingenberg [79]. Peter Mitchell [80], however, preferred to place ubiquinone between cytochromes *b* and c_1, that is to say somewhere in the region of the factor. According to current views, the quinone/quinol couple has the pool function envisaged by Klingenberg, but, in addition, two different species of ubisemiquinone function as intermediates in electron transfer between cytochromes *b* and c_1. However, neither is identical with the BAL-sensitive factor.

Work on the BAL-sensitive factor after 1962

After Deul and Thorn's important demonstration that the BAL-sensitive and antimycin-sensitive sites are not identical, no further experiments on the BAL-sensitive factor were carried out in the Amsterdam laboratory for nearly 20 years. Much attention was given to antimycin, which was experimentally a more rewarding inhibitor since it binds stoicheiometrically to the ubiquinol : cytochrome *c* oxidoreductase (1 molecule per enzyme monomer) and its binding can be followed by quenching of its fluorescence. This work culminated in the appearance in 1973 of a review [81] in which there is a brief reference to Thorn and Deul's experiment, but it is interpreted only as showing that the still not understood co-operative binding of antimycin to reduced particles requires an intact BAL-sensitive site. Neither in this review, nor in the earlier published paper by Mårten Wikström, then a post-doc in my laboratory, and my student Jan Berden [82], in which they suggested a very fruitful mechanism for explaining the oxidant-induced "extra" reduction of cytochrome *b*, is proper attention paid to the fact that Deul and Thorn had shown that BAL treatment prevents not only the "extra" reduction of cytochrome *b* in the presence of antimycin, but the total reduction. Wikström and Berden's scheme does not explain the "double kill" experiment of Deul and Thorn. I must take responsibility for this. In retrospect, I can only think that I had remembered Deul's original conclusion that it was the "extra" reduction that was involved and forgotten Deul and Thorn's extension of this finding.

In 1980, BAL returned to the scene again in three papers dealing with the effect of incubation with BAL on the reduction of cytochrome *b* in the presence of antimycin. Krenzenko and Konstantinov [83] confirmed Deul and Thorn's findings. Malviya and coworkers [84], on the other hand, found that the reduction of cytochrome *b*-566, but not that of cytochrome *b*-562, was inhibited by adding antimycin after BAL treatment (cf. Deul's original finding, in which *b*′ can now be equated with *b*-566). However, Qin-shi Zhu and Jan Berden, in unpublished observations, confirmed Deul and Thorn's and Krenzenka and Konstantinov's findings. The latter authors pointed out that neither the traditional scheme of linear arrangement of the respiratory carriers in the succinate-cytochrome *c* reductase span of the electron transfer chain nor the Wikström–Berden model provide a satisfactory explanation of the mode of BAL inhibitory action.

The third paper on BAL published in 1980 was the final identification of the BAL-labile factor, but first we must discuss the work of others that led up to this.

The Rieske Fe-S protein

In 1964, Rieske and co-workers published a paper with the title *Studies on the electron transfer system, LVIII. Properties of a new oxidation-reduction component of the respiratory chain as studied by electron paramagnetic resonance spectroscopy* [85], in which a new signal in the EPR spectrum with $g = 1.90$ was detected in what was called a subparticle constituting the ubiquinol : cytochrome *c* oxidoreductase. Although the exact nature of the cluster responsible for the signal was not at that time established, it was already known that a specific type of protein-bound non-haem iron is involved in similar signals with $g = 1.94$. The unique property of the compound giving the signal at $g = 1.90$ was its reducibility by ascorbate, that is that its potential is much higher than that of other compounds with a similar EPR spectrum.

Although Rieske and co-workers showed that the compound was reducible by succinate (via traces of succinate : Q oxidoreductase in the preparation) and oxidizable by cytochrome *c*, and that the reduction was partially inhibited by antimycin, they were cautious in their interpretations, stating that

> "its importance to the overall electron transfer process remains undetermined. To establish an obligate role for the $g = 1.90$ component, it must be shown that the component undergoes reversible oxidation-reduction at a rate comparable with that of the established components, and also that it is essential to the normal function of the oxidation-reduction system. ... It has been possible to separate the entity giving the EPR signal at $g = 1.90$ from the cytochrome components of Complex III only by employing procedures that are sufficiently drastic to destroy the enzymic activity of the complex".

Because of the effect of antimycin on the new compound and its (slightly) lower redox potential with respect to that of cytochrome c_1 Rieske and co-workers placed "tentatively between cytochrome and cytochrome c_1, if indeed it is an obligate electron carrier". Although this placed it precisely where I put the BAL-sensitive factor, Rieske and co-workers did not speculate on the possible identity of their new component with the BAL-sensitive factor. They were, however, able to exclude its identity with the antimycin-sensitive site.

I never knew what to do with the Rieske protein. In an accompanying paper, Rieske and co-workers placed it on a side-path from the main electron-transfer chain [86]. In any case, I was now more involved in the mechanism of oxidative phosphorylation. Until we moved into our new laboratory in the late 1960's we were not equipped to carry out the large-scale preparations of the respiratory enzymes pioneered by David Green's laboratory.

Even after the structure of the Fe–S clusters was established, the Rieske protein, which was later found to be a constituent of the corresponding enzymes in bacteria, chromatophores and chloroplasts, received comparatively little attention. In order to distinguish it from other proteins of the respiratory chain containing Fe–S clusters, it was (and is) always referred to as the Rieske Fe–S protein.

It is now a pleasure to re-read the paper by Rieske and co-workers, which is one of the classics in the history of the study of redox enzymes.

Oxidation factors (OxF)

In 1969, Ef Racker and co-workers [87] described the reconstitution of the succinate oxidase system from succinate dehydrogenase, cytochrome c_1, cytochrome c, cytochrome c oxidase and a particulate preparation containing cytochrome b. In 1970, Jan Berden and I [88] found that the reconstitutive activity of the particulate cytochrome b fraction correlated with the presence of residual cytochrome c_1 in this fraction. In 1972, in a paper entitled *Resolution and reconstitution of the mitochondrial electron transport system, III. Order of reconstitution and requirement of a new factor for respiration* [89], Racker and co-workers reported that by gel filtration on Sephadex a new protein factor could be separated from the preparation of cytochrome c_1 previously used in the reconstitution [87]. They found that this factor is required for the antimycin-sensitive reduction of cytochrome c by duroquinol and obtained spectral evidence that it is operative between cytochromes b and c_1. They did not relate this to the BAL-sensitive factor. Indications were also obtained for the requirements of a second component for the formation of an active complex. In 1977, my student Carla Marres [90] found that the major component separated from a crude preparation of cytochrome c_1 by this procedure is a 12-kDa protein which is also present as a subunit of the intact ubiquinol : cytochrome c oxidoreductase.

In 1979, Trumpower and Edwards [91] found that the active principle of a preparation of OxF isolated by a different procedure from that used by Racker and co-workers is not the 12-kDa subunit but is identical with the Rieske Fe–S protein. It should be pointed out, however, that it is by no means certain that the OxF of Trumpower and Edwards is the same as that of Racker and co-workers. All subunits of ubiquinol : cytochrome c oxidoreductase essential for enzymic activity could be called oxidation factors and it seems quite possible that the 12-kDa protein separated by Carla Marres [90] from cytochrome c_1 is also an oxidation factor. Furthermore, other persistent contaminants of cytochrome c_1 – the so-called hinge protein isolated by Tsoo King [30], which is usually missed in SDS gel electrophoresis because it stains so poorly with Coomassie Blue, and a 7-kDa subunit isolated by von Jagow [92] – are quite possibly oxidation factors.

However, it is unlikely that any of these oxidation factors act directly as an electron carrier and the demonstration by Trumpower and Edwards that the Rieske Fe–S protein has such a function was an important development, if not particularly surprising. What did excite me was a discussion comment made by Bernie Trumpower during a symposium held in honour of Ef Racker in Cranbrook on July 7, 1979, recalling that before he had identified the OxF with the Rieske Fe–S protein, he had reported that the factor is necessary for reduction of cytochrome b in the presence of antimycin. Since in the light of Deul and Thorn's experiment, this was exactly what was to be expected of the BAL-sensitive factor, I was now convinced that OxF = Rieske Fe–S protein = BAL-sensitive factor, and I said so during an after-dinner speech to that isolator of factors par excellence, Ef Racker. As a matter of fact I had seen this earlier paper by Trumpower, published in 1976 [93], but forgotten it. Maybe I had been put off by the title *Evidence for a protonmotive Q cycle mechanism of electron transfer through the cytochrome* b–c_1 *complex*, since I was not very sympathetic to the Q cycle at that time. In any case, neither Trumpower nor I had then realized that in its inhibitory effect on the reducibility of cytochrome *b* in the presence of antimycin, his OxF had the properties expected for the BAL-sensitive factor. It took three years for the penny to drop, so far as I was concerned.

However, it was still necessary to prove the identity of the BAL-sensitive factor with the Rieske Fe–S protein.

Identification of BAL-sensitive factor with the Rieske Fe–S protein

On returning to Amsterdam after the summer meetings in the U.S.A. and Canada in 1979, I asked one of my students to repeat the isolation of the Fe–S protein by the method of Trumpower and Edwards from both BAL-treated and untreated heart-muscle particles and to compare the activity of the two preparations in reconstitution. Unfortunately, he did not succeed in isolating an active "oxidation factor" fraction. However, one day another of my students, Simon de Vries, who was re-examining the Deul–Thorn "double-kill" effect, told me that he had measured the EPR spectrum of BAL-treated particles and found that the Rieske Fe–S protein had disappeared.

This is shown in Fig. 14, taken from a *Letter to Nature*, submitted on August 21, 1980 [94]. The EPR spectrum of heart-muscle particles reduced with ascorbate in the presence of tetramethyl-*p*-phenylenediamine, cytochrome *c* and cyanide shows the lines characteristic of the Rieske Fe–S protein at $g_z = 2.03, g_y = 1.89$ and $g_x = 1.80$. There is an additional line at about $g = 1.94$ derived

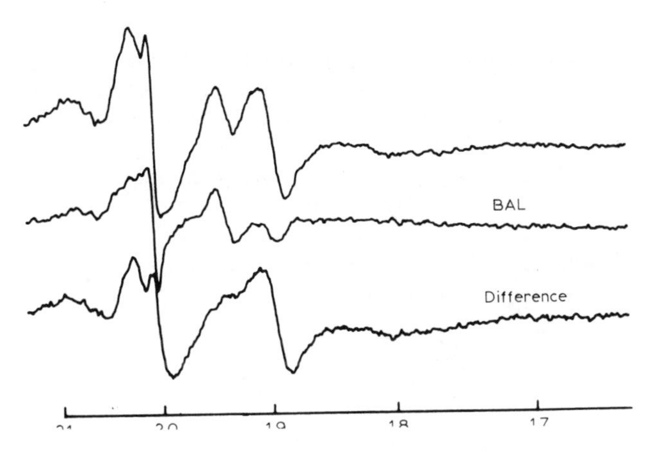

Fig. 14. Effect of treatment with BAL on EPR spectrum of Rieske Fe–S protein in submitochondrial particles, reduced with 5 mM ascorbate, 10 μM, *N,N,N',N'*-tetramethyl-*p*-phenylenediamine, 20 μM cytochrome *c* and 4 mM KCN. The top spectrum is that of untreated particles (70 mg/ml) and the middle spectrum that of particles shaken in air with 10 mM BAL at 36°C for 30 min. The bottom spectrum is the difference. The scale shows the *g* values. From [94].

from partial reduction of one of the Fe–S clusters associated with succinate dehydrogenase, and a free radical at about $g = 2.0$ derived from the semiquinone of ubiquinone and from ascorbate. (The broad line at about 2.1 is probably due to contaminating copper). After BAL treatment, the signals of the Rieske Fe–S cluster completely disappeared. The residual signals are due to the free radical and the g_y and g_x lines of the iron-sulphur cluster of the succinate dehydrogenase. The difference spectrum shows the three lines characteristic of the Rieske Fe–S protein (as well as some free radical, the concentration of which was different in the two spectra).

Spectra taken of NADH- and succinate-reduced particles showed that, although the mitochondrial inner membrane contains about 10 different Fe–S clusters, only that belonging to the Rieske protein is destroyed by BAL treatment of submitochondrial particles, thus confirming the specificity of BAL treatment. An Fe–S cluster present in outer membrane fragments contaminating submitochondrial particles is also destroyed by the BAL treatment.

TABLE III

Correlation between amount of residual Rieske [2Fe–2S] cluster, after destruction by different concentrations of BAL, with fraction of cytochrome *b* that is rapidly reduced by succinate in the presence of antimycin

Submitochondrial particles were inactivated by shaking with different concentrations of BAL for 45 min at 25°C. From [96]

BAL mM	[2Fe–2S] (relative)	Fraction of cytochrome *b* rapidly reduced (%)
0	100	89
1	51	54
2	20	20
4	0	0

It is rather remarkable that BAL treatment is so specific for this single 2Fe–2S cluster in the Rieske protein. Susceptibility to reducing agents in the presence of air is a rather common property of Fe–S clusters. The 4Fe–4S cluster of hydrogenase of *Chromatium vinosum* is, for example, converted to a 3Fe–xS cluster with concomitant loss of activity [95]. It is possible that, in submitochondrial particles, it is only the 2Fe–2S cluster in the Rieske Fe–S protein that is accessible to the very hydrophilic BAL. In this connection, it should be recalled that the phenomenon leading to the discovery of the BAL-sensitive factor was first revealed in experiments with glutathione instead of BAL. When, however, these experiments were repeated in Amsterdam, using glutathione obtained commercially, no effect was found. It is likely, then, that the inactivation by glutathione found in earlier experiments was due to an impurity in the preparation that I isolated from yeast. Possibly. an impurity present in some of Hopkins' earlier preparations of glutathione was the culprit.

The nature of the chemical reaction between BAL, oxygen and Fe–S protein is not established. It is possible that a disulphide bridge is formed between BAL and one of the cysteine molecules or inorganic sulphur that is bound to the iron atom in the Fe–S cluster.

In our recent work, BAL treatment has been found once again to be a useful technique for irreversibly and specifically eliminating the Rieske Fe–S Protein. Table III shows the correlation between the amount of residual Rieske Fe–S cluster and the amount of cytochrome *b* rapidly reducible by succinate in the presence of antimycin [96].

Structure of Rieske Fe–S protein

Since Leslie Grivell and his colleagues in our laboratory were engaged in isolating, cloning and sequencing the genes for the different subunits of yeast ubiquinol : cytochrome *c* oxidoreductase, Carla Marres joined his group in order to sequence the gene coding for the Rieske Fe–S protein [97]. The conventional technique of using antibodies to the protein translated by the messenger RNA transcribed by the gene was used to detect the encoded protein. The antibody for the detection of the Fe–S protein was the same as that used for studying the precursor–product relationship of this subunit [98]. Sequencing the gene responsible for the protein precipitated by this antibody was completed by Carla Marres by the summer of 1983 and I reported the sequence in an unpublished lecture that I gave to the FEBS Meeting in Brussels. Since it contains only one cysteine, we concluded that an unusual Fe–S cluster, in which at least 3 ligands are amino acids other than cysteine, is present in this protein. This was surprising but not impossible in view of unusual features of its EPR spectrum and its unusually high redox potential. Indeed, Fee [99] has recently proposed that a similar protein in *Thermus thermophilus* has non-cysteine ligands.

Later, however, it transpired that the gene product did not cross-react with other preparations of antibodies to the Rieske Fe–S protein, including that to the *Neurospora* enzyme and when Hanns Weiss informed us of the amino acid sequence of the latter enzyme, it was clear that the antibody used in Basle [98] and Amsterdam [97] had picked up another protein. Comparison with published sequences showed that this was the manganese-containing superoxide dismutase, which has about the same molecular weight as the Rieske Fe–S protein and had apparently contaminated the preparation of ubiquinol : cytochrome *c* oxidoreductase from which the antibody had been prepared. Thus, the antigen used to prepare the supposed antibody to the Fe–S protein was, in fact, superoxide dismutase [100]. That superoxide dismutase is associated with the ubiquinol : cytochrome *c* oxidoreductase is perhaps not surprising, since it would have a physiological function in destroying superoxide produced by auto-oxidation of one of the species of ubisemiquinone that is an intermediate in the reaction sequences (see below).

The amino acid sequence of the *Neurospora* enzyme is shown in Fig. 15. It contains 4 cysteines, but it remains to be determined whether these are the ligands.

Gly - Ser - Ser - Ser - Ser - Thr - Phe - Glu - Ser - Pro -
Phe - Lys - Gly - Glu - Ser - Lys - Ala - Ala - Lys - Val -
Pro - Asp - Phe - Gly - Lys - Tyr - Met - Ser - Lys - Ala -
Pro - Pro - Ser - Thr - Asn - Met - Leu - Phe - Ser - Tyr -
Phe - Met - Val - Gly - Thr - Met - Gly - Ala - Ile - Thr -
Ala - Ala - Gly - Ala - Lys - Ser - Thr - Ile - Gln - Glu -
Phe - Leu - Lys - Asn - Met - Ser - Ala - Ser - Ala - Asp -
Val - Leu - Ala - Met - Ala - Lys - Val - Glu - Val - Asp -
Leu - Asn - Ala - Ile - Pro - Glu - Gly - Lys - Asn - Val -
Ile - Ile - Lys - Trp - Arg - Gly - Lys - Pro - Val - Phe -
Ile - Arg - His - Arg - Thr - Pro - Ala - Glu - Ile - Glu -
Glu - Ala - Asn - Lys - Val - Asn - Val - Ala - Thr - Leu -
Arg - Asp - Pro - Glu - Thr - Asp - Ala - Asp - Arg - Val -
Lys - Lys - Pro - Glu - Trp - Leu - Val - Met - Leu - Gly -
Val - <u>Cys</u> - Thr - His - Leu - Gly - <u>Cys</u> - Val - Pro - Ile -
Gly - Glu - Ala - Gly - Asp - Tyr - Gly - Gly - Trp - Phe -
<u>Cys</u> - Pro - <u>Cys</u> - His - Gly - Ser - His - Tyr - Asp - Ile -
Ser - Gly - Arg - Ile - Arg - Lys - Gly - Pro - Ala - Pro -
Leu - Asn - Leu - Glu - Ile - Pro - Leu - Tyr - Glu - Phe -
Pro - Glu - Glu - Gly - Lys - Leu - Val - Ile - Gly

Fig. 15. The covalent structure of the Rieske Fe-S protein in *Neurospora crassa*. The four cysteine residues are underlined. From [108].

Function of the Rieske Fe–S protein

In 1976, Trumpower [93] showed that his OxF is necessary for the reduction of cytochrome *b* by succinate in the presence of antimycin and suggested that the direct reductant of cytochrome *b* is ubisemiquinone formed by the OxF-catalysed oxidation of ubiquinol by cytochrome c_1. With the identification of OxF with the electron carrier, the Rieske Fe–S protein, it was proposed that ubiquinol reacts with the latter. The proposed semiquinone would be expected to accumulate when the Fe–S protein is kept oxidized and the oxidation of cytochrome *b* is prevented by the addition of antimycin (Fig. 16) – the conditions of the oxidant-induced reduction of cytochrome *b*. In 1981, Simon de Vries identified a new species of ubisemiquinone under these conditions [101], clearly different

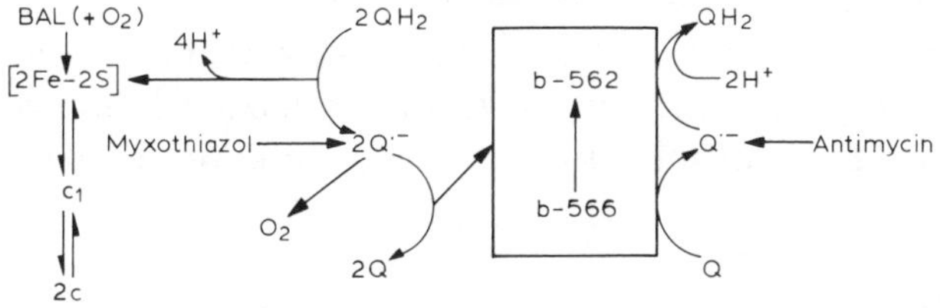

BAL (+ O_2)
4H$^+$
2QH$_2$
QH$_2$
[2Fe-2S]
b-562
2H$^+$
Myxothiazol
2Q$^{\cdot -}$
Q$^{\cdot -}$
Antimycin
c$_1$
b-566
O$_2$
2Q
Q
2c

Fig. 16. Simplified version of Q cycle. One molecule of QH_2 is oxidized by the [2Fe–2S] cluster, with subsequent electron transfer from the latter to cytochrome c_1 and cytochrome *c*. The semiquinone anion ($Q^{\cdot -}$) is oxidized by cytochrome *b*. The process is repeated with a second molecule of QH_2, after which 2 electrons from the 2 molecules of QH_2 are left in cytochrome *c* and 2 in the two-haem cytochrome *b* (the haems are designated *b*-562 and *b*-566, respectively). The two electrons in cytochrome *b* are transferred successively to Q with formation of one molecule of QH_2. The net result is $QH_2 + 2c^{3+} + 2H^+_b \rightarrow Q + 2c^{2+} + 4H^+_{fe}$ where the suffices b and fe indicate that the protons on the left-hand side of the equation come from the region around one of the haems of cytochrome *b* and those on the right-hand side are liberated in the region of the [2Fe–2S] cluster. The sites of action of BAL and antimycin and of the recently introduced inhibitor, myxothiazol [103], are indicated by broad arrows. The residual oxidation in the presence of antimycin is explained by reduction of O_2 to O_2^- by the species of $Q^{\cdot -}$ shown on the left-hand side of the diagram (cf. [49]). This pathway is sensitive to myxothiazol and to BAL treatment [50].

from the previously identified antimycin-sensitive ubisemiquinone [102]. In agreement with Trumpower's proposal, the new species of semiquinone was found not to be formed when particles were treated with BAL.

It is thus well established that the function of the Rieske Fe–S protein is to transfer a single electron from ubiquinol to cytochrome c_1. Whether the donor is free ubiquinol is less certain. It seems more likely that the ubiquinol is bound to a specific subunit of the enzyme and that the electron is transferred from bound ubiquinol to the 2Fe–2S cluster leaving the semiquinone firmly bound to this subunit [104].

The Deul–Thorn "double kill" experiment is accomodated by assuming that the oxidation of cytochrome *b* by the antimycin-sensitive ubisemiquinone is easily reversible. Thus, there are two pathways for the reduction of cytochrome *b*, one by the antimycin-insensitive ubisemiquinone and one by ubiquinol bound to the antimycin-binding site. The former pathway but not the latter is sensitive to BAL treatment or to myxothiazol which prevent the formation of the antimycin-insensitive ubisemiquinone.

Which factors are required for the reduction of cytochrome *c* by cytochrome *b*?

According to the widely accepted Q cycle (see Fig. 16), originally proposed but in a somewhat different form by Peter Mitchell in 1975 [105], the function of cytochrome *b* is to shunt electrons from one ubiquinone-binding site to another. It does not transfer electrons from substrate to cytochrome *c* as originally envisaged in a linear electron-transfer chain. Thus the question posed at the head of this section should be rewritten as: which factor are required for the oxidation of ubiquinol by cytochrome *c*? The following answer can be given, in chronological order of their discovery. (**1**) Cytochrome *b* [10], now known to contain two protohaems [106]; (**2**) cytochrome c_1 [31]; (**3**) the BAL-sensitive factor [18], now shown to be identical with the Rieske Fe–S protein [85] and with Trumpower's preparation of OxF [91]; (**4**) antimycin-sensitive ubisemiquinone [102], either bound to cytochrome *b* in the region of the b-562 haem [81] or to a separate antimycin- and ubisemiquinone-binding subunit; (**5**) antimycin-insensitive ubisemiquinone [101], either bound to the Fe–S protein or to a separate myxothiazol- and ubisemiquinone-binding subunit.

The quinone-binding sites are also the sites of action of all the inhibitors of the cycle, except BAL. Apart from those mentioned above, it now seems likely that these are also the site of action of the alkylhydroxynaphthoquinones studied by Eric Ball and co-workers as long ago as 1947 [26] and of the various ubiquinone analogues introduced by Karl Folkers. One of these, *n*-heptadecylmercapto-6-hydroxy-5,8-quinolinequinone, has been shown to bind to both semiquinone-binding sites, the antimycin-sensitive site binding slightly more weakly [107].

All 5 components listed above are directly involved in electron transfer. It seems likely that all the subunits of the enzyme are required, either with a structural role or specifically in energy conservation. However, this has still to be demonstrated.

Final remarks

It is more than 40 years since I first heard of BAL. The BAL-labile or -sensitive factor was discovered nearly as long ago, the Rieske Fe–S protein just 20 years ago, OxF about 15 years ago, and everything was tied together only a few years ago. The history of the research into the nature of the BAL-labile factor is, like that of most scientific investigations, a story of the slow accumulation of data, sifting of clues and putting them together until a coherent pattern emerges. There were few "leaps forward" and, as usual, very few moments of excitement. The first observations with Keilin's microspectroscope, Trumpower mentioning that removal of his OxF (= Rieske Fe–S protein) prevents reduction of cytochrome *b* in the presence of antimycin and seeing Simon de Vries' EPR spectra of BAL-treated particles were the "purple" moments. There were also the disappointments and mistakes – the failure to recognize the existence of cytochrome c_1, the unexpected finding of relatively large amounts of myoglobin in the Keilin and Hartree heart-muscle preparation and, most recently, the realization that the gene that Carla Marres had been given as coding for the Fe–S protein and which she had so nicely sequenced did not, in fact, code for this protein. There were also at least two occasions when I neglected to relate new findings to the result of Deul and Thorn's "double kill" experiment in 1962. Once when Mårten Wikström and Jan Berden were developing their explanation for the oxidant-induced reduction of cytochrome *b*, and once when Trumpower published the effect of removing OxF on the reduction of cytochrome *b* in the presence of antimycin. Maybe, the Deul and Thorn experiment was carried out too soon. Maybe there are complicated psychological reasons why the result of this experiment had retreated into my subconscious.

I have enjoyed writing this story. I have tried to present it honestly and objectively, if this is possible in an autobiography. It describes no great scientific achievement, but rather an ordinary run-of-the-mill scientific investigation. Despite the mistakes, however, I confess to be reasonably satisfied with the performance of the main character of the story!

One of the pleasures of writing an autobiographical piece is that, while so engaged, all one's friends with whom the story is associated pass before one's mind as if on a screen. First and foremost, my wife Marion waiting patiently in our Cambridge flat to cook the 1 shilling and threepence worth of meat, which was the weekly ration in 1946–1947, knowing that I would very likely be held up by an experiment or by a "talk with the Professor". The notebooks that I have consulted while writing this were typed by Marion. I promised her an emerald ring when I established the structure of the factor, a promise I have not kept. Those long talks with Professor Keilin, which so often kept me from coming home at an Australian time for dinner, were the most exciting intellectual experience of my life and have been decisive for my entire career. Ted Hartree, whom I think had difficulty with the Australian invasion of the Molteno Institute and whom I found difficult to get to know at the beginning, but who soon became a dear friend, as did all other members of the Molteno "Club". It would take too long to mention all my colleagues associated with this research, but two of my most recent students - Carla Marres and Simon de Vries - who have given their Professor the intellectual stimulus that he previously got from his, and in an equally pleasant uninhibited fashion, deserve special mention. The reversible interaction between teacher and student, where first one and then the other is recipient, is one of the great attractions of doing research in a University. That this relation was for so long disturbed in Universities all over the world in the late 1960's and early 1970's was a great disaster, from which we are now happily recovering.

For my part, I have always felt that both as student and as Professor, I have received much more than I have given - for which I can only feel grateful, if more than a little guilty.

REFERENCES

1 R.A. Peters, L.A. Stocken and R.H.S. Thompson, Nature, 156 (1945) 616-619.
2 F.G. Hopkins, E.J. Morgan and C. Lutwak-Mann, Biochem. J., 32 (1938) 1829-1848.
3 D. Keilin and E.F. Hartree, Biochem. J., 41 (1947) 500-502.
4 F.G. Hopkins, J. Biol. Chem., 84 (1929) 269-320.
5 E.C. Slater, Pasteur, 18 (1969) 90-92.
6 E.C. Slater, Biochem. J., 45 (1949) 130-142.
7 E.C. Webb and R. van Heyningen, Biochem. J., 41 (1947) 74-78.
8 E.S.G. Barron, Z.B. Miller and J. Meyer, Biochem. J., 41 (1947) 78-82.
9 D. Keilin, The History of Cell Respiration and Cytochrome, University Press, Cambridge, 1966, pp. 150-153.
10 D. Keilin, Proc. Roy. Soc. B, 98 (1925) 312-339.
11 D. Keilin and E.F. Hartree, Proc. Roy. Soc. B, 129 (1940) 277-306.
12 D. Keilin and E.F. Hartree, Biochem. J., 44 (1949) 205-218.
13 D.E. Green and Y. Hatefi, Science, 133 (1961) 13-19.
14 H. Laser and E. Friedmann, Nature, 156 (1945) 507.
15 E.C. Slater, Biochem. J., 45 (1949) 8-13.
16 W.D. Bonner Jr., Biochem. J., 56 (1954) 274-285.
17 R. Lemberg, J.W. Legge and W.H. Lockwood, Biochem. J., 35 (1941) 328-352.
18 E.C. Slater, Nature, 161 (1949) 405-406.
19 V.R. Potter and A.E. Reif, J. Biol. Chem., 194 (1952) 287-297.
20 E.C. Slater, Bull. Soc. Chim. Biol., 31 (1949) 176-179.
21 E.C. Slater, Biochem. J., 45 (1949) 14-30.
22 E.C. Slater, Biochem. J., 45 (1949) 1-7.
23 D. Keilin and E.F. Hartree, Proc. Roy. Soc. B, 125 (1938) 171-186.
24 E.C. Slater, Biochem. J., 44 (1949) 305-318.
25 F.B. Straub, Zeit. Physiol. Chem., 272 (1942) 219-226.
26 E.G. Ball, C.B. Anfinsen and O. Cooper, J. Biol. Chem., 168 (1947) 257-270.
27 E.M. Case and F. Dickens, Biochem. J., 42 (1947) 1.
28 K.W. Cleland and E.C. Slater, Biochem. J., 53 (1953) 547-556.
29 F.L. Crane, Y. Hatefi, R.L. Lester and C. Widmer, Biochim. Biophys. Acta, 25 (1957) 220-221.
30 T.E. King, Adv. Enzymol., 54 (1983) 367-366.
31 E. Yakushiji and K. Okunuki, Proc. Imp. Acad. Tokyo, 16 (1940) 299-302.
32 E.C. Slater, Nature, 163 (1949) 532.
33 F.J. Ogston and D.E. Green, Biochem. J., 29 (1935) 1983-2004.
34 D. Keilin and E.F. Hartree, Nature, 176 (1955) 200-206.
35 D. Keilin and E.F. Hartree, Nature, 164 (1949) 254-259.
36 E.C. Slater, Biochem. J., 44 (1949) xlviii.

37 F.B. Straub, Biochem. J., 33 (1939) 787-792.
38 V. Massey, Biochim. Biophys. Acta, 30 (1958) 205-206.
39 C. Veeger and V. Massey, Biochem. Biophys. Acta, 37 (1960) 181-183.
40 E.C. Slater, Biochem. J., 46 (1950) 484-499.
41 E.E. Lockhart and V.R. Potter, J. Biol. Chem., 137 (1941) 1-12.
42 E. Haas, B.L. Horecker and T.R. Hogness, J. Biol. Chem., 136 (1940) 747-774.
43 L.A. Heppel, Fed. Proc., 8 (1949) 205.
44 W.W. Wainio, S.J. Cooperstein, S. Kollen and B. Eichel, J. Biol. Chem., 173 (1948) 145-152.
45 L. Smith and E. Stotz, J. Biol. Chem., 209 (1954) 819-828.
46 E.C. Slater, Biochem. J., 46 (1950) 499-503.
47 P. Siekevitz, Trans. N.Y. Acad. Sci. (1983) 213-232.
48 E.C. Slater, Annu. Rev. Biochem., 22 (1953) 17-56.
49 A. Boveris and E. Cadenas, FEBS Lett., 54 (1975) 311-314.
50 M. Ksenzenko, A.A. Konstantinov, G.B. Khomutov, A.N. Tikhonov and E.K. Ruuge, FEBS Lett., 155 (1983) 19-24.
51 B.L. Horecker, J. Biol. Chem., 183 (1950) 593-605.
52 E.C. Slater, Biochem. Soc. Symp., 15 (1958) 76-102.
53 M.B. Thorn, Biochem. J., 63 (1956) 420-436.
54 D.H. Deul and M.B. Thorn, Biochim. Biophys. Acta, 59 (1962) 426-436.
55 E.C. Slater, Adv. Enzymol., 20 (1958) 147-199.
56 A.O.M. Stoppani and J.A. Brignone, Rev. Soc. Arg. Biol., 31 (1955) 282-289.
57 C. Widmer, H.W. Clark, H.A. Neufeld and E. Stotz, J. Biol. Chem., 210 (1954) 861-867.
58 D.H. Deul, Biochim. Biophys. Acta, 48 (1961) 242-252.
59 J.P. Colpa-Boonstra and K. Minnaert, Biochim. Biophys. Acta, 33 (1959) 527-534.
60 B. Chance, J. Biol. Chem., 233 (1958) 1223-1229.
61 E.C. Slater and J.P. Colpa-Boonstra, in J.E. Falk, R. Lemberg and R.K. Morton (Eds.), Haematin Enzymes, Vol. 2, Pergamon, Oxford, 1961, pp. 575-592.
62 B. Chance, in J.E. Falk, R. Lemberg and R.K. Morton (Eds.), Haematin Enzymes, Vol. 2, Pergamon, Oxford, 1961, pp. 593-594.
63 B. Chance, Nature, 169 (1952) 215-221.
64 D.E. Green, in C. Lièbecq (Ed.), Proc. 3rd Int. Congr. Biochem., 1955, pp. 281-284.
65 E.C. Slater, Symp. Soc. Exp. Biol., 10 (1955) 110-133.
66 C.L. Tsou and C.Y. Wu, Acta Physiol. Sinica, 20 (1956) 22-29.
67 O. Meyerhof, Arch. Ges. Physiol., 199 (1923) 531-536.
68 L.W. Wheeldon, Biochim. Biophys. Acta, 29 (1958) 321-332.
69 C. Martius and D. Nitz-Litzow, Biochim. Biophys. Acta, 12 (1953) 134-140.
70 M.J. Cormier and J.R. Totter, J. Am. Chem. Soc., 76 (1954) 4744-4745.
71 W.D. Wosilait and A. Nason, J. Biol. Chem., 208 (1954) 785-798.

72 J.A. Bouman and E.C. Slater, Biochim. Biophys. Acta, 26 (1957) 624–633.
73 C. Martius, Biochem. Z., 326 (1954) 26–27.
74 J.P. Colpa-Boonstra and E.C. Slater, Biochim. Biophys. Acta, 27 (1958) 122–133.
75 A.F. Brodie and P.J. Russell in Proc. 5th Int. Congr. Biochem. Moscow, 1961, Vol. 5, Pergamon, Oxford, 1963, pp. 89–100.
76 A. Nason and I.R. Lehman, J. Biol. Chem., 222 (1956) 511–530.
77 D.H. Deul, E.C. Slater and L. Veldstra, Biochim. Biophys. Acta, 27 (1958) 133–141.
78 E.C. Slater, 4th Int. Congr. Biochem., Vienna, 1958, Vol. 11, Pergamon, Oxford, 1958, pp. 316–344.
79 M. Klingenberg and A. Kröger, in E.C. Slater, Z. Kaninga and L. Wojtczak (Eds.), Biochemistry of Mitochondria, Academic Press, London, and PWN, Warsaw, 1966, pp. 11–27.
80 P. Mitchell, Biol. Rev., 41 (1966) 445–502.
81 E.C. Slater, Biochim. Biophys. Acta, 301 (1973) 129–154.
82 M.K.F. Wikström and J.A. Berden, Biochim. Biophys. Acta, 283 (1972) 403–420.
83 M.Y. Krenzenko and A.A. Konstantinov, Biokhimya, 45 (1980) 343–354.
84 A.N. Malviya, P. Nicholls and W.B. Elliott, Biochim. Biophys. Acta, 589 (1980) 137–149.
85 J.S. Rieske, R.E. Hansen and W.S. Zaugg, J. Biol. Chem., 239 (1964) 3017–3022.
86 J.S. Rieske, W.S. Zaugg and R.E. Hansen, J. Biol. Chem., 239 (1964) 3023–3030.
87 S. Yamashita and E. Racker, J. Biol. Chem., 244 (1969) 1220–1227.
88 J.A. Berden and E.C. Slater, Biochim. Biophys. Acta, 216 (1970) 237–249.
89 H. Nishibayashi-Yamashita, C. Cunningham and E. Racker, J. Biol. Chem., 247 (1972) 698–704.
90 C.A.M. Marres and E.C. Slater, Biochim. Biophys. Acta, 462 (1977) 531–548.
91 B.L. Trumpower and C.A. Edwards, FEBS Lett., 100 (1979) 13–16.
92 H. Schägger, G. von Jagow, U. Borchart and W. Machleidt, Z. Physiol. Chem., 364 (1983) 307–311.
93 B.L. Trumpower, Biochem. Biophys. Res. Commun., 70 (1976) 73–80.
94 E.C. Slater and S. de Vries, Nature, 288 (1980) 717–718.
95 S.P.J. Albracht, K.L. Albrecht-Ellmer, D.J.M. Schmedding and E.C. Slater, Biochim. Biophys. Acta, 861 (1982) 330–334.
96 Q.S. Zhu, J.A. Berden, S. de Vries and E.C. Slater, Biochim. Biophys. Acta, 680 (1982) 69–79.
97 A.P.G.M. van Loon, A.C. Maarse, H. Riezman and L.A. Grivell, Gene, 26 (1983) 261–272.
98 C. Côté, M. Solioz and G. Schatz, J. Biol. Chem., 254 (1979) 1437–1439.

99 J.A. Fee, K.L. Findling, T. Yoshida, R. Hille, G.E. Tarr, D.O. Hearshen, W.R. Dunham, E.P. Day, T.A. Kent and E. Münck, J. Biol. Chem., 259 (1984) 124–137.
100 C.A.M. Marres, A.P.G.M. van Loon, L.A. Grivell and E.C. Slater Eur. J. Biochem., 147 (1985) 153–161.
101 S. de Vries, S.P.J. Albracht, J.A. Berden and E.C. Slater, J. Biol. Chem., 256 (1981) 11996–11998.
102 A.A. Konstantinov and E.K. Ruuge, FEBS Lett., 81 (1977) 137–141.
103 G. Thierbach and H. Reichenbach, Biochim. Biophys. Acta, 638 (1981) 282–289.
104 Q.S. Zhu, J.A. Berden and E.C. Slater, Biochim. Biophys. Acta, 724 (1983) 184–190.
105 P. Mitchell, FEBS Lett., 56 (1975) 1–6.
106 E.C. Slater, in V.P. Skulachev and P. Hinkle (Eds.), Chemiosmotic Proton Circuits in Biological Membranes, Addison-Wesley, Reading, 1981, pp. 69–104.
107 Q.S. Zhu, J.A. Berden, S. de Vries, K. Folkers, T. Porter and E.C. Slater, Biochim. Biophys. Acta, 682 (1982) 160–167.
108 V. Harnisch, H. Weiss and W. Sebald, Eur. J. Biochem., 149 (1985) 95–99.

G. Semenza (Ed.) Selected Topics in the History of Biochemistry: Personal Recollections (Comprehensive Biochemistry Vol. 36) ©*1985 Elsevier Science Publishers*

Chapter 5

Experiences in Biochemistry

NATHAN O. KAPLAN

Department of Chemistry and Cancer Center, University of California, San Diego, La Jolla, CA 92093 (U.S.A.)

In reviewing my scientific career, I feel that there have been several factors which have played important roles in the direction which my research took and also in the success that developed from this work. One of the most important aspects, in my opinion, is the role of research teachers. These individuals play an important part in developing the form and style which a young investigator may take. A second factor, in my view, is to work in an environment where there are stimulating colleagues, postdoctoral fellows, and students. The students add an extra dimension not only because they carry out research based on the ideas of the faculty members, but they bring a sense of curiosity which is of value in consolidating an investigator's thoughts.

Chance plays an important role in producing a successful scientist. I say this because when one is carrying out experiments based on certain goals, and by chance finds an unexpected phenomenon, that can lead the investigator into new and more exciting areas of research. It is absolutely essential in order to have any type of success to be able to alter one's research direction into an area which has arisen during the course of ongoing experiments. Hence, an investigator must have insight and confidence in his ability to make changes in his research plans, even though he knows by con-

Plate 10. Nathan O. Kaplan.

tinuing along the same lines he might get solid results but findings which are not as interesting or important. It is unfortunate that at the present time many scientists in biochemistry and related fields carry on work which would meet the objectives of the research grant rather than explore new ideas which may have arisen during the course of their work.

Berkeley years

I began my research career as a graduate student at Berkeley. I had been an undergraduate at UCLA in Chemistry, and was certainly not one of the outstanding students. I must admit that I was not turned on by the Chemistry teaching at UCLA. It appeared to be much more textbook-oriented than creative. I did have a little experience in research doing some work, with the late Dr. Max Dunn, synthesizing some amino acids.

When I arrived at Berkeley in 1940, biochemistry was on the verge of becoming a field in which many new important developments would ensue. Fritz Lipmann's paper on the phosphate bond had not yet appeared, but there was interest in phosphate compounds [37]. Furthermore, the use of isotopes in biochemical studies had been initiated. The heavy isotopes by Schoenheimer, Rittenberg, and their colleagues at Columbia were being applied and the radioactive isotopes were beginning to be exploited by the group at Berkeley. Professor Greenberg suggested that with the availability of radioactive phosphate, I might look at the turnover of ATP and related compounds in the intact rat because of the availability of radioactive phosphate. Hence, I began reading some of the classical papers of Meyerhof, Warburg, Von Euler, Cori and their associates about phosphate metabolism. I was quite impressed by their contributions and also those of Fiske and Subbarow in their work with ATP and creatine phosphate. I thought that the understanding of the role of these phosphate compounds would be of importance in delineating metabolism in the whole animal.

When I look back upon the years that I was dealing with radioactive phosphate, I recognize the difficulty in evaluating results and in making any kinds of conclusions about the regulation of the phosphate compounds such as ATP and creatine phosphate without knowledge of the many enzymes in which these compounds react and form [17,32]. I am rather pleased that I made the conclusion that the insulin must have an effect on pyruvate oxidation. This was based on results obtained with the turnover of radioactive phosphate in ATP in the presence of the hormone [17,32]. In fact, I wrote a somewhat theoretical outline of this hypothesis. When I was in Lipmann's laboratory, he suggested that I submit the paper; however, the reviewers thought it was too speculative and that I should do more research on the problem. At any rate, I was very happy recently to read several papers which quoted some of these early studies of 40 years ago.

At Berkeley, just before the War, there were some unusual young investigators. I had the good fortune of taking a course on microbial metabolism given by H.A. Barker. He covered a number of microbial fermentations which indicated the existence of different pathways by which microorganisms obtained their energy. His course was most stimulating and probably was one of the primary reasons for my developing a lifelong interest in biochemistry. Little was known about intermediate metabolism, although the glycolytic pathway was being dissected and the intermediates isolated. It was at the time that I was taking Barker's course that Lipmann's article on the phosphate bond appeared. I recall Barker bringing me the first volume of *Advances in Enzymology* which contained Lipmann's paper and how excited I had been over its content, particularly since it explained a number of points which were discussed in the Barker course. It is of interest to note that Barker did not know anything about Lipmann, but once he read the paper, he decided immediately to take part of his sabbatical leave with Lipmann. It was through Barker that I eventually joined Lipmann after the War.

In addition to Barker, there was the late W.Z. Hassid who was at that time, I believe, an assistant professor in plant sciences; he was somewhat older than the other faculty members who had become interested in the new "biochemistry". Hassid was a carbohydrate chemist, and he had done some work with the starch phosphorylase in plants (potatoes and peas). It was through Hassid that I met Sam Ruben who was an assistant professor of chemistry and Martin Kamen who was working at the Radiation Laboratory and who became a lifetime friend. Kamen and Ruben had teamed up with Hassid to utilize the short-life ^{11}C to do some preliminary studies in photosynthesis. Ruben and Kamen became intensively interested and involved in studies relating to biochemistry because of the carbon isotope. When they later discovered ^{14}C, the field of biochemistry certainly was revolutionized. It was during this period, just prior to the War that I also met Mike Doudoroff, who was an assistant professor of bacteriology. It is of interest to note that Barker, Hassid, Ruben, Kamen and Doudoroff, who were not in the Biochemistry Department all saw the importance of trying these radioactive isotopes in the study of intermediate metabolism.

I was fortunate to have met these individuals at the time when they themselves were formulating new research interests. In a discussion with Doudoroff, he stated that he had an organism called *Pseudomonas saccrophilia* which utilized sucrose better than fructose. I had become the "phosphate expert" at Berkeley, so he asked me what I thought about the rapid utilization of sucrose by this bacteria. Since I had read the Cori's papers on muscle glycogen phosphorylase, I suggested that this organism contained an enzyme which would break down sucrose to glucose 1-phosphate and fructose, and that the glucose 1-phosphate would be handled more rapidly than either free glucose or fructose. Based on the analogy with the glycogen phosphorylase, we then started experiments with extracts of the organism to ascertain whether the hypothesis that a sucrose phosphorylase existed was correct. Indeed, it was found that, on dialysis, sucrose was not cleaved to glucose or fructose, but that, if phosphate was added to the dialyzed preparations, it was indeed broken down. We enlisted the help of Hassid who had isolated glucose 1-phosphate from the phosphorolysis of starch. We established then that glucose 1-phosphate could be formed from sucrose, the other product being fructose [11]. We then found that glucose 1-phosphate and fructose would yield sucrose with the liberation of inorganic phosphate. These were indeed interesting experiments for me because I had used some of what I had read and what I was doing, and applied it to a new problem successfully and I was greatly thrilled by the results. It is of interest to note that after we had submitted the paper to the *Journal of Biological Chemistry* a paper appeared in Russian describing a reaction of sucrose phosphorolysis analogous to what we had found in the *Pseudomonas* strain. The coincidence of similar work being carried out some 8000 miles away was quite sobering to me.

Looking back, it is my belief that the experience at Berkeley left a deep impression on me scientifically. It illustrated how a group of people coming from different disciplines can be excited by a new tool and can put it to use appropriately. I believe that the group I have described were the leaders and pioneers in applying radioactive isotopes for the solution of biological problems. This conclusion is certainly borne out by the immense amount of work that followed World War II.

While I was still at Berkeley, I joined the Manhattan Project, although I had not completed my degree. I was fortunate to be able to spend some time working on my thesis before leaving for Los Alamos. I am still amazed that I was able to carry out work for the Manhattan Project and at the same time complete the experimental work for my Doctoral Degree. I believe I must have worked at least 16 hours a day to achieve these goals, but in retrospect, those days were some of the most pleasant because one could concentrate on research even if you washed your own dishes.

Studies on diabetes

In the course of my association with the Manhattan Project, I was sent to Detroit for several months to carry out a special project. It was there that I met Dr. Maurice Franks who was an instructor in medicine at Wayne State Medical School. He was an internist who had a special interest in diabetes. I discussed my theory with him of insulin influencing pyruvate oxidation which would appear in diabetes as a lowering of the ATP levels and an increase in cellular inorganic phosphate. We decided to moonlight some experiments which dealt with producing diabetic rats by use of alloxan [31]. We were able to show that when the diabetic state developed, there was a marked fall in the ATP in the liver and an elevated inorganic phosphate. What was very striking was the large amount of inorganic phosphate that was excreted. This phosphate loss was particularly intriguing.

The studies of the rats suggested that it might be worthwhile to investigate what happens with phosphate in diabetic coma [15]. Since Dr. Franks had access to the Emergency Room at the Detroit City Hospital, we were able to study a number of patients who had been brought to the hospital in a state of diabetic coma. I was surprised by the number of patients in such a condition: the rate was usually about one per night. The problem of diabetic coma does not exist in such a magnitude at the present time.

We found that the most severely comatose patients had a lower serum inorganic phosphate which coincided with a marked increase in urinary phosphate. We then decided to administer inorganic phosphate as well as the other ingredients used in the treatment of a patient in a diabetic coma. In the presence of insulin, the phosphate levels became more normal and excretion decreased. We concluded that inorganic phosphate was helpful in the treatment of the coma. It is of some interest to note that, although this work was done almost 40 years ago, clinicians have paid little attention to the use of phosphate in the treatment of diabetic coma. Recently there have been a number of papers which have confirmed our early findings and have suggested that inorganic phosphate be used in treatment of both diabetic coma and acidosis.

Our results with phosphate were rewarding in the sense that a basic concept could be put to clinical use. It also showed that interactions between clinicians and basic scientists are important and should be encouraged. The association with Dr. Franks was a pleasant and stimulating one. To me, being able to do some research which was of interest to me overcame the dullness of the project which I had been sent to carry out. Dr. Franks and I have been close friends since our meeting in Detroit.

The Lipmann laboratory at the Massachusetts General Hospital

When the War ended and I could leave the Manhattan Project, I was most fortunate to be able to join Fritz Lipmann's laboratory at the end of 1945. When I arrived at Lipmann's laboratory, he had only one person working with him, Constance Tuttle, who was a technician. I had the good fortune of being the first professional to work with Lipmann since Barker was in his laboratory in 1941.

In my view Lipmann is one of the greatest biochemists of the century. He has an unequalled creativity and an insight into problems. His interests are broad and his thinking in research certainly represents much of the framework from which modern biochemistry and molecular biology were built.

I was particularly fortunate because Lipmann was able to give me a great deal of attention. We had many discussions and talks not only about the ongoing research efforts in the laboratory, but on nearly any subject. In looking back on those years, I am amazed at how accurate Lipmann was in predicting the future of biochemistry. The time spent with Lipmann was a great learning experience for me and strongly influenced my future research and thinking. Although I had some experience in research, I believe it was Lipmann who greatly affected my philosophical approach to science. I will be forever indebted to him for the stimulating and compassionate relationship between us at the Massachusetts General Hospital.

When I talked with Lipmann before officially arriving in Boston, we discussed the possibility of investigating photosynthetic problems. It is of interest that some of this was based on a paper published by Sam Ruben in the *Journal of the American Chemical Society* which stressed that the phosphate cycle was operative and important in photosynthesis. Lipmann also had ideas on the subject and we thought that this would be a rewarding area to investigate. However, upon my arrival, Lipmann had already begun the work on the acetylation of sulfanilamide. He had spent many hard years isolating and characterizing acetylphosphate as a product in the oxidation of pyruvate in bacteria. He had thought that acetylphosphate was the acetylating agent and he carried out studies to ascertain whether this compound could acetylate sulfanilamide in liver or choline in brain. In the acetylation of sulfanilamide studies, Lipmann had used an extract from pigeon liver acetone powder. The reaction was lost on dialysis and he could restore the activity by adding a boiled extract.

When I discussed the problem with him, he thought it might be worthwhile to ascertain what in the boiled extract might be responsible for the restoration of activity. In order to ascertain that a known compound was not involved, I tested many substances for activity. I recall obtaining less than a microgram of NAD (DPN) from Bernard Jandroff, who was at that time working in the Department of Biochemistry at Harvard. For the first month in the Lipmann laboratory I must have tried at least a hundred compounds to see if they would restore the activity in the manner which the boiled liver extract did. None of the compounds showed any activity.

After we were convinced that a new factor probably was responsible for the restoration of activity, Lipmann suggested that we attempt purification [38]. Not being certain how to approach this problem, I asked him what did he think we should do. He responded by saying,

"Read some of Warburg's papers on the isolation of NAD and NADP as well as FAD and FMM."

Perhaps from Warburg's procedures, one could develop a method of attack to purify the unknown factor. One must note here that at that time there were no chromatographic methods available for purification [33].

I did what Lipmann had suggested and followed some of Warburg's techniques. We used a combination of heavy metal and phosphotungstic acid precipitations and after about a year's time we were able to obtain about 150 mg of material from about one-half ton of beef liver [40].

During the course of the purification we had made several observations which were significant to the solution of the structure.

There was an enzyme present in the pigeon liver extract which cleaved the thioethanolamine from the pantothenic acid moiety part of the molecule as indicated in Fig. 18. Of course at that time, we did not know that the enzyme carried out this function, but we realized that there was an enzyme in the liver extract which could inactivate the factor. We also knew that diesterase and a phosphatase would inactivate the coenzyme. This is indicated also in the structure in the figure. During the course of the purification, it was noted that there was probably a sulfhydryl group present in the molecule. This possibility arose from smelling a heated solution, and then showing that an enzyme was activated in the presence of sulfhydryl groups.

When we first analyzed the purified compound, we found that it contained one molecule of adenine, one molecule of ribose, and

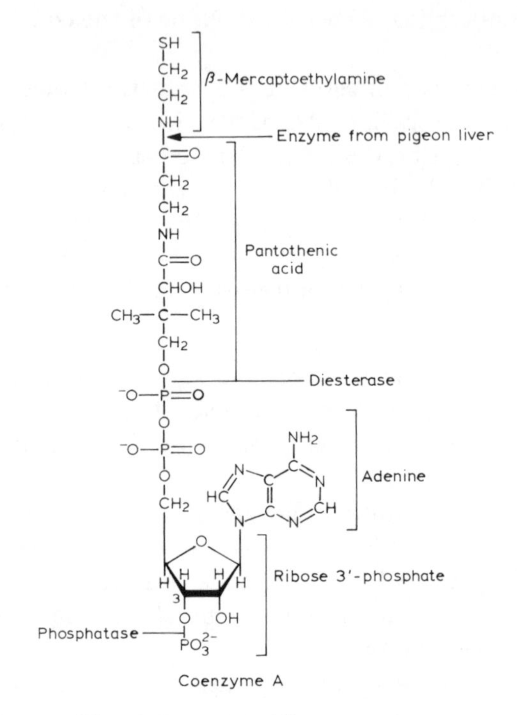

Fig. 18. Structure of Coenzyme A.

three phosphates. One of these phosphates was identified as a monoester phosphate since it was hydrolyzed by the prostate phosphatase which was graciously given to us at that time by Gerhardt Schmidt. However, we knew that there was an additional component present in the coenzyme. Lipmann and I had a number of discussions as to what the nature of this entity might be. We concluded that it must be pantothenic acid. This was based on the fact that the then known B vitamins had all been known to function as coenzymes such as NAD and FAD. Pantothenic acid was known to be present in relatively high concentrations in tissues. We felt that since there was no function reported for pantothenic acid that the unknown substance contained the vitamin. I would like to emphasize that we were quite convinced, even without any data, that pantothenic acid would be found as a component of coenzyme A.

At that time, David Novelli joined the group. We were interested in ascertaining whether pantothenic acid was a part of the coenzyme. There is no good chemical method for determining pantothenic acid. A microbiological assay was necessary and since Novelli had a background in microbiology, he was brought into the group.

Lipmann decided that what we should do was to send out the purified preparation to various groups who were involved in vitamin research and could do vitamin assays for us. At least six laboratories were given the material. All of them reported that there was no pantothenic acid or other vitamins present in the preparation. These results were quite depressing particularly when one group of "experts" wrote us that we really did not know what we were doing because apparently the compound was ATP since it contained 1 mol of adenine, 1 mol of ribose, plus 3 mol of phosphate. I recall that when Lipmann received this letter, he became so upset that he left the laboratory for the day. It looked like we had reached a dead end.

When Lipmann returned the next day, he still argued for the presence of pantothenic acid in the coenzyme. It was just not being liberated. He then suggested that, because of the lack of any ideas of how to go ahead, the best for me was to learn some infrared techniques from Elkan Blout. Perhaps that would give us a clue to the structure. The situation at that time seemed somewhat hopeless, but good fortune was with us. Lipmann had sent the preparation to R.J. Williams in Texas who was the discoverer of pantothenic acid. Williams had turned the preparation over to Beverly Guirard for pantothenic acid analysis. In agreement with the other groups, using a bacterial assay, she found no pantothenic acid. However, she had stored the preparation at 4°C for a month and then she decided to do another assay. On carrying out a second analysis for various vitamins, she found a trace of pantothenic acid. This aroused her curiosity and she hydrolyzed the preparation which would liberate β-alanine. There is a yeast mutant requiring β-alanine so that she was able to identify that β-alanine was present in the preparation, and she suggested that there indeed must be pantothenic acid in the molecule [39]. When we received this information, we were overjoyed. We then recognized that the pantothenic acid was not being liberated from the coenzyme so it would become available to the bacteria which required it. We then used a combination of the various enzymes that we knew inactivated the coenzyme. That is the diesterase, phosphatase, and the enzyme in the pigeon liver preparation. When we incubated the coenzyme with these three enzymes and then assayed for pantothenic acid with the bacteria, we indeed were able to obtain positive results, clearly indicating that pantothenic acid was a constituent of the molecule. We were then able to relate the pantothenic acid content to the units of activity [46,47].

I believe this was a remarkable example of how chance, circumstances, and good science can lead to a solution of a problem. Dr. Guirard was more curious than the other laboratories to whom we had sent the preparation. This was a stroke of luck for us because it was most fortunate that she assayed the coenzyme a second time. Her work is an example of how a hunch really pays off. I am even more impressed with the fact that we were able to identify pantothenic acid by collaborating with the right individual; this also indicates the role of luck and chance in carrying out scientific research. It also illustrates how different individuals deal with a problem — Dr. Guirard was curious enough to do a second assay after a month. It is highly unlikely that most people when sent samples will make a re-determination.

It was found by Lynen some years later that the active form was acetyl-CoA. The Lipmann laboratory had made an intensive search to look for this intermediate, but without success. I recall that during the purification of CoA itself from liver, I had encountered a fraction which promoted the acetylation of sulfanilamide without the addition of ATP. We thought at that time that the fraction was contaminated with ATP and that this was why no ATP was necessary. Later, after Lynen's discovery, this fraction was found to have substantial levels of acetyl-CoA. Although this is hindsight, it left an impression on me, because it stressed that when one obtains a result which is not expected, one should follow up on the finding. If we had been really imaginative, we probably would have isolated the acetyl-CoA 4 or 5 years before Lynen had achieved its purification and identification.

It is difficult for present-day students to realize that one could purify a substance without the use of some sort of column technique. Actually, Connie Tuttle and I worked up about half a ton of liver in order to get the first relatively pure CoA. The amount of work that went into this purification was immense, but the rewards were certainly great. Although there was a lot of hard work involved in this problem, my experience in the Lipmann laboratory from a scientific point of view was certainly most unusual, instructive, and rewarding.

I left the Lipmann laboratory to join the Biochemistry Faculty at the University of Illinois Medical School in Chicago as an assistant professor. I took this position because Sidney Colowick had already become a member of that department and there was promise of other young investigators joining the department. Our hopes for making a great and productive research department were not fulfilled. Due to a number of unusual personal problems, it became necessary for such people as Giulio Cantoni, Colowick, and myself to leave. My stay at Illinois was only 1 year, but despite all the negative aspects of that year, the one positive factor was to become friends with Sidney Colowick, whose career has been closely related to mine. Unfortunately, that year in Chicago, from a scientific point of view, was wasted. I attempted to do some studies relating pyruvate oxidation and CoA but the environment and situation did not allow any meaningful research to take place.

McCollum–Pratt Institute — Johns Hopkins University

After being at the University of Illinois for a few months, it became apparent to Colowick and myself that we had to leave. We both began to look for other positions. Fortunately for us, we were contacted by Bill McElroy who had just become Director of the McCollum–Pratt Institute, part of the Department of Biology at Johns Hopkins University. McElroy offered us an opportunity to carry out research in whatever we would like to, although we had some reservations about becoming involved in an institute which was set up to study trace metals. However, the atmosphere at Johns Hopkins was so different scientifically and socially from that at Illinois, that I thought we had just escaped from a concentration camp. The McCollum–Pratt was a new institute and the original members consisted of McElroy, Colowick, the late Al Nason, Henry Little, Bob Valentine, and myself. We were housed in a greenhouse. The laboratory facilities were not ideal, but we were able to obtain glassware and some equipment almost immediately and within a month or two after arriving, I was able to begin to do research again. This certainly was a happy feeling after what appeared to be a completely lost year at Illinois.

Also housed in the greenhouse was Elmer V. McCollum. He had retired as Chairman of the Department of Biochemistry in the School of Hygiene at Johns Hopkins. He was a friend of Mr. Pratt, who donated the resources for the new Institute. McCollum, in my opinion, was one of the great scientists of our century. He was the discoverer of Vitamins A and D, and was the driving force to put nutrition on a sound scientific basis. His contributions were of the greatest importance. I developed a strong friendship with him, spending many hours with him as he re-experienced his own career. Although he was in his seventies, he had an unusual insight. He was not intimately familiar with the new developments in biochemistry, but he strongly felt that the enzymatic approach would be most important in understanding biochemical problems. He was always a source of encouragement. I consider him as one of my teachers although I had no courses nor did any research with him. I believe that I was indeed fortunate to have contact with him. I shall have more to say about him and other associates in a planned future biography.

One of the projects which I started in Baltimore was the study of the mechanism of CoA action in pyruvate oxidation. From the work in Lipmann's laboratory, there was conclusive evidence that CoA was involved in the oxidation of the keto acid not only in microorganisms but also in plant and animal tissues [48]. During the course of these studies, I had added cyanide to inhibit respiration. I had also added an excess of NAD in order to measure the reducing equivalent generated from pyruvate dehydrogenase under these conditions. It was of interest that the first experiment showed that when pyruvate was oxidized NADH could be formed, but when the enzyme preparation was omitted, the same results were obtained. Then, in collaboration with Sidney Colowick, we found what was happening, i.e., that cyanide was adding to NAD to give an adduct which had absorption in the same region as NADH (340 nm). We then characterized the NAD cyanide compound and found it to be a good method for measuring the oxidized pyridine nucleotides. NADP also can be determined by the cyanide reaction. We later read that in the middle thirties, Meyerhof and his associates had also observed such a reaction. Our experiments showed that the cyanide adduct was a good quantitative measurement of the quaternary pyridine ring. We used this method for determination of the oxidized forms of the coenzyme in extracts of various organisms and tissues, and the procedure was used for determining whether the nicotinamide ribose bond is present when certain enzymes cleave the coenzyme [9,26].

In this connection, Nason was working with zinc-deficient *Neurospora*. The deficient *Neurospora* showed very little glycolytic activity and there was no alcohol dehydrogenase present in these cells whatsoever. No reduction of NAD could be demonstrated even when known dehydrogenases were added. In discussing the problem with Nason, I felt that there was a possibility that the NAD might be hydrolyzed and was unavailable for reactions in the deficient *Neurospora*. We then used the cyanide method to measure the NAD levels in the zinc-deficient *Neurospora*, and the amount found was negligible as compared to controls. We then added NAD and found that there was a rapid destruction of the coenzyme by use of the cyanide method. We therefore detected an enzyme which cleaved the NAD at the nicotinamide ribosidic linkage [27]. This enzyme is now termed NAD glycohydrolase. We called it NADase. The enzyme is very potent and present in relatively large amounts in zinc-deficient *Neurospora*. It is also found in high concentrations in the spores of *Neurospora*.

Hence, by coincidence and by the fact that Nason was present in the Institute, working with the deficient organisms was an important beginning in my lifetime interest in the pyridine nucleotides.

We were able to purify the *Neurospora* NADase partially and also isolate the product adenosine diphosphate ribose. During the course of these studies, we noted that the NAD hydrolysis by this enzyme was not complete as measured by the cyanide reaction. However, if we compared this to the enzymatic assay, that is, using alcohol dehydrogenase, all the reactive enzymatic coenzyme was destroyed. This indicated to us that there was a possibility that a molecule was present in the preparations which had the nicotinamide ribose linkage but which was not subject to attack by the NADase. We then decided to hydrolyze with the *Neurospora* enzyme large amounts of NAD and after all the biological activity had disappeared, we attempted isolation and purification of the compound which was resistant to the action of the NADase, but still maintained a reaction with cyanide. We were successful in this endeavor and the compound proved to be the α isomer of NAD which had very low or no activity with most dehydrogenases. It is of interest to note that the commercial preparations of NAD which were used at that time contained approx. 15% of the total NAD as the α derivative [25]. We also, using the *Neurospora* NADase, were able to cleave the NAD and obtain pure adenosine diphosphate ribose, a compound which is of extreme biological interest at the present time.

An enzyme, having activity similar to the *Neurospora* NADase had recently been described in a number of animal tissues. Colowick and I thought it might be of some interest to compare the properties of animal and *Neurospora* enzymes. We were able to follow the enzyme from animal sources which also hydrolyzes the NAD at the nicotinamide ribosidic link by the use of the cyanide assay. One can distinguish the hydrolytic cleavage at the glycosidic linkage from that of the pyrophosphate cleavage by comparing the cyanide reaction with the alcohol dehydrogenase reaction. If the cyanide reaction is maintained after treatment, this would indicate that the enzyme was a nucleotide pyrophosphatase, an enzyme which cleaves at the pyrophosphate linkage.

The animal enzyme has many different properties than that found with the *Neurospora* protein. We found in our early studies that the enzyme was largely present in the microsomes. However, there are traces of this enzyme in the nucleus. The nuclear enzyme apparently has turned out to be a distinct enzyme which has a role in adenosine diphosphate ribosylation. Very recent studies have shown that the NADase (glycohydrolase) is largely localized in the plasma membrane [34]. Since in those days we had no way of separating plasma membrane from the microsomes, we had concluded that it was a microsomal entity. The mammalian enzyme is a membrane protein and is difficult to solubilize.

There was another important distinction between the *Neurospora* NADase and the animal enzyme in that the animal system was inhibited by nicotinamide, whereas the *Neurospora* enzyme was quite insensitive to the free vitamin. In considering what the difference might be and why there was nicotinamide inhibition, I recollected the studies with the sucrose phosphorylase where one could get inhibition of sucrose phosphorolysis in the presence of excess fructose. It was found that labelled fructose could be incorporated into sucrose without any breakdown of the disaccharide. I speculated that the free nicotinamide could exchange with the bound nicotinamide with NAD and thereby inhibit the breakdown of the pyridine coenzyme. When Leonard Zatman joined the laboratory, we gave him the problem of determining whether radioactive nicotinamide could be incorporated into NAD, and indeed he found that an exchange reaction occurred. That is, ^{14}C-labelled nicotinamide could be found in NAD under conditions where there was no breakdown of the coenzyme. This is illustrated in the following reaction.

$$\text{NRPPRA} + \text{N} \rightarrow \text{NRPPRA} + \text{N} \; (\text{N} = {}^{14}\text{C-labelled nicotinamide})$$
(NAD)

The results of these studies clearly show that an exchange reaction was going on and that the nicotinamide was inhibiting because of the exchange reaction [56]. In contrast, the *Neurospora* enzyme does not appear to do so although Larry Grossman found in the laboratory some time later that, when ergothionine was added into the *Neurospora* system some exchange could be observed, but not of the magnitude which is promoted by the animal enzyme [16].

The fact that radioactive nicotinamide could be exchanged into NAD suggested the possibility that other pyridine compounds also undergo such a reaction and were being incorporated into NAD to promote an analogue of the coenzyme. This is illustrated below:

$$\text{NRPPRA} + \text{X} \rightarrow \text{XRPPRA} + \text{N}$$

X represents a pyridine or a pyridine-related compound.

In the early fifties isonicotinic acid hydrazide had been introduced as an antituberculosis agent which was found to be most effective. We thought the structure of the isonicotinic acid hydrazide was similar enough so that it might be an analogue of nicotinamide.

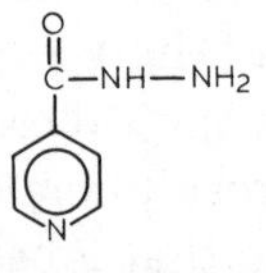

Indeed, using the enzyme from pig spleen glycohydrolase, we were able to demonstrate the formation of this isonicotinic acid hydrazide analogue of NAD. The compound was isolated and characterized [57,58]. What is not clear to this date is whether the isonicotinic acid hydrazide analogue of NAD is involved in the therapeutic action of the free base. It is of interest that isonicotinic acid hydra-

TABLE I

Sensitivity of NADases from different species to isonicotinic acid hydrazide

Sensitive	Insensitive
Beef	Horse
Sheep	Dog
Goat	Man
Deer	Pig
Buffalo	Frog
	Guinea pig
	Rat
	Chicken
	Mouse

zide can induce pellagra-like symptoms in patients. Hence it is usually given with additional nicotinic acid or nicotinamide.

It was indeed fortunate that we had used the pig brain NADase to show the formation of the isonicotinic acid hydrazide analogue, since we later tested the same enzyme from bovine and we found that it was strongly inhibited by this pyridine base. A later analysis showed that the enzymes from certain species were not inhibited by isonicotinic acid hydrazide and would form the corresponding coenzyme analogue. This variation in different skills is illustrated in Table I. The glycohydrolase from all tissues, for example, from the bovine or sheep tissues is inhibited by isonicotinic acid hydrazide, whereas the enzymes from all tissues, such as the pig, were not inhibited. Initially, this was a perplexing finding. We were confused by the observations summarized in Table I, but this turned out to be our first venture into comparative enzymology. When the table was originally consituted, we did not understand its significance, but one of my brighter graduate students scanned the list and suggested that the species that were inhibited by the isonicotinic acid hydrazide were all "kosher animals", whereas those showing no inhibition were non-kosher. He was correct; all ruminant species are sensitive to the inhibition by isonicotinic acid hydrazide. This is due to the fact that when the isonicotinic acid hydrazide analogue of NAD is formed, it markedly inhibits any further reaction. We showed this by adding the coenzyme analogue directly to a bovine NADase [21,22].

The fact that we were able to synthesize the isonicotinic acid hydrazide analogue by this exchange reaction suggested to me that we might be able to make other NAD analyses by such a method. In searching the literature, I was impressed with the observation that had been made in previous years with 3-acetyl pyridine.

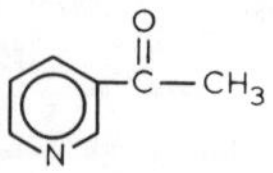

This compound was shown, in small quantities, to be converted to nicotinic acid or nicotinamide by a mechanism which has yet to be elucidated. The compound, however, in high levels is quite toxic to animals. In dogs, it could produce a nicotinamide deficiency syndrome which is known as black tongue. We postulated that perhaps the 3-acetyl pyridine would be exchanged for the nicotinamide to make the coenzyme analogue according to the following reaction.

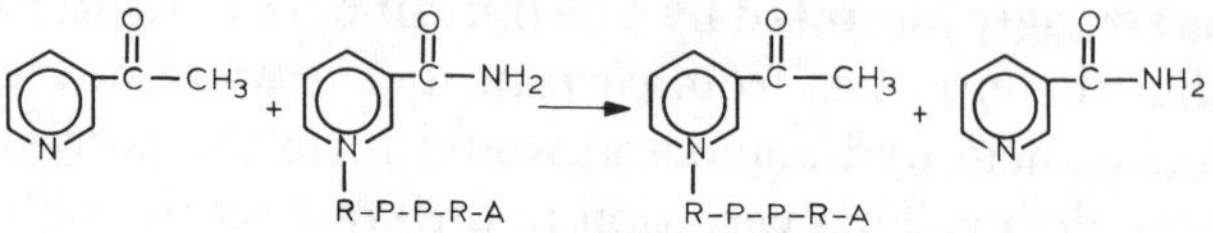

We, therefore, again used the pig brain system and this analogue was readily formed. Its generation could be easily measured by the fact that it has an optimum absorption in the reduced form at 375 nm as compared to 340 nm for NADH. Alcohol, lactic, and a number of other dehydrogenases reduce this coenzyme analogue [19,24].

It was of interest that this analogue, which we made extensive studies with, was the compound that led us into the isozyme research. We had purified for certain reasons the beef heart and beef muscle lactic dehydrogenases. Earlier, Straub had obtained pure enzyme from beef heart and Racker had obtained the homogeneous rabbit skeletal muscle LDH. When we originally compared the beef heart enzyme activity with the 3-acetyl pyridine analogue of NAD, it reacted quite differently from the rabbit skeletal muscle. We thought this was due to the fact that there was a species difference, but when we found that the beef heart and beef muscle enzyme varied in their reaction to the analogue, we then believed that the two enzymes were quite different and this led us into an extensive study of the heart-muscle enzymes which will be described later when I had moved to Brandeis.

The exchange reaction has been widely used. Approximately 150–200 of these coenzyme analogues have been synthesized by the enzymatic procedure. They have been widely used in many enzymatic studies of pyridine-nucleotide-linked enzymes and they have been found to be useful in many different types of studies [1]. We will describe some of these later in the paper.

It is of interest that the nicotinamide exchange reaction led us into the coenzyme analogue synthesis which then was the basis of our studies on isozymes.

During my stay at Johns Hopkins, I became interested in studying oxidative phosphorylation in bacteria. I had taken up Lipmann's philosophy that if you cannot get anywhere in studying animal tissues, go to bacteria which are simpler for resolving the problem; hence, I began an investigation, in collaboration with Sidney Colowick, on oxidative phosphorylations. We selected as an organism to study *Pseudomonas aeruginosa* a well-known microbe which has a potent oxidative capacity. We thought that by growing this organism on citrate we would have a citric acid cycle which then could be probed for oxidative phosphorylation reactions. In order to ascertain whether the citric acid cycle was operating in this organism, we prepared extracts from the bacteria. It had recently been reported by Arthur Kornberg that the yeast contained not only NADP isocitrate dehydrogenase, but also a second NAD enzyme which was specific for NAD. We thought it would be of value to assay the *Pseudomonas* extracts for both enzymes. As indicated in Fig. 19, the first study showed that indeed there was an NADP enzyme reacting with isocitrate to give NADPH; however, there was no activity with NAD. In order to check all components of this latter assay, I added NADP to another cuvette with NAD and the kinetics of the enzyme was similar to adding NADP alone

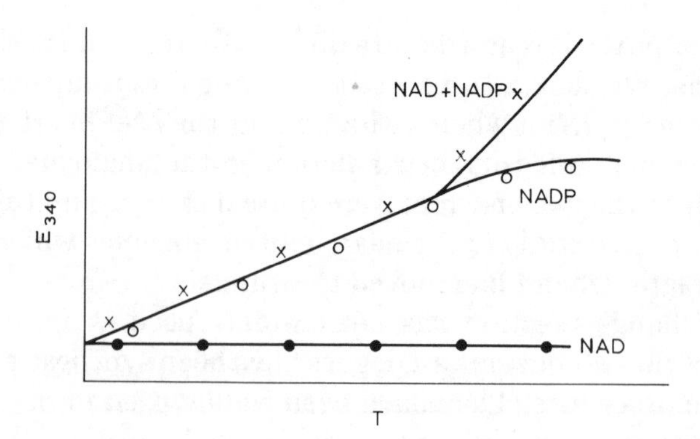

Fig. 19. Initial data schematically indicating the presence of pyridine nucleotide transhydrogenase. T, time (arbitrary units). Isocitrate was the substrate.

in the initial stages of the reaction. At this time, McElroy suggested that we go to lunch, so I turned off the Beckman and proceeded to the Johns Hopkins Faculty Club for a leisurely luncheon and some pool playing. When we returned, I took another look at what was happening in the cuvettes. To my amazement, the cuvette that contained NADP and NAD had a much higher absorbance than the cuvettes containing only NADP and yet there was no reaction with NAD. We checked this out and it certainly was indicated that in the presence of NADP, isocitrate was reducing NAD. We, therefore, postulated that there must have been a transfer or a reduction of the NAD presumably by the NADPH generated by the NADP isocitrate dehydrogenase. This is illustrated in Fig. 19 [10,28].

After doing the necessary controls, we were convinced that we had an enzyme which catalyzed the following reaction.

$$NADPH + NAD^+ \rightarrow NADP^+ + NADH$$

What astounded us was the fact that the reaction seemed to go from left to right very proficiently, and there was no reaction between NADH and NADP. This reaction was shown by Dr. Antony San Pietro, when he came to our laboratory as a post-doctoral fellow, to be a direct hydride transfer [50]. The fact that the reaction appeared to be irreversible was confusing since the redox potential of the two pyridine coenzymes is close to unity. However, one should emphasize that the reaction does not defy the laws of thermodynamics, since it will go from right to left, but at a very slow rate. What Elizabeth Neufeld found when she was working with the enzyme was that 2′-adenylic acid stimulated the reverse reaction; that is, the oxidation of NADH by NADP [29,45]. It appeared that the 2′-AMP had a regulatory effect, but we were not smart enough to recognize its significance at that time. In fact, later when Philip Cohn worked on the enzyme at Brandeis for his thesis, he scolded me for not emphasizing this point. He suggested that we had found the first allosteric enzyme and had not been smart enough to recognize it. This enzyme has gone through intensive study by a number of students and post-doctoral fellows. It has been shown through the years that it can be isolated as an aggregate whose molecular weight may be greater than one hundred million. The enzyme observed in the electron microscope appears as a helical rod of about 200–2000 Å with an M_r as great as 100 million, but when 2′-adenylic acid is added, it changes into a circular molecule which appears to be an octagonal structure with 36 subunits of about 54 000 each or a total M_r of one and a half million [7,8,43,53]. The dramatic effect in the electron microscope paralleled many physical, chemical, and catalytic studies which pointed to the fact that there was a great alteration in structure when the nucleotide was added.

It is of interest that the *Pseudomonas* transhydrogenase can be followed directly by the use of analogues. We now routinely use the thionicotinamide analogue of NAD. This analogue has an absorption maximum of around 400 nm and is actually yellowish, and one can observe reduction with this analogue visually without the use of the spectrophotometer. We followed the reaction usually by measuring the increase in absorption at 400 nm according to the following equation:

$$\text{NADPH} + \text{(TN)AD}^+ \rightarrow \text{(TN)ADH} + \text{NADP}^+$$

Once we had established that the bacteria had a pyridine nucleotide transhydrogenase, we began searching animal tissues for such an enzyme. We found that the enzyme was localized in the mitochondria, and we were able to solubilize and partially purify the catalyst [30]. It differed greatly from the *Pseudomonas* enzyme in that 2′-adenylic acid was not required and the reaction proceeded in both directions after extraction from the mitochondria. Later, a number of groups showed that in the mitochondrial system ATP would drive the reaction of NADH and NADP beyond its equilibrium point [14]. That is, higher levels of NADPH were generated than could be predicted from the equilibrium of the reaction. The mechanism by which this occurs is still not clear. We confirmed their findings and showed that the mitochondrial enzyme which was purified is also responsible for the overshoot when ATP is added. There is no energy effect on the purified animal enzyme [18].

Summarizing the work that I was involved in at Hopkins, I am somewhat impressed that a casual finding led to the main research focus of my career. The work with the pyridine coenzymes and the enzymes that they are involved with certainly was not my goal when I began to do research at the McCollum–Pratt Institute. Those years at Hopkins were indeed productive and greatly rewarding.

When I moved to Brandeis, I was most excited by the challenges of setting up a new department. Johns Hopkins had been very good to me in allowing me to develop my potential. When I left Hopkins, I went with the blessings not only of Bill McElroy and Dr. McCollum, but also of President Milton Eisenhower. I shall not forget President Eisenhower's letter stressing how proud he and the Hopkins community were of me in going to a fledgling institution to set up a new department built in the Hopkins tradition.

Brandeis

When I moved to Brandeis, as Chairman of the Department, I believe it was not only a satisfaction to my ego to set up an excellent department, but it was also of great value to my research. I state this because I believe that one must continuously be bombarded with new types of stimuli from different individuals. I found that at Brandeis I was able to rejoin Martin Kamen. We had brought into the Department at the beginning a group of very promising younger scientists. These included Bill Jencks, Laurie Levine, Larry Grossman, Mary Ellen Jones, Stanley Mills, and a short time later, Helen Van Vunakis, Bob Abeles, Gordon Sato, Gerry Fasman, and Serge Timasheff. All of these individuals were somewhat different in their scientific outlook from those at Hopkins. Even though the Hopkins environment was very enthusiastic and stimulating, the opportunity to be with new people brought new ideas and new approaches to my own research.

Perhaps the most important aspect of the research at Brandeis was the beginning of the investigations of what we now call isozymes. We had used the coenzyme analogues to show that the heart and muscle lactate dehydrogenases of a given species were different. We found that the heart enzyme of one species was much more closely related to the heart enzyme of another species as compared to the muscle enzyme of the same species. We also made the same observation that there was similarity with the muscle lactic dehydrogenases of different species. We had examined a number of other tissues with some of the analogues and were amazed to find that their activity was somewhat between the heart and the muscle types of lactic dehydrogenases (LDHs) as measured by the reactivity with a number of the coenzyme analogues. We were puzzled by this finding when Markert and associates, as well as Weiland and Pfleiderer, reported that multiple bands of LDH could be observed on electrophoresis. We then, in collaboration first with Levine, made antibodies to the heart and muscle types. We showed that there was no cross reaction of the antibodies to the heart enzymes with the muscle enzyme. As expected from the data with the coenzyme analogues, the heart enzymes or muscle enzymes reacted with the specific antibodies to a degree related to their classification. (We will return to this below.)

Since we had these antibodies, we began discussions with the late Edgar Zwilling who was an embryologist in the Biology Department. He had a student, Robert Cahn, who had become interested in the LDH during development of the chick. He began to look at the types of LDH that occurred in the embryonic chick breast muscle. To our surprise, he did not find the now known M_4 type but actually the heart-type LDH. This he was able to determine by immunological methods and later by electrophoretic technique. We observed the changes in M_4 LDH during development where the M-type was being expressed at an increased rate, and it became the principal type at the time of hatching. Immediately after hatching, there was a great increase in the LDH so that the breast muscle of the chicken is largely M_4 enzyme with traces of H_4 being present. In contrast, the embryonic chicken heart also contains pure H_4 enzyme. There is no change in the type of enzyme, during development although the concentration of the H_4 enzyme does rise somewhat during development [5]. Realising that the findings of Markert showed that LDH was a tetramer, our results strongly supported a view that the five forms of LDH consisted of two parent types, H and M, which occur as H_4 and M_4 with three intermediate hybrids which migrated predictably in between the H and M forms.

Enzyme evolution

Perhaps one of the most interesting aspects of our work at Brandeis was a development of interest in comparative enzymology and evolution. The fact that we had antibodies to the two homopolymers of LDH led to the idea that we could show the relationship by using the immunological techniques; at that time, we were utilizing the microcomplement fixation assay which was being exploited by Levine. The whole program was stimulated by the arrival of Allan Wilson, as a Postdoctoral Fellow. He had recently obtained a Ph.D. in biochemistry at Berkeley, but had a zoological background. He wanted to become active in molecular evolution, and saw the potential of the LDH system.

Wilson used antibodies to various LDHs and by this mechanism was able to classify various organisms [54]. In general, this classification agreed with that carried out by the taxonomist. However, there were some divergences from the classical finding that proved to be of interest. The use of enzymes in following evolution appears to be now a field of great interest and is actively being pursued by a number of investigators, both at the DNA and protein levels. Wilson has remained at the forefront of this work.

In discussing relationships among animals, I always think of the "kosher" NADases and that certain catalytic properties are characteristic of an enzyme of a given group. I believe some of these catalytic differences are important because they relate to the operation of the enzyme in a particular environment. For example, the flat fish (halibut, flounder, and sole) have only the anaerobic type of LDH (M_4) in their tissues as adults. Since they bury themselves in the sand and live in an anaerobic environment, one can see why the M enzyme is present even in the heart of these fish. It is of interest when these flat fish young are hatched, they are free swimming forms. They have one eye on each side. At the time when one of the eyes moves from one side to the other, there is also a change in the lactic dehydrogenase. The young fish have the H type present in heart and other tissues and at the time of the eye movement, there is a change to the M form.

A biochemist who knows little about classification in zoology can sometimes run into difficulties which turn out to be educational and somewhat exhilarating. When we examined some of the invertebrates, we found that the arthropods, particularly the lobsters and crabs, had a very unusual LDH [22]. In the reaction, lactate oxidation by NAD did not occur at a significant rate; however, this group of arthropods had high activity with the acetyl pyridine analogue of NAD and lactate. The contrast of reactivity with the analogue to the natural coenzyme was so great that we thought that the analogue was the actual nucleotide inherent in the tissues of lobsters and crabs. However, we were able to show that NAD was present in the muscles of these species. We have studied the enzyme from lobster and it is of interest that it is probably the only allosteric LDH present in nature. The reason for this is still not clear, but may be related to some physiological parameters in which regulation is important. At that time, we also began a study of the enzyme in the horseshoe crab (*Limmulus*). We thought this organism was a crab and when we examined it, we found to our dismay that the LDH reacted very well with the natural coenzyme in contrast to the other crabs. Margaret Ciotti was working with us at that time. She read an elementary zoology textbook and found that the *Limmulus* was indeed not a crab, but belonged to the spider family. We then examined spiders, scorpions, tarantulas, and found that they all had enzymes like the horseshoe crab. So, it was evident that the horseshoe crab was related to the spider species and not to the lobster–crab group. This was of interest to me because by using a biochemical test and not knowing what the classification of the horseshoe crab was, we had innocently found by a biochemical method the relationship of this animal to other members of the arthropod phylum. Another surprise with the horseshoe crab and its relatives was that the enzyme turned out to be a D-LDH. This was an interesting observation and it came about accidentally. George Long, one of our graduate students, had purified the *Limmulus* enzyme. When he had achieved this, I suggested that he now obtain accurate measurements K_m, V_{max} and other properties so that he should use as substrate pure L-lactate instead of the DL mixture which he had used to assay the enzyme during the purification. When he attempted to use L-lactate with the purified enzyme, no activity was observed. We were somewhat distressed by this finding, but he found it was still active with the DL mixture of lactate. When he tried the D lactate, it reacted like the racemic mixture. The *Limmulus* and its relatives have a D-LDH, which is a dimer. This was a surprise finding since most animal LDHs are L enzymes and tetramers, the D enzyme is also found in a number of worms. Our investigations showed that a given species may have either D- or L-LDHs, but only one is found in a given species [41,42]. Usually two types of the D enzyme are found in a given species which are equivalent to the H and M of the animals with the L-LDH. We believe that the D and L enzymes have evolved from a common gene. We are now beginning studies to show this relationship at the gene level.

One interesting anecdote that occurred during the course of these comparative studies on LDH was about the enzymes from the two fish, haddock and cod. These are two closely related species. After we had purified the haddock muscle lactate dehydrogenase, we received a call from the Bureau of Fisheries asking if we could determine the difference between the cod and haddock enzymes. I asked them why they wanted to know. They responded by saying that they were having a problem with frozen filets of haddock which were being sold in supermarkets and that they suspected these filets contained a large proportion of cod rather than haddock. Since haddock sold at four times the cost of cod, they suggested that there might be fraud with respect to the consumer. We received from the Fisheries some cod, haddock, and frozen "haddock" filets. I asked Allan Wilson and Barrie Kitto to look into the problem. Indeed, they found the haddock and cod muscle LDHs migrated distinctly different on various gels and also could be distinguished immunologically [55]. When several brands of the frozen filets were tested, they all turned out to contain almost totally the cod enzyme. The data unquestionably showed that the frozen filets were not the material marked on the labels. Armed with this information, the Bureau of Fisheries as well as Food and Drug Administration obtained an injunction and confiscated most of the frozen filets which were on the market. The suppliers all claimed no contest and said that they had purchased the fish from foreign fishermen. The Bureau of Fisheries was so delighted over the results that they had the findings put in the Congressional Record as an illustration of how molecular biology could be of value to the average consumer. Today the filets are monitored by determining the LDH either electrophoretically or immunologically and the chances are now when you buy frozen filets of haddock, it is really haddock. My coworkers and myself were really pleased with the outcome, since it showed the power of molecular techniques to resolve problems which could not be resolved by morphological methods.

During the years at Brandeis we attempted to understand the mechanism of lactate dehydrogenases and other dehydrogenases. It is of interest to note that in our comparative studies, we had crystallized a number of lactate dehydrogenases. I remember clearly the day William Lipscomb of Harvard brought Michael Rossman to the laboratory. Rossman was interested in solving the structure of an LDH by X-ray crystallography. Since he knew we had a number of LDHs crystallized from different sources, he asked if he could look through our "library" to see if any of the crystals would be suitable for study. After an hour of examination, he appeared excitedly in my office and asked what kind of an animal an "Oscar" was. I was perplexed by the question, but after looking at the tube with the crystals, we found that it was labelled "Oscar" since it was a preparation of Oscar Chilson's who had been carrying out studies on the dogfish M enzyme. So, "Oscar" turned out to be the dogfish M_4 enzyme and the crystals were ideal for Rossman's study. Fortunately we had a large supply of dogfish muscle and Rossman was able to carry out his outstanding investigation on the structure of the enzyme.

When we moved to La Jolla several years later, we crystallized the M_4 LDH from the West Coast dogfish muscle. The West Coast fish is a different species, and interestingly enough, we could never prepare crystals which were of the same quality as the East Coast dogfish and was not usable for X-ray work. It was indeed fortunate that we were on the East Coast when we were doing the initial LDH work. This is an example of the importance of comparative enzymology in the area of structure determination.

Later at Brandeis, we expanded our work on isozymes to a number of other enzymes, besides lactate dehydrogenase. We made studies on various forms of creatine kinase, malate dehydrogenase, the glutamic aspartic transaminase, glycerol phosphate, dehydrogenase, as well as other systems. We did extensive purification and characterization studies. In this work, we were impressed by the fact that the various isozyme forms had different catalytic properties. This was particularly manifested in the LDHs. At Brandeis, and later at University of California, San Diego (UCSD) in La Jolla, we obtained evidence indicating the molecular reason why isozymes should exist. In addition, we continued studies on NADases and transhydrogenases. This occupied a good deal of our time both at Brandeis and at UCSD. Involvement in this work led us to look at new techniques of protein modification and into newly developing physical approaches utilizing fluorescence, circular dichroism methods, and nuclear magnetic resonance.

I will not describe the many studies at great length. As I look back on them now, we were actually largely expanding on what we had found initially at Johns Hopkins. The work was rewarding and had to be done, but I think even though it was much more sophisticated, it was not as exciting as when the enzymes were first discovered.

Many of my former postdoctoral fellows and students have continued on a number of aspects of this work that was initiated at Hopkins and Brandeis and I must say that their work, as well as my own, certainly has given a good understanding on how pyridine nucleotide enzymes work [12].

One of the areas in which I became interested in my later years at Brandeis and continued at UCSD, was the application of nuclear magnetic resonance to an understanding of the structure of the pyridine nucleotides. First with the relatively low frequency nuclear magnetic resonance (NMR), we began to study the structures of the nucleotides. This was followed in a collaborative study with Bill Phillips at DuPont with the high frequency NMR and later when we arrived in La Jolla, the NMR approach was emphasized. From this work, I think there was some new insight into the structure of the co-enzymes which became helpful to many in understanding the mechanism of action of dehydrogenases with which the pyridine coenzymes interact.

I believe that the years spent at Brandeis were most exciting and fruitful, not only because we became involved in a number of new and important fields, such as the isozymes and molecular evolution, but that these areas introduced us to new approaches to enzymology which I believe was probably due to the stimulus of the other faculty members at Brandeis.

UCSD years

In 1968 I moved from Brandeis to the UCSD. This move came as a result of considerable deliberation. During my years at Brandeis, I became more and more involved with University politics. Although I do believe this is very important for faculty members to engage in, particularly senior faculty, the conditions at Brandeis began to require more and more time for this type of activity. There were very few problems in the Biochemistry Department itself and the

Department operated with great efficiency and camaraderie. What I believe was happening was that more and more of my time was going into University activities which seemed to be non-productive. When I look back now at the time spent on these activities and what has occurred over the years, I now feel considerably more comfortable that that time was not spent in vain.

Nevertheless, the attraction of UCSD was a great one. It was a new institution just beginning to develop, and I thought that this would offer new challenges. I also was attracted to the institution because of my association with old friends such as Martin Kamen. I have been at UCSD for 16 years. The experiences I have had at UCSD were quite different from that at Brandeis and Hopkins.

When I arrived at UCSD, we began the sequencing of the LDHs since Rossman was now well on his way to establish the crystal structure of the dogfish M_4 enzyme. In this endeavor, William Allison and Susan Taylor, both of whom are still here at UCSD, were involved. We thought that by understanding the structure, we could explain more of the mechanisms. Of interest to me was the question why the M and H LDHs had such different properties. I do not believe yet that we had enough knowledge to tell why the physical and catalytic properties of the two forms are different. But the mechanism of the reaction of the two forms appears to be quite similar; they differ quantitatively.

We had initiated studies to determine when the two forms of dehydrogenase became established [13]. It certainly appears reasonable that they originated from a common gene. In this connection, it is of interest that a third LDH called LDH-C which was first isolated from the sperm was reported. It is evident now that this form from the spermatozoa is not related as closely to the M and H forms as these two main types are. Comparative sequence data has now been reported indicating that the M and H forms arose from a common gene, but the question of the origin of the C-type is still unclear [36]. One should emphasize that the C-type has very different specificity characteristics. It will react with α-ketobutyrate as well as or better than pyruvate.

As discussed above, a good deal of work had been carried out on the structure of the pyridine coenzymes themselves. We continued this work and we were particularly pleased to find that the ternary complex consisting of LDH, NAD, and pyruvate, could be demonstrated utilizing NMR techniques. This complex has a stronger tendency to be formed with the H-enzyme than the M-form. We were able, by the NMR techniques to show how the pyruvate adduct of NAD formed chemically was related to the adduct formed with the enzyme.

These and other results gave us greater insight into the structure and properties of the pyridine nucleotides. We were able to show more about the formation of alpha NAD and its characteristics. What was called the NADH-X structure was also resolved by NMR work and the mechanism by which it was originated was elucidated. In these studies, two graduate students, Norman Oppenheimer and Lyle Arnold, played an important part along with several postdoctoral fellows such as Ramaswamy Sarma, C-Y Lee, and Fred Evans. I believe it is appropriate at this time to acknowledge the contributions of John Wright who kept the 220 MHz NMR in operation in the early days at UCSD. Much of this work has been summarized in the volume edited by Everse, Anderson and You.

Some 25 years ago, I spent a number of months at the Weizmann Institute in Israel. During this period, I had the opportunity to discuss with Professor Ephraim Katzir a number of potential applications of immobilizations of enzymes. One such problem we thought would be worthwhile pursuing was to immobilize enzymes which were sequential in metabolic pathways. We also thought we might duplicate an organelle function such as the mitochondria by immobilizing the enzymes from this cellular compound. It is of interest for whatever reason, we did not carry out these studies. It was very rewarding when Klaus Mosbach showed a number of years later that intermediates were in a sense not free in solution as we had always thought, but were in close position with the enzymes that produced them and those that utilized them [44]. I believe that from a fundamental point of view this has been one of the most important contributions to immobilization of enzymes. Certainly their practical use to produce products is of great significance, but it has now been shown quite convincingly that immobilization techniques make it possible to more thoroughly understand mechanisms of enzyme reaction as well as its flow rate of metabolites in a cell.

At the time we were first carrying out the studies on immobilized enzymes, a new graduate student, Craig Venter, became interested in metabolic functions of the embryonic heart cells. In the course of his work, he became interested in the effects of catecholamines. I thought it might be of interest to see if immobilized catecholamines are active since these compounds act at the membrane level. I suggested this problem to Jack Dixon who was doing the enzyme immobilization studies, and he and Venter set out to determine whether this was a realistic possibility. Indeed, the results turned out to be very interesting. Venter was able to obtain activity with the hormone on the beats, and this he could show both by an increase in the rate of beating of various heart preparations as well as an elevated synthesis of cyclic AMP [51,52]. This became an exciting potentiality; that is, studying the effects of an agent acting on the membrane without its being destroyed by the enzymes on the exterior or interior of the cell. Later studies show that it was equivocal whether the catecholamine was still on the bead or whether they dissociated giving a high localized concentration which would induce the activity, since the amount of immobilized catecholamine was less than that required to produce a physiological response. This is a sort of semantic argument, but for whatever reason, it certainly indicates that the immobilization technique can be helpful in studying the action of ligands and specific receptors in the membrane.

During those years we were involved with immobilized systems, I thought we did make some interesting findings which were of value in the application of immobilized enzymes or drugs for use for different purposes. We became interested in affinity chromatography, particularly using different types of ligands for separation of different dehydrogenases. This certainly appeared to be a new development in the purification of enzymes. We were able, by the use of the appropriate nucleotide column, to adsorb enzymes and then elute them with specific binary complexes which we knew were specific for different dehydrogenases. Other methods of elution were also worked out. One can get purification of an enzyme which exists in a very small quantity in serum by this type of procedure [35]. (The method appears to be most useful in fishing out enzymes which are present in a small concentration.) It is not clear as yet whether techniques of affinity chromatography will work for the purification of large amounts of enzymes. It is possible that classical techniques such as column fractionation and precipitation will be of more value in scaling of protein purification.

However, the affinity chromatography now plays an important role in the development of the new biotechnology [6]. If one wishes to clone enzymes which are present in small amounts, the most suitable way to obtain the enzyme is by affinity methods. Consequently, the new mass sequencing techniques allow for the obtaining of partial sequences. The partial sequences are of value in synthesizing DNA probes for possible hybridization studies with mRNA. Once a DNA clone can be formed from the mRNA, then theoretically the clone can be inserted into bacteria so large amounts of the enzyme can be produced. We are now pursuing this type of technology to obtain enzymes which previously we had not been able to obtain because of the presence of only very small amounts of catalysts in the microorganism or tissue.

It is evident that enzymology in recent years has altered its course. It is closely associated with molecular biology. Not only can one get adequate enzymes for practical purposes, but one can by the recombinant technology obtain enzymes which would not have been possible previously in enough quantity to do catalytic and physical studies. The recent interest in site-specific mutagenesis allows for the making of specific enzyme analogues which I am sure will give precise information as to mechanisms and active sites. I believe that the advances in molecular biology have made enzymology a more interesting field. We are pursuing such areas in our present studies.

In connection with the above, it is of interest that obtaining the genes, for example such as the different forms of LDH will give us a good understanding of the evolution of these enzymes and certainly more precise information of phylogeny than the protein sequences themselves have yielded.

Finally, I would like to discuss briefly an area which we became involved in quite accidentally. As our studies on receptors and immobilized hormones became more a part of our work, we engaged more and more in tissue culture studies. Since we were not experts in this field, we renewed our collaboration with Dr. Gordon Sato who had also moved to La Jolla from Brandeis.

Sato had obtained several athymic mice and was planning studies with these animals. These mice, since they lack a thymus, do not reject human tumors and as a result, human tumor xenografts can grow in this animal. I was impressed by the potentiality of using this animal for studies in chemotherapy as well as metabolism of human tumors. I had always thought that it was important in order to develop chemotherapeutic agents against human tumors, that the human system itself should be investigated. The athymic mouse made it possible to look at agents before trials on man are initiated.

I have always been interested in the problems of chemotherapy. This came from our work with the anti-nicotinamide compounds [23]. At that time we felt that if one could produce an analogue of NAD in vivo it might interfere with cellular metabolism. This does happen. However, in the early 50's when we first brought some of these drugs to the clinic, the results were not promising. These compounds were effective against experimental mouse tumors, but they were quite toxic. We had learned how to decrease the toxicity and side effects, but the compounds were not received by the oncologists with any great enthusiasm and so we dropped the program. This work had been done largely with Dr. Abraham Goldin and his associates at the NIH.

Sato and I decided to try to develop an athymic mouse colony, and with the help of the University we were given a small building to house his laboratories as well as my own. The building was oriented also for tissue culture work as well as for the athymic mice. We started our studies with bacterial and plant toxins, seeing if we could make conjugates with the A-chain, which is a toxic entity and an antibody. These are now called immunotoxins. Whether they will be of use remains to be determined.

After a number of years with the athymic mouse colony, it is still too early to determine whether this model might be useful in developing agents for the treatment of human cancer. It certainly has the potential to do so, but obtaining the essential mentioned data is a slow process.

However, I would like to point out that if one is a careful observer, there is always new information to be found. For a number of reasons, we had thought that our colony had genetic drift since it was found that some of the animals had some capacity to form antibodies. We could not understand why a few animals had this potential whereas others did not. A careful examination by Gillian Beattie in our laboratory showed that those animals producing antibodies were infected with pinworms. Hence the explanation that these variations were due to the degree of parasites present. Following this observation, it was also noted that those animals with pinworm infection developed lymphomas; these were of the mouse type [3,4]. The connection between lymphomagenesis and the pinworm was not clear, but later it was observed that animals with human xenografts also had a very high incidence of lymphoma and the induction occurred over a very short period of time. These observations led us to attempt to find what was inducing the lymphoma. We found that the serum of the animals which were developing lymphoma had a high gp70 concentration. gp70 is a capsule protein of the leukemic virus. More careful examination showed that this protein was being synthesized in human xenografts in the athymic mouse since this apparently occurred as a result of infection of the human tumor by a mouse virus. The gene responsible for the synthesis of this protein was apparently incorporated into the DNA of the human tumor that was growing in the athymic mouse. About 10% of the membrane protein of the human tumor membrane was found to be gp70, a protein of viral origin [2].

We have not yet been able to demonstrate clearly what the role of the gp70 is in the induction of lymphoma. It is a system certainly on which further work should be carried out. During the course of these studies we examined normal mice for this viral glycoprotein, and we found very low concentrations in all tissues of the mouse except for the pancreas where it is present in extremely high levels. The protein was traced to islet cells and by electron microscopy we found it to be present in the granules of the islets. At this time we do not know the significance of this finding.

I have discussed the gp70 in the athymic mice here because I feel it is an example when one is searching for one type of phenomenon, one might become involved in a completely different field.

If one makes a conscious decision to enter a new area, many changes become manifested such as development of techniques, and in our case, immunology, virology, electron microscopy, and cellular biology. This also requires different types of people, so during the course of this change, the nature of my group has been altered. I believe change is for the good when one enters a new field, but there are always the jaundiced eyes of study sections questioning whether one has the appropriate background to do so. This has become a great fallacy since experience has shown that people can alter fields and still make notable contributions. What I am afraid is happening is that one must prove himself in a field before money is allotted to do work in a new venture.

In conclusion, I can say that I have had a very interesting, stimulating and exciting career. I feel that the years at the Massachusetts General Hospital, Hopkins, and Brandeis were particularly rewarding. I do not have such a positive feeling toward UCSD. It is not that the University was not good to me, but the fact that the stimulation was difficult to obtain. I do not mean that there were no adequate seminars or high-quality, helpful scientists present on the campus. I am suggesting that there was a lack of interaction between investigators. This is characteristic of an institution which is large and has people of the same scientific interest scattered through many buildings. There is not the day-by-day interaction which proved to be so successful at Brandeis and Hopkins. I think that one of the problems which institutions must face in the future is largeness. I believe that after a certain size is reached, the congeniality begins to decrease. There are no longer special relationships between members of the faculty. In a way, what I am saying is that we should try to maintain the smaller institutions. They are of great importance. I recognize that doing significant research at the present time in the biological sciences requires large groups, but the smaller universities can play an important role because of the close-knitness of the faculty who are interested in each other's work.

I would also like to make a few comments about the future of biochemistry. When I entered the field about 45 years ago, there was actually very little known about the chemistry of life. Through the years, many of the major pathways of metabolism have been elucidated and much is known about the mechanism and structure of enzymes. Biochemistry has been established as a distinct discipline; it is taught to undergraduates, in contrast to the time when I was student - it was a graduate undertaking. Biochemistry uses the principles of chemistry, but the approaches and technology are quite different. Biochemistry has become the language of biology. The growth of cellular and molecular biology has come from the principles developed in biochemistry. The biochemist is changing; he is leaning more towards biological problems. This is an admirable turn since the new fields are challenging and exciting. Many believe that biochemistry of the past decades is "old hat", but it should be remembered that the principles of biochemistry are what have led to the advances in biology. Students should not lose sight of the eloquence of the experiments of Warburg because it is the same eloquence which is inherent in the isolation, characterization and manipulation of genes.

I recognize that this narrative has not included all of my research activities. I have noted only some of the work. I have not mentioned much significant work performed by numerous postdoctoral fellows and graduate students. Their studies were not selected on the basis of quality or importance since this paper was written randomly. However, I do wish to take this opportunity to thank all my colleagues and friends who made possible any success that I might have had.

REFERENCES

1 B.M. Anderson, in J. Everse, B. Anderson, K.-S. You (Eds.), The Pyridine Nucleotide Coenzymes, Academic Press, New York, 1982, pp. 91-126.
2 G.M. Beattie, R.A. Lannom, S.M. Baird, E.V. Heksell III, F.C. Jensen, J.F. Leis and N.O. Kaplan, Cancer Res., 43 (1983) 4349-4354.
3 G. Beattie, S.M. Baird, J.S. Lipsick, R.A. Lannom and N.O. Kaplan, Cancer Res., 41 (1981) 2322-2327.
4 G. Beattie, S. Baird, R. Lannom, S. Slimmer, F.C. Jensen and N.O. Kaplan, Proc. Natl. Acad. Sci. USA, 77 (1980) 4971-4974.
5 R.D. Cahn, N.O. Kaplan, L. Levine and E. Zwilling, Science, 136 (1962) 962-972.
6 I.M. Chaiken, M. Wilchek and Indu Parikh, Affinity Chromatography and Biological Recognition, Academic Press, New York, 1983.
7 P.T. Cohen and N.O. Kaplan, J. Biol. Chem., 245 (1970) 4666-4672.
8 P.T. Cohen and N.O. Kaplan, J. Biol. Chem., 245 (1970) 2825-2836.
9 S.P. Colowick, N.O. Kaplan and M.M. Ciotti, J. Biol. Chem., 191 (1951) 447-459.
10 S.P. Colowick, N.O. Kaplan, E.F. Neufeld and M.M. Ciotti, J. Biol. Chem., 195 (1952) 95-119.
11 M. Doudoroff, N.O. Kaplan and W.Z. Hassid, J. Biol. Chem., 148 (1943) 67-75.
12 J. Everse, B. Anderson and K. You, The Pyridine Nucleotide Coenzymes, Academic Press, New York, 1982, pp. 1-390.
13 J. Everse and N.O. Kaplan, Adv. Enzymol., 37 (1973) 61-133.
14 R.R. Fisher and S.R. Earle, in J. Everse, B. Anderson, K.-S. You (Eds.), The Pyridine Nucleotide Coenzymes, Academic Press, New York, 1982, pp. 280-321.
15 M. Franks, R.F. Berris, N.O. Kaplan and G.B. Myers, Arch. Int. Med., 81 (1948) 42-55.
16 L. Grossman and N.O. Kaplan, J. Am. Chem. Soc., 78 (1956) 4175-4178.
17 N.O. Kaplan, Studies with P of the Changes in the Acid Soluble Phosphates in the Liver Coincident to Alterations in Carbohydrate Metabolism. Summary of the dissertation submitted in partial satisfaction of the requirements for the degree of Ph.D. University of California, 1943.
18 N.O. Kaplan, The Harvey Lectures. Series 66, Academic Press, New York, 1972, pp. 105-133.
19 N.O. Kaplan and M.M. Ciotti, J. Am. Chem. Soc., 76 (1954) 1713-1715.
20 N.O. Kaplan and J. Everse, in G. Weber (Ed.), Advances in Enzyme Regulations, Vol. 10, Pergamon, Oxford, 1972, pp. 323-336.
21 N.O. Kaplan, M.M. Ciotti, J.van Eys and R. Burton, J. Biol. Chem., 234 (1959) 134-138.
22 N.O. Kaplan, M.M. Ciotti, M. Hamolsky and R.E. Bieber, Science, 131 (1960) 392-397.

23 N.O. Kaplan, A. Goldin, S.R. Humphreys, M.M. Ciotti and J.M. Venditti, Science, 120 (1954) 437-441.
24 N.O. Kaplan, M.M. Ciotti and F.E. Stolzenbach, J. Biol. Chem., 221 (1956) 833-839.
25 N.O. Kaplan, M.M. Ciotti, F.E. Stolzenbach and N.R. Bachur, J. Am. Chem. Soc., 77 (1955) 815-818.
26 N.O. Kaplan, S.P. Colowick and C.C. Barnes, J. Biol. Chem., 191 (1951) 461-472.
27 N.O. Kaplan, S.P. Colowick and A. Nason, J. Biol. Chem., 191 (1951) 473-483.
28 N.O. Kaplan, S.P. Colowick and E.F. Neufeld, J. Biol. Chem., 195 (1952) 107-119.
29 N.O. Kaplan, S.P. Colowick and E.F. Neufeld, J. Biol. Chem., 205 (1953) 1-15.
30 N.O. Kaplan, S.P. Colowick, E.F. Neufeld and M.M. Ciotti, J. Biol. Chem., 205 (1953) 17-29.
31 N.O. Kaplan, M. Franks and C.E. Friedgood, Science, 102 (1945) 447-448.
32 N.O. Kaplan and D.M. Greenberg, J. Biol. Chem., 156 (1944) 553-558.
33 N.O. Kaplan and F. Lipmann, J. Biol. Chem., 174 (1948) 37-44.
34 A.F. Knowles, J.F. Leis and N.O. Kaplan, Cancer Res., 41 (1981) 4031-4037.
35 C.Y. Lee and N.O. Kaplan, J. Macromol. Sci. Chem., A10 (1/2) (1976) 15-52.
36 S.S. Li, W.M. Fitch, Y.E. Pan and F.S. Sharief, J. Biol. Chem., 258 (1983) 7017-7028.
37 F. Lipmann, Adv. Enzymol. I (1941) 99-162.
38 F. Lipmann and N.O. Kaplan, J. Biol. Chem., 162 (1946) 743-744.
39 F. Lipmann, N.O. Kaplan, G.D. Novelli, L.C. Tuttle and B.M. Guirard, Coenzyme for Acetylation, a Panthothenic Acid Derivative, 167 (1947) 869-870.
40 F. Lipmann, N.O. Kaplan, G.D. Novelli, L.C. Tuttle and B.M. Guirard, J. Biol. Chem., 186 (1950) 235-243.
41 G.L. Long and N.O. Kaplan, Arch. Biochem. Biophys., 154 (1973) 711-725.
42 G.L. Long and N.O. Kaplan, Arch. Biochem. Biophys., 154 (1973) 696-710.
43 D.D. Louie, N.O. Kaplan and J. McLean, J. Mol. Biol., 70 (1972) 651-654.
44 K. Mosbach, Sci. Amer., 224 (1971) 26-33.
45 E.F. Neufeld, N.O. Kaplan and S.P. Colowick, Biochim. Biophys. Acta, 17 (1955) 525-535.
46 G.D. Novelli, N.O. Kaplan and F. Lipmann, J. Biol. Chem., 177 (1949) 97-107.
47 G.D. Novelli, N.O. Kaplan and F. Lipmann, Fed. Proc., 9 (1950) 209.
48 R.E. Olson and N.O. Kaplan, J. Biol. Chem., 175 (1948) 515-529.
49 N.J. Oppenheimer and N.O. Kaplan, Biochemistry, 23 (1974) 4685-4693.
50 A. San Pietro, N.O. Kaplan and S.P. Colowick, J. Biol. Chem., 212 (1955) 941-950.
51 J.C. Venter, J.E. Dixon, P.R. Maroko and N.O. Kaplan, Proc. Natl. Acad. Sci. USA, 69 (1972) 1141-1145.

52 J.C. Venter, J. Ross Jr., J.E. Dixon, S.E. Mayer and N.O. Kaplan, Proc. Natl. Acad. Sci. USA, 70 (1973) 1214–1217.
53 B. Wermuth and N.O. Kaplan, Arch. Biochem. Biophys., 176 (1976) 136–143.
54 A.C. Wilson, N.O. Kaplan, L. Levine, A. Pesce, M. Reichlin and W. Allison, Fed. Proc., 23 (1964) 1258–1269.
55 A.C. Wilson, G.B. Kitto and N.O. Kaplan, Science, 157 (1967) 82–85.
56 L.J. Zatman, N.O. Kaplan and S.P. Colowick, J. Biol. Chem., 200 (1953) 197–212.
57 L.J. Zatman, N.O. Kaplan, S.P. Colowick and M.M. Ciotti, J. Biol. Chem., 209 (1954) 453–466.
58 L.J. Zatman, N.O. Kaplan, S.P. Colowick and M.M. Ciotti, J. Biol. Chem., 209 (1954) 467–479.

G. Semenza (Ed.) Selected Topics in the History of Biochemistry: Personal Recollections (Comprehensive Biochemistry Vol. 36) ©1985 Elsevier Science Publishers

Chapter 6

A half Century of Biochemistry

HENRY A. LARDY

Institute for Enzyme Research, University of Wisconsin, Madison, WI 53705 (U.S.A.)

My scientific career began at South Dakota State University, Brookings, in 1935. As a freshman I needed employment to pay tuition and living costs and found it in the Dairy Department caring for cows and rats on vitamin D experiments. Dr. G.C. Wallis produced the first vitamin D deficiencies in mature cows and I reported to him that when the deficiency became severe the cows shed their teeth. My first scientific paper, with Dr. Wallis, was published in the Proceedings of the South Dakota Academy of Science in 1938. In my second college year I began to think of majoring in chemistry and events carried me in that direction. Dr. Eugene Burr, the professor who taught quantitative analysis approached me one day in the laboratory and suggested I go to the Experiment Station Chemistry Laboratory to speak to Alvin Moxon, the head chemist. The laboratory was engaged in research on selenium (Se) toxicity and employed 3rd and 4th year students to conduct much of the research. It was the goal of most chemistry majors to be employed by the Experiment Station Laboratory for the facilities were the best on the campus, students were coauthors on published papers and it was a good preparation for graduate study at better known universities. I soon learned that the "quant" instructor was the recruiter for the Laboratory and I was employed part time during my junior and senior years.

Plate 11. Henry A. Lardy (1967).

The Experiment Station Laboratory had become known among scientists because of its director, Dr. Kurt Walter Franke. While conducting his experiments on Se toxicity he had invented the metal battery cage and the efficient conical feed cups for experimental animals. He had established that Se in grains grown on seleniferous soil west of the Missouri River was found in the protein fraction. Franke died during my first year at Brookings and his younger colleague Alvin Moxon became head chemist and continued the research programs at the Laboratory.

An advantage of working there was that chemical and biochemical journals were first delivered to the Laboratory before they were deposited and available in the college library. I had read the papers of Stekol dealing with mercapturic acid synthesis in animals and wondered whether the selenium in liver proteins might conjugate with bromobenzene to form mercapturic acids. To investigate this possibility I fed two dogs a ration containing 10 ppm of Se in the form of seleniferous corn. After some weeks on this diet they showed the typical symptoms of Se poisoning in dogs, the most notable of which is a greatly distended abdomen from which one could tap a liter of ascites fluid every few days. I gave each dog a ml of bromobenzene in a capsule and the following morning I found two dogs, thin as greyhounds, in a very wet kennel. Bromobenzene caused the dogs to excrete greatly increased amounts of Se, presumably in the form of selenium cysteine conjugated with bromobenzene and acetylated on the amino group. I attempted to isolate the Se-containing conjugate by following procedures for isolating mercapturic acids. Initial fractions contained Se but as purification proceeded the Se was lost. Others in the laboratory then fed bromobenzene to steers grazing on seleniferous soils and Se excretion in the urine was enhanced while blood Se decreased to zero over a period of 2 weeks [1].

In the summer of 1938 Moxon drove to Madison, WI with a group of his student chemists to acquaint them with possible opportunities for graduate study. On the way we stopped at the Mayo laboratories in Rochester, MN. The facilities there were impressive and what I remember most vividly is the large steel vessels full of adrenal glands and immersed in huge tanks of steaming water. Kendall was isolating steroid hormones!

We arrived in Madison about 10 p.m. and Moxon took us directly to the animal rooms of the Biochemistry building. At that late hour the place was humming. Bob Madden, who with Elvehjem, Strong and Wooley, had discovered nicotinic acid as the cure for pellagra, was feeding his dogs. Harry Waisman, Ed Hove, Olaf Michelson and others were doing the variety of chores that were part of the research enterprise when students did everything from planning the experiments, to cleaning the cages, to writing the papers.

Of the six students on that trip, all became graduate students at Wisconsin. Four completed the Ph.D. degree and two took jobs after completing the M.S. degree. Moxon himself returned to Madison in a few years to complete his doctorate.

Graduate school at Wisconsin

In 1939 a biological problem with practical implications for agriculture interested scientists in many parts of the world. Artificial insemination (AI) of livestock was being studied with the hope that semen might be preserved so as to retain its fertility for long periods of time. This would permit the insemination of several females from a single ejaculate and thus a valuable male might sire many more offspring than he could with natural mating.

Russian investigators were especially active in this area of endeavor. Their biochemists and animal husbandmen had developed equipment and techniques for collecting semen and had initiated studies of sperm metabolism.

At Madison, WI, funds were made available to Biochemistry Professor Paul H. Phillips to study sperm preservation and he offered me a Research Assistantship to work on that problem. I arrived in Madison late in the afternoon of a June day and found Professor Phillips still at work. He suggested I be prepared to start work at 8 o'clock the next morning. We worked together and finished our first joint experiments at midnight. Phillips suggested that spermatozoa would require a "pabulum" to supply energy for motility and that we should try materials available to farmers as a source of the goodies that would keep sperm fertile. Empirically we tested a wide variety of materials and within a few weeks found that fresh egg yolk diluted with an equal volume of phosphate buffer (pH 6.8) provided a medium in which bull sperm, stored at 5 to 10° C, maintained motility for 7 to 10 days, and in some cases for as long as 2 weeks. Egg yolk is especially effective in preventing cold-shock damage to the sperm plasma membrane. Ram, stallion, and turkey spermatozoa also retained motility for several days when diluted with buffered egg yolk [2]. Almost immediately cows in the University dairy herd were inseminated with stored semen and the retention of fertility was demonstrated to parallel retention of motility. Cooperative Dairy Bull Studs for the application of AI were started at several locations in Wisconsin and in other states where dairying was an important industry. The extensive use of purebred bulls with the genetic potential for high milk productivity has more than doubled the yearly yield of milk per cow in most countries of the world. In Denmark, Israel and Russia nearly 100% of the cows are bred artificially and in some countries this applies to sheep as well. AI is a multibillion dollar industry.

Some years after our development of the yolk-dilutor, Polge and his colleagues in England found that sperm could be frozen if suspended in a yolk buffer mixture containing glycerol. Motility and fertility are regained after storage for as long as 25 years at the temperature of liquid nitrogen.

With the practical problem solved I was free to do other research as long as it involved spermatozoa. I had been collecting semen from several of the University herd sires and after observing my prowess (acquired on the show circuit) in handling them, Prof. Ed Heizer, Chairman of the Dairy Science Department, gave me permission to use Pabst Comet, a vicious beast that was previously handled only by the University Herdsman. The elevation of my rank in the University barn was the equivalent of a freshman being assigned the quarterback position on the varsity team.

With a supply of spermatozoa assured, I pursued my interest in intermediary metabolism and enzymology with cells whose chemical processes could be correlated with a function — motility [3,4]. Conrad Elvehjem had brought Barcroft manometers and flasks to the Department from England after his post-doctorate year at

Cambridge so I studied the respiration and glycolysis of both epididymal and ejaculated spermatozoa of the bull [3], and of several other species. Evidence was found that phospholipids were the endogenous energy source in spermatozoa [4]. Striking differences in these metabolic parameters were found between epididymal and ejaculated cells [5] and differences are still being found in other functions [6,7]. I was interested in the mode of action of the thyroid hormone and compared some of its effects with those of 2,4-dinitrophenol (DNP), a known stimulant of cell respiration. This work permitted me to state that the effect of DNP is "an interference of the energy coupling mechanism with the result that oxidation and glycolysis run rampant while the energy is lost as heat rather than as work" [8]. I elaborated on that theme in a review written in 1944 [9].

While investigating sperm storage and its application in AI we were made aware of the problem of infertility in both male and female cattle. We ran hundreds of analyses for ascorbic acid in the blood of these animals, for Phillips had a hunch that a deficiency of vitamin C might be responsible. Some animals did have lower than normal amounts of ascorbate in their blood or semen and injections of ascorbic acid were found to improve fertility in both bulls [10] and cows [11]. These results were greeted with skepticism in some dairy circles but workers at Michigan State repeatedly confirmed the effectiveness of ascorbate in a stallion of low fertility and the relation of ascorbate status to estrogen production and ovulation is now well documented [12–14].

One source of disappointment occurred during my first year of graduate study. An examination question in a nutrition course asked for an explanation of the thiamine-sparing effect of dietary fat. I proposed that the β-oxidation of fatty acids elucidated by Knoop did not require thiamine whereas it was well known that pyruvate oxidation did. I went on to elaborate a theory that the 2-carbon unit derived from fat must be similar to that from pyruvate and that fat was oxidized via the Krebs tricarboxylic acid cycle. So far as I know that had never been proposed before. The novelty of the idea apparently missed the professor for he did not comment and did not return the exam paper. A few years later

Breusch [15] proposed that fat is oxidized via the citric acid cycle and we finally published data [16,17] showing that acetate as well as the endogenous lipid in spermatozoa is oxidized in a system that is blocked by malonate.

Graduate study in biochemistry at Wisconsin was an exciting experience. Conrad Elvehjem, the Chairman of the Department was one of the world's outstanding nutritionists; Harry Steenbock and Karl Paul Link were equally famous. Though working in the College of Agriculture, each had discovered effective treatments for serious human ailments — pellagra, rickets, and coronary heart disease, respectively. Marvin Johnson, first an instructor and later a member of the professorial staff was an exceptional teacher and a source of ideas for examining any problem experimentally. Seminars were lively, lectures by visitors were frequent and interaction between departments thrived. As a graduate student I shared authorship on papers with professors from five departments other than Biochemistry.

A symposium on Respiratory Enzymes held in our building in 1941 was an inspiration to many of us. Carl Cori, Herman Kalckar, Fritz Lipmann, Otto Meyerhof, Severo Ochoa, Harland Wood and several from our own campus, all in the prime of their careers, were among the speakers.

Coworkers are also a stimulus to creativity and productivity. There were some 50 to 75 graduate students in the Department and my closest associate was Paul Boyer [18]. We both began graduate study in 1939 and finished our doctorates in the spring of 1943.

In January of that year Annrita Dresselhuys and I were married in New York City, and then began a personal life, now in its fourth decade, of such joy and richness with her and our children, Nicholas, Diana, Jeffrey, and Michael, that it eclipses professional rewards.

Post-doctoral research

I remained in Phillips' laboratory for a year on a post-doctorate appointment and in the autumn of 1944 went to the University of

Toronto as a National Research Council Fellow in Chemistry. The laboratory of Professor Herman O.L. Fischer in the Banting Institute was again an inspiration. John Sowden, later Professor of Chemistry at Washington University and John Grosheintz were using nitromethane to make a variety of new sugar and inositol derivatives. Professor Eric Baer was beginning his sojourn into lipid and phospholipid syntheses and his student Leon Rubin (later Director of Research at Canada Packers) was synthesizing selachyl alcohol. Morris Kates was a diligent young undergraduate who showed promise of becoming the outstanding lipid chemist he is today. During the year I was able to demonstrate why Meyerhof thought the pyruvate kinase reaction to be irreversible [19], I developed a chemical synthesis of glucose 6-phosphate [20] and, with Leon Rubin, proved the configuration of the butanediol enantiomers by syntheses from D- and L-mannitol [21].

Return to Wisconsin

While at Toronto I received an offer from Elvehjem to return to the Biochemistry Department as an Assistant Professor. I protested mildly that the salary offered would be less, after taxes, than I was earning as an NRC Fellow but was careful to imply that I wanted to accept the offer. Elvehjem compromised by putting me on the payroll a month before my arrival. University protocol would not allow that today.

My 5 years in the Biochemistry Department involved teaching in three different courses, directing the research of several graduate students and collaborating with other departments on a practical problem in agriculture — the transmission of DDT from canning crops to the feed, flesh and butterfat of dairy cows. My students developed the synthesis of some rare sugars by the *Acetobacter suboxydans* oxidation of appropriate precursors [22,23] and tagatose 6-phosphate and 1,6-bisphosphate were prepared [23]. The latter is a substrate for muscle aldolase [23a] and both are normal intermediates in the metabolism of D-galactose by *Staphylococcus aureus* [23b].

While in Fischer's laboratory I had had many interesting conversations with Bruno Mendel who described his use of levorotatory glyceraldehyde as an inhibitor of tumor glycolysis. This compound had been used extensively by Joseph Needham and his colleagues at Cambridge to document their hypothetical non-phosphorylating glycolysis in embryonic tissue. This pathway was antithetical to the concepts of Meyerhof [24] whose work and ideas I revered, so I decided to study the mechanism of the inhibition. Virgil Wiebelhaus and I found [25] that the L-sorbose 1-phosphate formed by aldolase-catalyzed condensation of dihydroxyacetone phosphate with L-glyceraldehyde inhibited hexokinase and we pointed out the similarity in structure between L-sorbose 1-phosphate and D-glucose 6-phosphate. Coincident with the studies of glyceraldehyde effects on glycolysis, Rex Mann had undertaken the synthesis of phosphate esters of sorbose for completely unrelated reasons, but his synthetic L-sorbose 1-phosphate [26] proved the identity of the inhibitory ester prepared enzymically. Not long after that the inhibition of hexokinase by glucose 6-phosphate was reported [27].

Our biochemical interests were catholic and one's journal reading covered broader areas in the 1940's than is possible now. The papers of Stokes et al. [28] dealing with biotin requirements of *Lactobacillus arabinosus* revealed that the vitamin could be replaced by the combination of aspartic acid and oleic acid. I had no idea what biotin might be doing to promote the formation of oleate, but I was certain that in an organism fermenting glucose, biotin must participate in the conversion of pyruvate to aspartate by condensing pyruvate and CO_2 to form oxalacetate — the Wood–Werkman reaction. Richard Potter, a student of Conrad Elvehjem, was doing microbial assays so we collaborated in a study [29] that revealed that biotin functioned in the CO_2 fixation process as predicted. Using $[^{14}C]HCO_3^-$ we documented the role of biotin in incorporating HCO_3^- into aspartate and demonstrated that L-aspartate specifically prevented the synthesis of labeled aspartate [30]. We thus described a feed-back inhibition long before the celebrated announcement of this phenomenon [31] but also long after it had been reported by Dische [32].

My student Patricia MacLeod studied CO_2 incorporation into tissue amino acids in normal and biotin-deficient rats and demonstrated that the decreased incorporation into arginine in the latter group could be accounted for by defective synthesis of citrulline from ornithine [33]. Gerhard Plaut, Joseph Betheil and George Drysdale established the role of folic acid in the incorporation of formate carbon into serine, glycine, glutamate, aspartate, arginine, heme, CO_2 and purines [34,35]. Gerhard's studies demonstrated the stoichiometric incorporation of CO_2 into the carboxyl group of acetoacetate from isovalerate in rat liver homogenate [36].

The Institute for Enzyme Research

The Institute for Enzyme Research at Wisconsin had been planned for four research teams. David Green came to Wisconsin in 1948 as the first team leader housed in temporary quarters until the new building was completed in 1950. I was then invited to be the second team leader. After 5 years in the Biochemistry Department, my graduate students, former student Gerhard Plaut, and I began work in the Institute in July of 1950. I retained a part-time appointment in Biochemistry involving the supervision of graduate students but a lighter classroom teaching load.

One of the first studies undertaken at the Institute dealt with the regulation of mitochondrial oxygen utilization. The dramatic influence of inorganic phosphate and phosphate acceptor concentration on the rate of substrate oxidation was reported in April of 1951 [37], was discussed at the first Johns Hopkins symposium on Phosphorus Metabolism in June of that year, and was published in early 1952 [38].

The mechanisms involved in respiratory control have been the subject of myriad papers. Our conclusion that "In general, rates of respiration vary inversely with the 'P potential' against which the oxidative systems must work" ([38], p.222) still seems valid. My prediction that reductive synthesis of fatty acids, for example, would be favored when respiration was limited by lack of phosphate acceptors ([39], p. 1006) was borne out when it was found that pyridine nucleotides are reduced in state 4 and become more oxidized on addition of a phosphate acceptor [40].

At the Michigan symposium [37] I also presented the theory that an organism could generate more ATP per unit time at high phosphate potential by partially uncoupling a bottle-neck step. This concept has appeared in far more sophisticated format in the writings of Stucki [41].

My ideas on respiratory control were summarized in a lecture at the Third International Congress of Biochemistry in Brussels [42]. For me, a high point of the meeting was a prolonged discussion with Fitzi Lynen — he in German and I in English — during which I explained the importance of control at phosphofructokinase by ATP [43] for the regulation of glucose phosphorylation via glucose 6-phosphate inhibition of hexokinase.

Simultaneously with the work on respiratory control, John Copenhaver, a postdoctorate fellow, did quantitative studies of oxidative phosphorylation in which oxygen consumption, phosphate esterification, substrate disappearance and product formation were measured [44]. The jury is still out as to whether whole numbers or fractions apply to the three phosphorylation steps associated with the electron transport chain. Our whole numbers have been confirmed in some of the most careful investigations [45] and are still favored in our own prejudice. Our contention that the oxidation of α-oxoglutarate to succinate yielded P/O values in excess of 3.0 was challenged by Slater and Holton [46] but our arguments [47] in favor of a value approaching 4 have not been repudiated.

When studying the role of phosphate acceptors on mitochondrial respiration rates I had prepared partially purified creatine kinase by fractionation with ammonium sulfate. My student Stephen Kuby, a perfectionist, felt strongly that only pure kinase should be used for this purpose and therefore he and Lafayette Noda undertook its purification [48,49]. Thus was initiated a program of enzyme purification that resulted in the first pure — and in almost every case, crystalline — creatine kinase, adenylate kinase, nucleoside diphosphokinase, 5'-adenylic deaminase, malic enzyme, glucose 6-phosphate dehydrogenase, phosphofructokinase and liver aldolase. When Robert Peanasky purified the latter enzyme he found that, contrary to muscle aldolase, the liver enzyme has relatively high activity toward fructose 1-phosphate and thus is capable of participating in the liver's metabolism of fructose [50].

My next attempt at enzyme purification was aimed at mitochondrial ATPase. Harlene Wellman, a technician, and I were studying the stimulation by 2,4-dinitrophenol of ATP hydrolysis in liver mitochondria and extracted ATPase activity from an acetone powder of the mitochondria [51]. Over an extended period of time we repeatedly failed to retain the activity during fractionation despite working in the cold to avoid denaturation. Success in purifying the heart enzyme was achieved by Efraim Racker and his coworkers who were astute enough to realize that the enzyme is cold-labile. Much later David Lambeth purified the liver ATPase [52] and we found that the nucleotide binding sites could be differentiated into two types, regulatory and catalytic, based on the response to nucleotides with different purine bases as well as to nucleotide analogs [53,54].

Berton Pressman studied the role of potassium ions in mitochondrial oxidative phosphorylation for his Ph.D. thesis [55]. He found that the microsomal fraction of rat liver increased the respiratory response to K^+, and that an acetone extract of the microsomes produced similar effects. This led to the discovery that free fatty acids are uncoupling agents with maximum activity displayed by myristate, and *cis* unsaturation greatly enhancing the activity of longer chains [56]. These results have been widely confirmed and have implications for thermogenesis, especially in brown fat.

Our work on biotin's role in fixing HCO_3^- stimulated me to think about propionate metabolism. It was known that propionate is antiketogenic and a precursor of carbohydrate via gluconeogenesis. Lorber et al. [57] had demonstrated that it was not oxidized directly to lactate, for the α and β carbon atoms of propionate were randomized in the 3-carbon molecules that formed glucose. I found that $^{14}CO_2$ was incorporated into non-volatile organic acids by extracts of rat liver mitochondria by a process dependent on both propionate and ATP [39]. The product was characterized as succinate [58,59]; later Katz and Chaikoff [60] and Flavin [61] found methyl malonyl-CoA to be an intermediate in the production of succinate, but free methyl malonate was not produced by soluble extracts of rat liver [61].

For his Ph.D. thesis research, George Drysdale undertook a study of fatty acid oxidation by soluble enzymes from liver mitochondria. He confirmed the requirement for ATP that had been demonstrated in particulate systems, demonstrated the formation of acyl-CoA intermediates, and found NAD to be required in dialyzed preparation [62,63].

While investigating the kinetic behavior of the several kinases we had purified, a possible general mechanism of metabolic regulation became apparent. In the final paper at a symposium in Detroit on *Enzymes: Units of Biological Structure and Function*, I presented data showing that a variety of enzymes that use Mg ATP as a substrate are inhibited by excess free ATP. These included fructokinase, creatine kinase, phosphofructokinase and the system that fixes CO_2 onto propionyl-CoA in the synthesis of succinate [43]. Others have extended the list of enzymes responsive to such regulation and have documented the interaction of other allosteric regulators with the effects of ATP [64,65].

Any experimental scientist can look back at his/her old data and regret that apparently aberrant results were not studied more carefully. Frank Maley, a graduate student in our group who demonstrated ability in both organic chemistry and enzymology, synthesized a number of sugar phosphates [66,67] and with his wife Gladys studied their conversion to uridine diphosphosugar derivatives. Glucose 1-phosphate was used as a control in comparison with glucosamine 1-phosphate and *N*-acetylglucosamine 1-phosphate. It was found that liver cytosol "rapidly breaks down UDPG to . . . UDP" [68]. The reaction was probably a manifestation of the glycogen synthesis pathway discovered the following year by Leloir and Cardini [69] but we were not thinking of glycogen at the time.

Another example is found in the thesis research of Julius Adler [70]. While investigating the pathway of itaconic acid metabolism by liver mitochondria he found that the molecule must first be converted to itaconyl-CoA by succinic thiokinase and that it was then cleaved to pyruvate and acetyl-CoA. Extracts of acetone-dehydrated liver mitochondria were found to fix $^{14}CO_2$ in the presence of itaconate and to produce radioactive glutamate, malate, citrate and α-ketoglutarate.

"In the presence of Mg^{2+}, ATP, and these extracts, pyruvate or phosphopyruvate fixes carbon dioxide one-half as fast as does itaconate, but the addition of acetyl-CoA increases the rate to 4 times that for itaconate" ([70], p. 874).

We assumed that pyruvate was carboxylated to oxalacetate and that acetyl-CoA stimulated "by trapping the oxalacetate as citrate". Some 3 years later Utter and Keech [71] purified pyruvate carboxylase and demonstrated the highly specific, allosteric activation of the enzyme by acetyl-CoA.

The thyroid hormone has intrigued me since my early graduate student days [8]. Later in collaboration with my own students we pursued various lines of research in an effort to clarify its mode of action. I was impressed with clinical studies that demonstrated the thyroid hormone decreased the metabolic efficiency of work performance [72–74]. Rats rendered hyperthyroid by feeding desiccated thyroid yielded liver mitochondria with elevated rates of state 4 oxygen consumption (L-glutamate as substrate) and decreased efficiency of phosphorylation [75]. Thyroxine and triiodothyronine in greater than physiological doses were found by us and several other groups to diminish phosphorylation efficiency but that effect is probably not relevant to the normal function of the hormone.

Following the discovery of triiodothyronine by Gross and Pitt-Rivers we championed the idea that thyroxine (T_4) probably needed to be converted to triiodothyronine (T_3) to become the active hormone. A visiting scientist, Dr. Kenkichi Tomita and I collaborated with Drs. Larson and Albright of our Department of Medicine in studies of the in vitro conversion of T_4 to T_3 and of both to their corresponding analogs in which the side chain is degraded to an acetic acid substituent [76].

Richard Doisy [77] discovered a dramatic reversal, by estrogenic hormones, of the elevated basal metabolic rate caused by T_4 or T_3. Corticosteroids were not effective and the antagonism was not mediated via pituitary, adrenal or thyroid glands. It appeared to be exerted at the cellular level. Dr. Ya Pin Lee made an equally dramatic finding in the great influence of thyroid hormone on the mitochondrial α-glycerophosphate oxidase [78,79]. The enzyme was nearly totally absent from heart, liver, kidney, and adipose tissue of thyroidectomized rats and feeding 2% desiccate thyroid for a week to 10 days increased the enzyme's activity 5- to 20-fold above the euthyroid level in some tissues. More recently we have demonstrated a significant role for this enzyme in the transport of electrons from liver cytosol to mitochondria under the stimulating influence of catecholamine hormones [80–82]. The increased cytosolic free Ca^{2+} brought about by the α_1 adrenergic receptor activation increases the affinity of α-glycerophosphate oxidase for its substrate sufficiently to permit rapid oxidation at the glycerophosphate concentration present in liver [81]. The substrate is not removed from the cytosol because the product, dihydroxyacetone phosphate is reduced by the NADH in the cytosol via the soluble dehydrogenase. The combination of thyroid hormone affecting the amount of α-glycerophosphate oxidase, and catecholamines affecting the enzyme's K_m probably accounts in part for the synergistic effect of these two hormones on metabolic rates [83].

While near the subject of dihydroxyacetone phosphate it is appropriate to discuss an "orphan" compound that we synthesized some years ago in an attempt to find a function for glyoxalase. The usual substrate for this enzyme, methylglyoxal does not occur naturally in amounts sufficient to justify the relative abundance of glyoxalase. Furthermore, the lactic acid product of glyoxalase action on methylglyoxal has the unnatural D configuration. I speculated that D-3-phosphoglycerate could conceivably be produced by this enzyme and Dr. Robert Weaver undertook the synthesis of its hypothetical precursor, phosphohydroxypyruvic aldehyde (PHPA). He oxidized dihydroxyacetone phosphate with cupric acetate, isolated and rigorously characterized the product as PHPA, demonstrated that it was a good substrate for glyoxalase I and that the glutathione derivative formed by this enzyme is hydrolyzed by glyoxalase II to form 3-phosphoglycerate possessing the natural D-configuration [84]. The "orphan" now has a parent.

The complex of aldolase with dihydroxyacetone phosphate is subject to oxidation by a number of electron acceptors to produce PHPA [85] and the latter has been found to occur in erythrocytes [86]. Currently we are searching for possible dehydrogenases that might oxidize the aldolase-substrate complex to produce PHPA. Such a system would comprise a "futile pathway" of glycolysis without supporting phosphorylation at the triose phosphate oxidation step.

Antibiotics as tools for the biochemist

Warburg's experiments with cyanide and narcotics, Lundsgaard's with iodoacetate, and Embden's with fluoride demonstrated the usefulness of enzyme inhibitors for the study of metabolic reactions. In the early 1950s the need for specific inhibitors of oxidative phosphorylation was obvious because the conventional approaches of enzyme chemists were not fruitful. In a Festschrift for Professor H.O.L. Fischer [87] we outlined our reasons for seeking useful inhibitors among the toxic antibiotics. We had collected antibiotics from several industrial firms and academic sources and screened for effects on mitochondrial oxidation and phosphorylation, ATPase, and in early experiments, glycolysis. Among more than 200 compounds screened we documented the mode of action of 16 inhibitors of ATP synthesis - usnic acid, oligomycin (5 homologs and analogs), aurovertin, 2 venturicidins, venturicidin aglycone, ossamycin, efrapeptin, leucinostatin, A20668 B and C, and bongkrekic acid [87–89].

Of the antibiotics we examined, several that influenced mitochondrial function were found to be ionophores. These included nonactin, monactin, dinactin, trinactin, enniatins A and B, monazomycin, nigericin, dianemycin, stendomycin, monensin, beauvericin, A204, and A23187 [87,90,91]. The last-mentioned has been employed in experiments reported in thousands of research papers.

Research with spermatozoa

Our original interest in sperm metabolism and function has been retained despite major efforts in other areas. I tried always to have at least one graduate student working with spermatozoa and frequent collaboration with Professor Neal First of the Animal Science Department has been fruitful. Bruce Morton devised systems for measuring oxidative phosphorylation efficiency in spermatozoa by blocking glycolysis with fluoride and using 2-deoxyglucose as phosphate acceptor [92]. Phosphorylation efficiencies approaching those measurable with isolated mitochondria from somatic cells were achieved by rendering the plasma membrane permeable to substrates and nucleotides. This was accomplished by shaking sperm with 75-μm glass beads, by exposure to hypotonic suspension media, or dissolving membrane cholesterol with the polyene antibiotic, filipin [93].

In 1970 Sutherland's group reported the presence of adenylate- and guanylate-cyclases in sea urchin sperm [94], and Casillas and Hoskins found adenylate cyclase in monkey sperm [95]. David Garbers, a student working jointly with Neal First, undertook a study of the role of cAMP in mammalian spermatozoa [96]. Phosphodiesterase inhibitors such as caffeine induced cAMP and cGMP formation, and stimulated motility and respiration. Caffeine stimulation of endogenous substrate oxidation is transient and appears associated with the utilization of acetyl carnitine [97]. In the presence of substrates that yield acetyl groups (lactate, pyruvate, acetate, β-hydroxybutyrate) respiration is enhanced 2- to 4-fold [96,98].

The effect of caffeine or of cyclic nucleotides is exerted on the motility mechanism and the respiratory increase results from the increased energy utilization [99]. The enhanced motility was characterized as being of the "whiplash" type observed in sperm capacitated under natural conditions in utero. There is reason to believe that during the capacitation process in the female reproductive tract, some natural secretory product activates motility in the manner of caffeine, thus aiding sperm to penetrate the cumulus oophorus and zona pellucida in order to reach the egg.

It is a strange fact that many of the enzymes in spermatozoa are different from those in somatic cells. Cytochrome *c* in mouse sperm differs from that in somatic cells at 13 of the 104 amino acids [100]. Sperm contain a lactic dehydrogenase that is not only a unique protein, it is also distributed in an unconventional manner — in both cytosol and mitochondrial matrix spaces [101]. The presence of the dehydrogenase in mitochondria permits an unusual metabolism of pyruvate in these cells [102,103]. In the presence of rotenone each molecule of pyruvate oxidized to CO_2 is balanced by 4 molecules being reduced to lactate. Motility can be supported by this pyruvate dismutation even in the presence of oligomycin because of the generation of GTP at the succinic thiokinase step. This dismutation system is very likely utilized when the sperm are deposited in the female reproductive tract where the oxygen supply is limited.

It has been known since the work of Jacques Loeb [103a] that fertilization of sea urchin eggs does not occur in the absence of calcium, and several investigators have shown that calcium must be present to achieve in vitro fertilization of mammalian eggs. Calcium is especially important for the acrosome reaction. The acrosome is a lysosome-like compartment on the head of the sperm that contains hyaluronidase and a proteolytic enzyme, acrosin. During the process of capacitation the acrosomal membrane is discomposed, the enzymes are released and can aid in establishing a path through the investments that surround the egg, thus permitting the spermatozoan to reach the vitellin membrane which it penetrates readily. Yanagamachi and Usui [104] found conditions for obtaining the acrosome reaction in mammalian sperm during prolonged incubation in vitro and one of the requirements was the presence of calcium in the suspending medium. Some investigators believe the sperm must be "capacitated" before the acrosome reaction can occur. However, my student, Jai-Pal Singh found that the acrosome reaction could be produced within minutes by the addition of minute quantities of the calcium ionophore A23187 to guinea pig sperm suspended in a medium containing 1 mM $CaCl_2$ [105]. This rules out the necessity for a slow process that must occur before the sperm are capable of undergoing the acrosome reaction. It supports Chang's contention [105a] that the term "capacitation" should include all the processes that occur in the female reproductive tract to make the sperm capable of fertilizing eggs. The acrosome reaction is a part of the capacitation process [105a].

It is quite likely that the role of calcium in the acrosome reaction is to activate phospholipase A_2. The formation of lysolecithin from phosphatidyl choline in the acrosomal membrane would obviously destroy the integrity of the acrosomal membrane structure.

My colleague Donner Babcock found a striking difference between epididymal and ejaculated sperm with respect to their permeability to Ca^{2+} [6]. Epididymal bovine spermatozoa contain 7 nmol of exchangeable Ca per 10^8 cells. On incubation in media containing 0.2 mM $CaCl_2$ they take up 4 to 8 times as much calcium. In contrast, ejaculated sperm, bathed in seminal fluid containing 9 mM calcium, contain no more calcium than epididymal sperm and when washed free of seminal plasma they still take up no calcium. We found that the explanation for these phenomena is the presence in seminal fluid of a protein that binds to the plasma membranes and inhibits calcium transport. This protein, which we have named caltrin, has been isolated and found to behave on gels as if its M_r were about 10 000 [106]. More recent studies of the protein's sequence indicate an M_r of 5400 [106a].

Plasma membrane vesicles prepared from epididymal spermatozoa take up calcium much more rapidly and in greater amount than vesicles prepared from ejaculated spermatozoa [7]. The flux of Ca^{2+} requires a counter flow of Na^+ and is enhanced by the presence of K^+. Thus the classic 3 Na^+ exchange for each Ca^{2+} may, in the case of bovine spermatozoa, be altered to a 3 $Na^+/Ca^{2+}+K^+$ electroneutral transport.

We believe the function of caltrin is to delay calcium uptake and thus prevent premature development of the acrosome reaction. The delay permits the acrosome reaction to occur at about the time that sperm encounter the egg. Were the acrosome reaction to occur much earlier, the hydrolytic enzymes required to aid penetration of the egg investments would be wiped away as the sperm swim up the reproductive tract. The possibility is being investigated that the female tract contains enzymes that slowly destroy caltrin thus permitting calcium uptake. Decapacitating factors in seminal fluid have been reported by others beginning with Chang [107]. However, the reported size differs from that of caltrin and there is disagreement about the significance of the factors [108]. The physiological significance of caltrin is a problem of major interest to our research group at present.

Gluconeogenesis

Because of an odd circumstance, the major effort in our laboratory shifted in the early 1960s from mitochondrial functions to studies of carbohydrate synthesis. I was interested in the mechanism by which 2,4-dinitrophenol enhanced the synthesis of phosphoenolpyruvate in isolated mitochondria as had been demonstrated by Mudge et al. [109]. In repetitive experiments I failed to demonstrate the formation of phosphopyruvate despite using conditions similar to those employed by Mudge et al. [109]. It finally dawned on me that I was using rat liver mitochondria, whereas Mudge had used rabbit. Furthermore, Bandurski and Lipmann [110] who had also demonstrated phosphoenolpyruvate formation had used guinea pig liver mitochondria and on rereading their paper, I learned that they had found little or no phosphoenolpyruvate carboxykinase (PEPCK) activity in rat liver mitochondria. At the time, Robert Nordlie, a post-doctoral fellow, had just completed his studies of inorganic pyrophosphatase [111] and I asked him to look into the species differences that might influence phosphopyruvate formation. He found that Utter's carboxykinase was present in the cytosol of rats, mice, and hamsters [112] with little or none in the mitochondria where it is found in chicken liver [113]. The pigeon and guinea pig have this enzyme in both cytosol and mitochondria [114] and the rabbit's liver, which has little or none in the cytosol when the animal is fed, develops relatively large amounts of the cytosolic enzyme during prolonged fasting [115]. These findings raised new questions, for, if pyruvate is converted to oxalacetate in liver mitochondria [70,71] and oxalacetate is converted to phosphopyruvate in the cytosol, the transport of oxalacetate from one compartment to another needed clarification. We established that oxalacetate formed from pyruvate was not transported as such but was either transaminated to form aspartate or reduced to form malate and that these two compounds, together with some citrate and glutamate, moved out of the mitochondria [116]. In the cytosol, aspartate was postulated to be transaminated and malate would be oxidized to oxalacetate providing the substrate for PEPCK. These findings thus provided a pathway for gluconeogenesis from lactate and pyruvate [116]. Earl Shrago, Bob Nordlie and David Foster [117] demonstrated the rapid adaptation of PEPCK to the need for glucose synthesis. Fasting 12–48 h more than doubles the activity of PEPCK in liver; refeeding rapidly returns the activity to the normal range; diabetes may increase the activity 5- to 10-fold and insulin rapidly brings the activity back to normal or below; glucagon elevates the activity, as do glucocorticoids, but adrenalectomy does not alter the response to fasting [117–119].

All of these responses are the result of new enzyme synthesis [119]. When fasted animals are refed diets devoid of carbohydrate, the activity of PEPCK increases instead of declining [120]. This finding prompted a study of the effect of feeding amino acids on PEPCK activity and thus was discovered the dramatic effect of tryptophan [121]. Tryptophan administered orally or by any parenteral route doubles the enzyme's activity in 30 min and the effect is not abolished by acetoxycycloheximide or puromycin [122]. It was first thought that tryptophan's effect on PEPCK activity might enhance gluconeogenesis and we were amazed that, instead, it inhibited carbohydrate formation [121], and astonished to find that the inhibition was exerted at the step catalyzed by PEPCK [123]. Manganese, which enhances the activity of PEPCK in normal liver cytosol has no effect on the elevated activity of the rat given tryptophan thus implicating metal involvement in the effect of tryptophan [122].

There are many pathways of tryptophan metabolism and a search of each [121,124] disclosed that the effect of tryptophan is mediated by its metabolite, quinolinate, an intermediate in the biosynthesis of NAD. Quinolinate blocks PEPCK by garnering Fe^{2+} from tissue stores to form a tetradentate complex of 2 quinolinates and 1 Fe^{2+} [125,126] which is the effective inhibitor. There is, as yet, no firm evidence that the amounts of tryptophan in normal diets play a role in the regulation of PEPCK via quinolinate formation but the study of this phenomenon has made us aware of the possible regulatory role of Fe^{2+}. In addition to Fe^{2+} and Mn^{2+}, Co^{2+} and Cd^{2+} enhance PEPCK activity but the latter two are not sufficiently abundant in liver to participate. The fact that quinolinate inhibits the enzyme in vivo or in the isolated perfused rat liver indicates the availability of Fe^{2+} in the liver. Carlo Veneziale found that the inhibition by quinolinate can be reversed by the addition of $MnCl_2$ to the perfusion fluid [124] and Roy Snoke found that Mn-quinolinate is not an inhibitor [125].

When highly purified PEPCK was used in studies of Fe^{2+} activation it was found that the enzyme was not stimulated even though it was still capable of responding to Mn^{2+} [127]. This led us to search for the reason that the enzyme in unfractionated cytosol could respond to Fe^{2+} and a protein was found that permits the pure enzyme to respond.

We named the protein "ferroactivator" [127], developed a radioimmunoassay for it and demonstrated that the liver content of this protein was increased in diabetic rats [128]. The unusual properties of PEPCK have given rise to evidence of unorthodox catalysis and stability [129–131]. Reynolds [132] presented evidence that ferroactivator functions by protecting the enzyme from damage by Fe^{2+}. We were well aware of the many highly active products that can be generated from oxygen in the presence of iron but felt that Reynolds' explanation was not correct, for in many experiments [133–135] our control values for PEPCK activity in the presence of Fe^{2+} but no ferroactivator were identical with the activity in the absence of Fe^{2+}. However, Reynolds was correct, for we have found that purified ferroactivator is glutathione peroxidase. Our assay conditions led to a combination of activation, by Fe^{2+}, of surviving PEPCK sufficient to compensate for the enzyme that had been destroyed by reactive oxygen species [136]. PEPCK displays far greater sensitivity to the effects of reactive oxygen species than do a number of other Fe^{2+}-requiring enzymes in keeping with the presence on PEPCK of a highly active thiol that is essential for activity [137] and sheltered by Mn^{2+} and ITP [130]. Currently some interesting relationships between Se, glutathione peroxidase and gluconeogenesis are being investigated.

On and off the campus

The academic environment at Wisconsin has traditionally been inspiring, challenging and pleasurable. Until 1976 University policies were established democratically at monthly faculty meetings. Since then representative democracy prevails in the form of an elected senate.

For all the years I have been here the Dean of the graduate school has been a scientist sympathetic to the needs of young faculty members; Presidents or Chancellors, more often than not, have come from the science faculties. A system of Divisional committees (Physical Science, Biological Science, Humanities, Social Studies) screens departmental recommendations for promotion and productive individuals are not held back by an arbitrary time schedule.

The University has a long record of liberal thought and action. During the grim period of McCarthyism, President E.B. Fred resisted efforts of many organizations and of the Board of Regents to impose a loyalty oath on the University faculty and staff. In 1952 I helped organize and served as President of "Citizens vs. McCarthy". We publicized the falseness of Senator Joseph McCarthy's claims in ads sponsored by distinguished professional, business, and academic people. At the peak of his national influence McCarthy came to the university to deliver a lecture and was laughed off the stage. However, we failed to defeat him in his campaign for re-election to the senate and he continued to damage, irreparably, our experts on China, the Voice of America and gifted writers and actors. The Army-McCarthy hearings in 1954 changed fear to opprobrium and he died 3 years later of alcoholic cirrhosis.

Among the many rewards of an academic life, the friendships developed with students are perhaps the greatest. Of my graduate students, 56 have completed their Ph.D. degrees to date and that number should be 60 at the time of my retirement. They have all been talented and a pleasure to work with. Planning for the future has me torn between continuing research beyond the retirement age of 70 or spending more than weekends in beautiful Sauk County on our farm with its forests, a lake and stream, tennis court, and elegant Arabian horses. I wonder which plan will win out.

REFERENCES

1 A.L. Moxon, A. Schaeffer, H.A. Lardy, K. DuBois and O.E. Olson, J. Biol. Chem., 132 (1940) 785-786.
2 H.A. Lardy and P.H. Phillips, Am. Soc. Animal Prod., 32 (1939) 219-221.
3 H.A. Lardy and P.H. Phillips, Am. J. Physiol., 134 (1941) 542-548.
4 H.A. Lardy and P.H. Phillips, Am. J. Physiol., 133 (1941) 602-609.
5 H.A. Lardy, R.G. Hansen and P.H. Phillips, Arch. Biochem., 6 (1945) 41-51.
6 D. Babcock, J.P. Singh and H.A. Lardy, Develop. Biol., 69 (1979) 85-93.
7 G. Rufo, P. Schoff and H.A. Lardy, J. Biol. Chem., 259 (1984) 2547-2552.
8 H.A. Lardy and P.H. Phillips, J. Biol. Chem., 149 (1943) 177-182.
9 H.A. Lardy and C. Elvehjem, Annu. Rev. Biochem., 14 (1945) 1-30.
10 P.H. Phillips, H.A. Lardy, E. Heizer and I. Rupel, J. Dairy Sci., 23 (1940) 873-878.
11 P.H. Phillips, H.A. Lardy, P.D. Boyer and G. Werner, J. Dairy Sci., 24 (1941) 153-158.
12 G. Rona and C. Chappel, Endocrinology, 72 (1963) 1-10.
13 C. Deb and A. Chatterjee, Endocrinology, 72 (1963) 159-160.
14 M. Igarashi, Int. J. Fert., 22 (1977) 168-173.
15 F.L. Breusch, Science, 97 (1943) 490-492.
16 H.A. Lardy and P.H. Phillips, Nature, 153 (1944) 168-169.
17 H.A. Lardy and P.H. Phillips, J. Biol. Chem., 148 (1943) 333-341.
18 H.A. Lardy and P.E. Hughes, Curr. Topics Cell. Reg., 24 (1984) 171-179.
19 H.A. Lardy and J. Ziegler, J. Biol. Chem., 159 (1945) 343-351.
20 H.A. Lardy and H.O.L. Fischer, J. Biol. Chem., 164 (1946) 513-519.
21 L. Rubin, H.A. Lardy and H.O.L. Fischer, J. Am. Chem. Soc., 74 (1952) 425-428.
22 L. Anderson and H.A. Lardy, J. Am. Chem. Soc., 70 (1948) 594-597.
23 E. Totten and H.A. Lardy, J. Am. Chem. Soc., 71 (1949) 3076-3078; J. Biol. Chem., 181 (1949) 701-706.
23a T.C. Tung, K.H. Ling, W.L. Byrne and H.A. Lardy, Biochim. Biophys. Acta, 14 (1954) 488-494.
23b D.L. Bissett, W.C. Wenger and R.L. Anderson, J. Biol. Chem., 255 (1980) 8740-8744; 8745-8749; 8750-8755.
24 O. Meyerhof and E. Perdigon, C. R. Soc. Biol., 132 (1939) 186-190; Enzymologia, 8 (1940) 353-362.
25 H.A. Lardy, V. Wiebelhaus and K.M. Mann, J. Biol. Chem., 187 (1950) 325-337.
26 K.M. Mann and H.A. Lardy, J. Biol. Chem., 187 (1950) 339-348.
27 H. Weil-Malherbe and A.D. Bone, Biochem. J., 49 (1951) 339-347; R.K. Crane and A. Sols, J. Biol. Chem., 203 (1953) 273-292.
28 J. Stokes, H. Larsen and M. Gunness, J. Biol. Chem., 167 (1947) 613-614.
29 H.A. Lardy, R. Potter and C.A. Elvehjem, J. Biol. Chem., 169 (1947) 451-452.
30 H.A. Lardy, R. Potter and R.H. Burris, J. Biol. Chem., 179 (1949) 721-731.

31 H. Umbarger, Science, 123 (1956) 848; R. Yates and A. Pardee, J. Biol. Chem., 221 (1956) 757-770.
32 Z. Dische, Bull. Soc. Chim. Biol., 23 (1941) 1140.
33 P.R. MacLeod, S. Grisolia, P.P. Cohen and H.A. Lardy, J. Biol. Chem., 180 (1949) 1003-1011.
34 G.W.E. Plaut, J.J. Betheil and H.A. Lardy, J. Biol. Chem., 184 (1950) 795-805.
35 G.R. Drysdale, G.W.E. Plaut and H.A. Lardy, J. Biol. Chem., 193 (1951) 533-538.
36 G.W.E. Plaut and H.A. Lardy, J. Biol. Chem., 192 (1951) 435-445.
37 H.A. Lardy, in L.F. Wolterink (Ed.), The Biology of Phosphorus, Michigan State College Press, 1952, pp. 131-147.
38 H.A. Lardy and H. Wellman, J. Biol. Chem., 195 (1952) 215-224.
39 H.A. Lardy, Proc. Natl. Acad. Sci. USA, 38 (1952) 1003-1013.
40 B. Chance and G.R. Williams, J. Biol. Chem., 217 (1955) 383-393.
41 J.W. Stucki, Eur. J. Biochem., 109 (1980) 269-283.
42 H.A. Lardy, Proc. 3rd Int. Congress of Biochem., Academic Press, New York, 1956, pp. 287-294.
43 H.A. Lardy and R.E. Parks Jr., in O.H. Gaebler (Ed.), Enzymes: Units of Biological Structure and Function, Academic Press, New York, 1956, pp. 584-588.
44 J.H. Copenhaver Jr. and H.A. Lardy, J. Biol. Chem., 195 (1952) 225-238.
45 N.G. Forman and D.F. Wilson, J. Biol. Chem., 257 (1982) 12908-12915.
46 E.C. Slater and F.A. Holton, Biochem. J., 56 (1954) 28-40.
47 H.A. Lardy and J.H. Copenhaver Jr., Nature, 174 (1954) 231-233.
48 S.A. Kuby, L. Noda and H.A. Lardy, J. Biol. Chem., 209 (1954) 191-201.
49 S.A. Kuby, L. Noda and H.A. Lardy, J. Biol. Chem., 210 (1954) 65-82.
50 R. Peanasky and H. Lardy, J. Biol. Chem., 233 (1958) 365-373.
51 H.A. Lardy and H. Wellman, J. Biol. Chem., 201 (1953) 357-370.
52 D.O. Lambeth and H.A. Lardy, Eur. J. Biochem., 22 (1971) 355-363.
53 H.A. Lardy, S.M. Schuster and R.E. Ebel, J. Supramol. Struct., 3 (1975) 214-221.
54 S.M. Schuster, R.E. Ebel and H.A. Lardy, J. Biol. Chem., 250 (1975) 7848-7853.
55 B.C. Pressman and H.A. Lardy, J. Biol. Chem., 197 (1952) 547-556.
56 B.C. Pressman and H.A. Lardy, Biochim. Biophys. Acta, 21 (1956) 458-466.
57 V. Lorber, N. Lifson, W. Sakami and H. Wood, J. Biol. Chem., 183 (1950) 531-538.
58 H.A. Lardy and R. Peanasky, Physiol. Rev., 33 (1953) 560-565.
59 H.A. Lardy and J. Adler, J. Biol. Chem., 219 (1956) 933-942.
60 J. Katz and I. Chaikoff, J. Am. Chem. Soc., 77 (1955) 2659-2660.
61 M. Flavin, Fed. Proc., 14 (1955) 211.
62 G. Drysdale and H.A. Lardy, Phosphorus Metabolism, II (1952) 281-285.
63 G. Drysdale and H.A. Lardy, J. Biol. Chem., 202 (1953) 119-136.

64 J.V. Passoneau and O.H. Lowry, Biochem. Biophys. Res. Commun., 7 (1962) 10–15.
65 H.-G. Hers, in G. Semenza (Ed.), Comprehensive Biochemistry, Vol. 35, Elsevier, Amsterdam, 1983, pp. 71–101.
66 F. Maley and H. Lardy, J. Am. Chem. Soc., 78 (1956) 1393–1397.
67 F. Maley, G.F. Maley and H. Lardy, J. Am. Chem. Soc., 78 (1956) 5303–5307.
68 F. Maley, G.F. Maley and H. Lardy, Science, 124 (1956) 1207–1208.
69 L.F. Leloir and C.E. Cardini, J. Am. Chem. Soc., 79 (1957) 6340–6341.
70 J. Adler, S.F. Wang and H.A. Lardy, J. Biol. Chem., 229 (1957) 865–879.
71 M.F. Utter and D.B. Keech, J. Biol. Chem., 235 (1960) PC 17–PC 18.
72 J.H. Smith, Arch. Int. Med., 42 (1928) 47–52.
73 S.P. Briard, J.T. McClintock and C.W. Baldridge, Arch. Int. Med., 56 (1935) 30–37.
74 H. Bruch, Jahrb. Kinderheilk., 121 (1928) 7–28.
75 H. Lardy, Brookhaven Symp. Quant. Biol., 7 (1953) 90–101.
76 H. Lardy, K. Tomita, F.C. Larson and E.C. Albright, Ciba Found. Colloquia Endocrinol., 10 (1957) 156–164.
77 R. Doisy and H. Lardy, Am. J. Physiol., 190 (1957) 142–146.
78 Y.-P. Lee, A.E. Takemori and H. Lardy, J. Biol. Chem., 234 (1959) 3051–3054.
79 Y.-P. Lee and H. Lardy, J. Biol. Chem., 240 (1965) 1427–1436.
80 B.B. Yip and H. Lardy, Arch. Biochem. Biophys., 212 (1981) 370–377.
81 M.E. Wernette, R.S. Ochs and H. Lardy, J. Biol. Chem., 256 (1981) 12767–12771.
82 R.S. Ochs and H. Lardy, FEBS Lett., 131 (1981) 119–121.
83 W.R. Brewster Jr., J.P. Isaacs, P.F. Osgood, T.L. King and A.B. King, Circulation, 13 (1956) 1–20.
84 W.R. Weaver and H. Lardy, J. Biol. Chem., 236 (1961) 313–317.
85 M.J. Healy and P. Christen, Biochemistry, 12 (1973) 35–41.
86 M. Cogoli-Greuter and P. Christen, J. Biol. Chem., 256 (1981) 5708–5711.
87 H. Lardy, D. Johnson and W.C. McMurray, Arch. Biochem. Biophys., 78 (1958) 587–597.
88 H. Lardy, J. Connelly and D. Johnson, Biochemistry, 3 (1964) 1961–1968.
89 H. Lardy, Pharmacol. Ther., 11 (1980) 649–660.
90 H. Lardy, Fed. Proc., 27 (1968) 1278–1282.
91 D. Wong, J.-S. Horng, R. Hamill and H. Lardy, Biochem. Pharmacol., 20 (1971) 3169–3177.
92 B.E. Morton and H.A. Lardy, Biochemistry, 6 (1967) 43–49.
93 B.E. Morton and H.A. Lardy, Biochemistry, 6 (1967) 50–56; 57–61.
94 J.P. Gray, J.G. Hardman, T. Bibring and E.W. Sutherland, Fed. Proc., 29 (1970) 608.
95 E.R. Casillas and D.D. Hoskins, Biochem. Biophys. Res. Commun., 40 (1970) 255–262.

96 D.L. Garbers, W.D. Lust, N.L. First and H.A. Lardy, Biochemistry, 10 (1971) 1825-1831.
97 A. Milkowski, D.F. Babcock and H.A. Lardy, Arch. Biochem. Biophys., 176 (1976) 250-256.
98 D.L. Garbers, N.L. First, J.J. Sullivan and H.A. Lardy, Biol. Reprod., 5 (1971) 336-339.
99 D.L. Garbers, N.L. First and H.A. Lardy, Biol. Reprod., 8 (1973) 589-598; 599-606.
100 B. Hennig, Eur. J. Biochem., 55 (1975) 167-183.
101 C. Goldberg, Curr. Topics Biol. Med. Res., 1 (1977) 79.
102 C. van Dop, S.M. Hutson and H.A. Lardy, J. Biol. Chem., 252 (1977) 1303-1308.
103 S.M. Hutson, C. van Dop and H.A. Lardy, J. Biol. Chem., 252 (1977) 1309-1315.
103a. J. Loeb, J. Exp. Zool., 17 (1915) 123.
104 R. Yanagamachi and N. Usui, Exp. Cell. Res., 89 (1974) 161-174.
105 J.P. Singh, D.F. Babcock and H.A. Lardy, Biochem. J., 172 (1978) 549-556.
105a M.C. Chang, J. Andol., 5 (1984) 45-50.
106 G.A. Rufo, J.P. Singh, D.F. Babcock and H.A. Lardy, J. Biol. Chem., 257 (1982) 4627-4632.
106a R.V. Lewis, J. San Agustin and H.A. Lardy, Proc. Natl. Acad. Sci. USA, 82 (1985) in press.
107 M.C. Chang, Nature, 179 (1957) 258-259.
108 R.A. McRorie and W.L. Williams, Arch. Biochem. Biophys., 43 (1974) 777-803.
109 G.H. Mudge, H.W. Neuberg and S.W. Stanbury, J. Biol. Chem., 210 (1954) 965-979.
110 R.S. Bandurski and F. Lipmann, J. Biol. Chem., 219 (1956) 741-752.
111 R.C. Nordlie and H.A. Lardy, Biochim. Biophys. Acta, 53 (1961) 309-323.
112 R.C. Nordlie and H.A. Lardy, J. Biol. Chem., 238 (1963) 2259-2263.
113 M.F. Utter and K. Kurahashi, J. Biol. Chem., 207 (1954) 787-802.
114 H.A. Lardy, Harvey Lectures, 60 (1966) 261-278.
115 D.C. Johnson, C.A. Ebert and P.D. Ray, Biochem. Biophys. Res. Commun., 39 (1970) 750-756.
116 H.A. Lardy, V. Paetkau and P. Walter, Proc. Natl. Acad. Sci. USA, 53 (1965) 1410-1415.
117 E. Shrago, H.A. Lardy, R.C. Nordlie and D.O. Foster, J. Biol. Chem., 238 (1963) 3188-3192.
118 H.A. Lardy, D.O. Foster, E. Shrago and P.D. Ray, Adv. Enzyme Reg., 2 (1964) 39-47.
119 D.O. Foster, P.D. Ray and H.A. Lardy, Biochemistry, 5 (1966) 555-562.
120 J.W. Young, E. Shrago and H.A. Lardy, Biochemistry, 3 (1964) 1687-1692.
121 D.O. Foster, P.D. Ray and H.A. Lardy, Biochemistry, 5 (1966) 563-569.
122 D.O. Foster, H.A. Lardy, P.D. Ray and J.B. Johnston, Biochemistry, 6 (1967) 2120-2128.

123 P.D. Ray, D.O. Foster and H.A. Lardy, J. Biol. Chem., 241 (1966) 3904-3908.
124 C.M. Veneziale, P. Walter, N. Kneer and H.A. Lardy, Biochemistry, 6 (1967) 2129-2138.
125 R.E. Snoke, J.B. Johnston and H.A. Lardy, Eur. J. Biochem., 24 (1971) 342-346.
126 M.J. MacDonald, Biochem. Biophys. Res. Commun., 90 (1979) 741-749.
127 L.A. Bentle, R.E. Snoke and H.A. Lardy, J. Biol. Chem., 251 (1976) 2922-2928.
128 M.J. MacDonald, L.A. Bentle and H.A. Lardy, J. Biol. Chem., 253 (1978) 116-124.
129 R.I. Brinkwoth, R.W. Hanson, F.A. Fullin and V.L. Schramm, J. Biol. Chem., 256 (1981) 10795-10802.
130 G. Colombo, G.M. Carlson and H.A. Lardy, Biochem. J., 176 (1978) 495-504.
131 M.F. Utter and H.M. Kolenbrander, The Enzymes, 6 (1972) 117-168.
132 C.H. Reynolds, Biochem. J., 185 (1980) 451-454.
133 L.A. Bentle and H.A. Lardy, J. Biol. Chem., 251 (1976) 2916-2921.
134 L.A. Bentle and H.A. Lardy, J. Biol. Chem., 252 (1977) 1431-1440.
135 M.J. MacDonald and H.A. Lardy, J. Biol. Chem., 253 (1978) 2300-2307.
136 N.S. Punekar and H.A. Lardy, Unpublished.
137 G.M. Carlson, G. Colombo and H.A. Lardy, Biochemistry, 17 (1978) 5329-5338.

G. Semenza (Ed.) Selected Topics in the History of Biochemistry: Personal Recollections (Comprehensive Biochemistry Vol. 36) © *1986 Elsevier Science Publishers*

Chapter 7

A Biochemist's View of his Struggle for Knowledge Review of Forty Years Service to Science

M. KLINGENBERG

Institute for Physical Biochemistry, University of Munich, Goethestrasse 33, 8000 Munich 2 (F.R.G.)

My prehistoric years in science

At the beginning, I want to apologize for the imperfections of the English in this personal review, since English is not my mother tongue. Still more so than when writing a technical paper, I realize how deficient my English is in expressing the more general themes pertinent to this review. Although a considerable portion of energy on my part is being diverted into the language struggle, I realize that the outcome is far from satisfying. Therefore, the reader is kindly asked to approach this review with patience.

In every scientist's life, there is a "prehistoric" period which is not recorded in his publications or thesis. This is the nebulous past in which the foundations for his scientific career were laid down. The introduction may permit me to provide a few glimpses into those pristine years.

My interest in science started with astronomy at the age of 13. In the next 2 years I began to get hold of popular and even serious literature, particularly the writings of A.S. Eddington. It was astro-

Plate 12. Martin Klingenberg.

physics which caught my fancy and I was following astrophysical calculations without really understanding them. After we had lost our dwellings and most of our belongings in the firestorms which deluged Berlin in November 1943, I came to the smaller town of Schwerin in Mecklenburg, where I became friends with a gifted, but polio-stricken boy who introduced me to chemistry. Shortly afterwards he died and I inherited his small amateur laboratory. This event lured me to chemistry into which I plunged now with all my energy, however, not without some farsighted regrets about leaving the more esoteric and eclectic world of astronomy and astrophysics. My path into chemistry was finally settled when at the end of the war after several years of deprivations, I was one of the few students to be admitted to the study of Chemistry at the Humboldt University in Berlin. By that time I had worked through some textbooks of chemistry, and I had, in long-hand, copied a large part of Pauling's *Nature of the Chemical Bond*, possibly one of the first copies in Germany, which was accessible in the American Information Center. My first encounter in those years with Biochemistry was through a reprint of Warburg's monography *Schwermetalle — Wirkungsgruppen von Fermenten*, which was one of the very few books printed in those impoverished times. In late 1948 I moved to the University of Heidelberg, where I developed an interest for physical chemistry. I graduated here in the field of gas kinetics; my diploma thesis was concerned with "Heat conduction through porous materials" and my doctoral thesis was about the "Determination of accommodation coefficient of various gases", and was actually published in the *Zeitschrift für Elektrochemie* [1]. Due to my keen interest in broader aspects of science, I spent extra time reading literature also on the borderlines of physical biochemistry. The work of Werner Kuhn on the kidney readsorption mechanism and his models on muscle contractions together with Katchalsky, the new mechanisms on nerve conduction, Lynen's work on the Pasteur effect and Otto Warburg's studies on photosynthesis — all of these intrigued me in those early 1950s and led me to ponder whether I might not work with one of these groups for my doctoral thesis or as a post-doc. However, for my Ph.D. work, I still stayed in physical chemistry before I moved into the biological field.

Particularly the work on the kinetics of catalase by Britton Chance made a decisive impression on me [2]. I avidly looked into new issues of the *Archives of Biochemistry* for new publications by Chance which struck me by their novel approach, reaching beyond the usual studies of enzyme kinetics. Feodor Lynen suggested to me that Britton Chance's spectrophotometric methods might be the best approach to oxidative phosphorylation. This strengthened my resolution to attempt to spend a post-doc period with Britton Chance and through recommendations by Richard Kuhn on his visits to Philadelphia, this could finally be settled. On the 14th of July 1954 the final doctoral exams took place and I finished my university studies. Before I left for Philadelphia, I met Theodor Bücher through the help of Benno Hess whom I knew from my study period in Heidelberg. It was during this encounter that the foundations for my work after my return to Germany were laid.

Philadelphia 1954 to 1956

In September 1954 I came to the Johnson Foundation for Medical Physics of the University of Pennsylvania on the basis of a fellowship administered through the National Academy of Sciences which was provided to young European Bioscientists. Britton Chance accepted me kindly despite my shortcomings in language, in knowledge of biochemistry and in experience of working in a crowded, very active laboratory. At that time he was actively pursuing, together with Ron Williams, the interaction of oxidative phosphorylation with the respiratory components of mitochondria. Among Britton Chance's suggestions I chose one less connected to mitochondria, namely the study of electron transport function in rat liver microsomes. These could be recovered as a byproduct from the supernatant of the mitochondrial preparations still obtained with archaic centrifugal conditions. To me it seemed to be a much simpler system of electron transport. In the beginning the components in the system were surveyed using the scanning "split beam"-photometer. With this I could obtain absorption difference spectra in turbid suspensions with high sensitivity, and in the next few months I was eagerly applying all the tricks of difference and double-difference spectra in order to unscramble single signals from the mixture of nature's still untangled systems. CO was available to identify terminal oxidases. When bubbling CO through the microsome suspension in which cytochrome b_5 was reduced by DPNH to my great surprise a large band at 450 nm appeared in the difference spectrum. It was clearly set apart from the γ-band of cytochrome b_5. The CO-band was much enlarged by reduction with dithionite. TPNH*, which later turned out to be the major natural reductant, was not available at that time.

Although the 450-nm peak was near the 445-nm CO-band of the oxidase in mitochondria, it did not seem to be a hemoprotein since no α-band was found. I also noted that on aging of the microsomes or on addition of cholate the CO-band shifted to 420 nm whereas the cytochrome b_5 reduction remained unchanged. By the hemochromogen method I showed that there was about 2.8 times more heme in the microsomes than could be accounted for by cytochrome b_5. With all these data at hand the conclusion seems inevitable that the 450-nm CO-pigment is a hemoprotein.

When I attempted to solubilize and purify cytochrome b_5 and the 450-pigment using the then available pancreatic enzyme mixture, David Garfinkel joined me, at the Johnson Foundation, in this work which, however, was bound to fail with the imperfect methods of those times. Only a few years later Sato and Omura cut the "Gordian knot" and concluded that the CO-pigment is a cytochrome, henceforth called cytochrome P-450. Ryo Sato also spent several months at the Johnson Foundation while I was there but worked at that time on cytochrome b_1 from *Escherichia coli*. During this period I also determined the flavin content in the microsomes as one FAD per cytochrome b_5 and it was concluded that the DPNH-cytochrome b_5 reductase was a flavoprotein. This work, performed in 1954 and early 1955, was published only in 1958 [1a].

* The term pyridine nucleotides (DPN, TPN, etc.) is used here in accordance with the publications referred to until approx. 1965. After that the nomenclature nicotinamide-adenine dinucleotides (NAD, NADP, etc.) was applied.

My main interest then turned to the operation of this apparently simple electron transport chain, in particular, to the question of how two electrons from DPNH and $FADH_2$ are transferred to the one-electron acceptor cytochrome b_5. Briefly, I collaborated with the late Enzo Boeri from Padova, then an exuberant scientist who was visiting the Johnson Foundation for a few weeks and worked with a cytochrome b_2 preparation from yeast. Just as in the liver microsomal system, a flavoprotein–cytochrome *b* complex seemed to exist here, too. Also, there was the question of whether two-electron type acceptors such as methylene blue, bypass the cytochrome *b* by reacting directly with the FAD and whether one-electron acceptors such as ferricyanide and cytochrome *c* have to use the cytochrome b_5 [3].

To solve these questions I looked directly at the oxido-reduction kinetics of cytochrome b_5. Using an early double-beam instrument belonging to Chance, the electron "sinks" in microsomes were titrated with DPNH, ferricyanide or cytochrome *c*, and the sequence of the electron transport components was analyzed from DPNH to flavin, to cytochrome b_5, and the b_5-oxidase. Since large amounts of microsomes were available I could use the newly constructed regenerative stopped-flow machine for determination of the single-step rate constants. Many nights were spent in mastering the complex instrumentation consisting of elaborate optics, the impressive but not infallible vacuum tube electronics and in dealing with the flow systems and the rapid recordings. The interaction of b_5 with ferricyanide or cytochrome *c* was so fast I had to lower the temperature down to −20° C with mixtures of methanol and glycol. Thus kinetic constants for electron transfer between iron pigments were determined [4].

A nonlinearity in the concentration dependence for NADH and cytochrome *c* was observed. This could be explained only by assuming at least two intermediary electron carriers between DPNH and cytochrome *c*. On this basis I derived a general treatment for a donor–acceptor binding system with flow schemes using knots. From here, using matrix systems, the rate equations were easily derived. Possibly due to my imperfect English, a drafted manuscript of this work and its application to the microsomal electron transport did not find the approval of Britton Chance and therefore was never published. Several years later, others by developing general enzyme steady-state kinetics used a similar formalism. I also derived the non-steady-state kinetic equation for the time sequence of up to 3 intermediates. All this experimental and theoretical kinetic work on which I spent 1955 and much of 1956 never saw the light of publication. This was partly due to my solitary way of working and a never corrected tendency to postpone writing down the results until they are pushed into second rank by the current work.

In working on these problems I learnt several things important for my later research. First, to deal with highly sophisticated instrumentation which was home-built and subject to continuous adaption and improvement, depending on one's own inventiveness. Second, following Britton Chance's example, if necessary, to push the measurements to the limits of sensitivity in order to enter new grounds. Another benefit from my stay at the Johnson Foundation was that I encountered many leading scientists passing through this very active place. Besides, I was working in direct contact with many good scientists such as Ronald Estabrook, Lucile Smith, Ronald Williams and the Baltscheffky's.

Marburg 1956 to 1967

In June 1956 I arrived in Marburg where Theodor Bücher was just setting up the Institute of Physiological Chemistry at the University. At that time research and research positions were still poorly funded in Germany. Bücher had obtained some funds from the Rockefeller Foundation to buy modern instrumentation and was also subsidized by the Deutsche Forschungsgemeinschaft because of his growing reputation. He had developed new procedures for purifying and crystallizing glycolytic enzymes and was now embarking on studies concerning their metabolism, regulation and compartmentation in animal cells. He developed new concepts about the flow of metabolites, in particular of reducing equivalents, by emphasizing the thermodynamic relationships in glycolytic systems. He called DPNH/DPN and TPNH/TPN hydrogen-transferring metabolites and ATP/ADP phosphate-transferring metabolites, regarding them as pools and pipelines for the transduction of reducing and phosphorylating equivalents. For this purpose he developed to perfection procedures for instantaneous quenching of a tissue and for precise optical assays based on coupling to the DPNH or TPNH system. Bücher had the idea of determining the redox potential of the DPN and TPN systems by measuring the contents of the equilibrating metabolites (see below). A favorite and particularly exciting tissue was the flight muscle of insects, particularly from locusta, which can produce two radically different metabolic states — at rest and in flight. The changes of the metabolic level were dramatic and readily showed the restriction of the metabolic flow at regulatory sites. Realizing the great value of these pioneering concepts and methods I soon became infected by the enthusiasm generated by Bücher.

Instrument building, the dual beam, oxygen electrode

When I left Philadelphia to go to Marburg, it was on the understanding with Bücher that I would use my experience to construct a double-beam instrument for developing the research on these new concepts, particularly on the mitochondrial metabolism. At that time vacuum tube electronics and suitable DC-amplifiers were not yet available and there were hardly any dealers in Germany representing the manufacturers of the needed US electronic equipment. So I had to go to the trouble of importing on my own many parts, for example operational amplifiers, stable DC-power supplies and so on, spending valuable time with the chores of building an instrument from scratch. The optics were constructed at the institute workshop following the basic Chance design; however, I used mirrors instead of lenses and high-intensity quartz monochromators. Part of the electronics were kindly built in the workshop of the Eppendorf Company in Hamburg according to my design, the electronic parts being supplied by me. The difficulties in getting the optics and electronics tuned were greater than anticipated and I solved them in several months of day-and-night work in 1957. In fact, my working day had 15 hours and my working week 7 days in those years. The performance of the double beam was then tested by optical assays on micromolar concentrations of DPNH, both in clear and highly turbid solutions. Only a small (8%) attenuation by the turbidity was observed and the absorbance resolution was a proud 10^{-5}. cm^{-1} [5]. Bücher was quite relieved, as he had become sceptical about the outcome of the investments, financed from his research grants.

Based on my experience with the "split beam" spectrophotometer I adapted the first commercial Beckman DK-1 recording spectrophotometer to record difference spectra of mitochondria, with a moderate performance. In adapting optics and electronics of the DK-1, I obtained the help of a service engineer from Beckman in Frankfurt, Werner Rudow, who later was to become president of Kontron AG in Zürich. I was also occupied with the construction of an oxygen electrode setup, making my own electrodes from platinum wire and experimenting with different amperometric arrangements and amplifiers. This was the first platinum oxygen electrode used in Europe and still based on the "open" form of the Johnson Foundation.

Insect flight muscle mitochondria, the "ideal" mitochondria

I entered the "mitochondriology" with the task of isolating flight muscle mitochondria with respiratory control. So far several laboratories had failed to obtain "coupled" insect mitochondria which led them to suggest that these mitochondria might not have this control which was known from liver and heart mitochondria. After successfully isolating coupled mitochondria from locust flight muscle and in addition to having built the first dual-beam spectrophotometer in the "old world", my status at the small institute in Marburg increased considerably [5]. Flight muscle mitochondria turned out to be a most suitable material for the double-beam technique since they surpass mitochondria from other sources because of their extreme specialization in energy production. The content of electron transport pigments is unusually high and therefore the absorption signals were strong and clear in relation to the noise. A new assessment was made of the wavelength differences suitable for recording the individual cytochromes, flavin and TPNH, determining the mutual interference factor and molar difference extinction coefficients [5]. These formed the basis for the later quantitation of the respiratory-chain components.

An extremely active respiratory substrate in insect muscle mitochondria was glycerolphosphate, as discovered first by Bücher and Zebe. Electron pressure from this substrate was so extreme that the pipe was filled to a very high degree of reduction of the cytochromes, facilitating the study of the cross-over points on switching between the "active" and "controlled" states. Encountering communicative difficulties with the nomenclature "state 3" and "state 4", etc., I introduced the more descriptive terms "active" (a) for the phosphorylating state and "controlled" (c) for the coupled state. Whereas with glycerolphosphate the cross-over was between c and a, the much weaker but still well-coupled DPN-linked substrates like pyruvate and malate had cross-overs between NAD and flavoprotein [6]. A most striking observation was that at the DPNH wavelength the signal was much larger with glycerolphosphate than with DPN-linked substrate [7,8]. Most important in the active and uncoupled state this signal disappeared, indicative of a high-energy storage state. These observations, supplemented by the enzymatic analysis of the DPNH/DPN led to the discovery of reversed electron transfer with which I will deal further below.

The intramitochondrial pyridine nucleotides, energy-linked transhydrogenase and reversed hydrogen transfer

Early on as a standard material for checking the performance of mitochondrial preparations and the equipment, I used mitochondria from rat liver. A special interest in liver mitochondria was expressed by H. Hübener, from the University of Frankfurt, who was studying the influence of gluco- and mineral corticoids on metabolism. Mitochondria seemed to be a possible target since their involvement in gluconeogenesis was also just becoming known. Indeed, when screening the many corticoid derivatives developed at that time, Hübener and I observed an increase of the mitochondrial DPNH absorption parallel to partial inhibition of respiration. These facts seemed to be well correlated to the glucocorticoid efficiency. Unfortunately Hübener suddenly died and these studies were never published. But they raised the question whether more TPNH than DPNH is formed, since mitochondria just had been found to contain TPN too.

With the double beam as an excellent instrument for highly sensitive enzymatic assays, I started with a medical student, Werner Slenczka, to determine the content of DPNH and TPNH in extracts from these mitochondria [9]. In order to maintain the reduced forms the alkaline extraction procedure of mitochondria was perfect for our enzymatic measurements of DPNH and TPNH. Less problems were found in assaying DPN and TPN in the parallel acid extracts. Several determinations of DPNH and TPNH reported in the literature suffered from serious errors. The variability of the sum of DPNH plus DPN suggested to Bill Slater's group the existence of DPN-X as an energized chemical intermediate of oxidative phosphorylation. In our assays the constancy of this sum, despite wide variations in the ratio DPNH/DPN, was a criterium for the performance of the whole procedure. Samples were taken during the continuous recording of the DPNH/DPN absorption changes of the mitochondria on addition of various substrates and of ADP, phosphate, inhibitors, etc. When assays were properly performed, the sum of DPNH + DPN content was actually found to be constant and the existence of a third form (DPN-X) was shown to be erroneous.

This delicate work done together with Werner Slenczka produced a number of new discoveries [9]: in liver mitochondria there is about twice as much TPNH as DPNH with the consequence that more than 60% of the absorbance signal is due to TPNH. DPNH is reduced maximally to 50–60% in the "controlled" and "anaerobic state", whereas TPNH reaches 80–90%. Previously the DPNH-system had been assumed to be 99% reduced in the controlled state by Chance and Williams [10] and the results were interpreted to reflect an energy-rich "DPNH-I". In the active state (state 3) DPN was found to be reduced to only 5–20% and TPN to 40–60%, depending on the substrate.

Most important was the finding that the ratio TPNH/TPN was in all states higher than that of DPNH/DPN. In the various states (viz. 1 to 5), controlled or active state, and with different substrates, a displacement of the TPN redox potential by −30 to −60 mV as compared to the DPN system was found. In the uncoupled state this difference collapses, both systems now have an equal redox potential. Thus it was shown for the first time that in mitochondria the transhydrogenation is energy-dependent.

Another great surprise was the finding that DPN was most extensively reduced with succinate. The preferential reduction of DPN by flavin-linked substrates was a striking finding based primarily on the extraction procedure for DPNH, etc. and the absorbance changes. Particularly strong was the effect in mitochondria from flight muscle, heart muscle and kidney. Here the TPN content is lower, in flight muscle it nearly vanishes. Correspondingly the absorbance changes due to DPNH between the energized controlled (state 4) and the active state (state 3) are much more dramatic. In insect muscle glycerolphosphate could reduce DPN to about 80%, whereas DPN-linked substrates only reached 10–30% reduction. This high level of reduction was very dependent on tight coupling and thus the "quality" of the mitochondria. This exciting finding led me to conclude that the DPN receives reducing equivalents from the flavin-linked substrates "uphill" by energy transfer.

The strong reduction by these very active electron donors indicated that the reduction pressure generated at the entry point in the respiratory chain from the flavin-linked substrate, later to be identified as the ubiquinone pool, was a more critical factor than the redox potential of these equivalents, provided that enough energy was generated. In flight muscle mitochondria the case was most drastic. Here, using glycerolphosphate, DPN was reduced about 10 times more than by pyruvate, etc. [8,11]. Moreover the reversed hydrogen transfer was less equivocal than with succinate since the oxidized product dihydroxyacetonphosphate is not a

DPN-reducing substrate, such as malate, generated from succinate. These two highly unpredictable findings were later to become the main indications for the existence of a phosphorylated energy pool in mitochondria.

Glycerolphosphate shuttle

The discovery by Bücher of the high glycerolphosphate oxidase in flight muscle mitochondria and his previous characterization of the cytosolic soluble glycerolphosphate dehydrogenase, led him to the concept of the "glycerolphosphate shuttle" [8,12]. Reducing equivalents from glycolysis enter the mitochondria via glycerolphosphate and the membrane-bound oxidase. The prerequisite for this shuttle was the impermeability of mitochondria to DPNH. In order to verify the distribution of the shuttle, I measured the glycerolphosphate oxidase in mitochondria isolated from various organs. In mammalian tissues (from rat) this oxidase is low, it is relatively active in brain, less in liver and least in heart. The significance of the shuttle seemed to follow this pattern of the glycerolphosphate activity. Somewhat later the malate-aspartate shuttle was conceived by Borst and others, which can out-perform the glycerolphosphate cycle in many mammalian cells.

Here I should mention a wide-ranging publication by Bücher and myself entitled *Hydrogen pathways of the living organization in the cell*, first published in 1958 in German and then translated into English [8]. The concepts on the organization of hydrogen transfer in the glycolytic system and the idea of a high-pressure TPNH system and low-pressure DPNH system were elaborated on the basis of experimental data and a derivation of the thermodynamic relations. Bücher introduced the idea of determining the redox potentials of the DPN- and TPN-systems in the cell by measuring the redox ratio of the metabolites in equilibrium with these coenzymes. The ramifications of this method were elaborated also in this paper. A smaller section was devoted to my results on the mitochondrial redox system, stressing the difference between the hydrogen pipelines in the two compartments. In mitochondria the

DPN-system has a more negative redox potential and most important, its reduction is geared to the phosphorylation potential in the mitochondria, whereas in the cytosol it is linked to the glycolytic phosphorylation and inversely proportional to the phosphorylation potential. Therefore, reducing equivalents from the cytosol cannot funnel hydrogen into the mitochondrial DPN system but close to the more positive flavin level for example via the glycerol phosphate shuttle. However, against its exclusive role argued the low activity of glycerolphosphate oxidase in most mammalian mitochondria.

This very concentrated publication laid a foundation for the quantitative analysis of metabolite systems in several laboratories. It soon became a "citation classic" and some noted scientists have remarked that they never read a paper more often and with more difficulties than this one.

Mitochondrial components, enzyme activities and stoichiometric relations

I now started several projects in parallel, possibly too many, so that a considerable portion of the results was never published. In order to understand more closely metabolism and its specific expression in various organs, Bücher proposed to measure enzyme distribution in terms of their activity. In this context a group of five glycolytic enzymes were found to occur in "constant proportion" in all organs, reflecting probably one common operon. My task was to determine some of these mitochondrial enzymes and components. Dirk Pette had developed a gradual extraction procedure which permitted differentiation between cytosolic and mitochondrial soluble and bound enzymes.

My first contribution was to develop a method for determining the content of the cytochromes. The contents of cytochromes a, a_3 and c rendered the relative respiratory capacities of various tissues [13]. Another important outcome was the possibility to use cytochromes as a quantitative parameter for the content of mitochondrial components in tissues. By relating tissue cytochrome content to that of isolated mitochondria one obtains the "cytochrome factor" = (cyt a, a_3/protein)$_{tissue}$/(cyt a, a_3/protein)$_{mit.}$ = mitochondria/tissue protein. This useful parameter shows that in rat liver, for example, mitochondria account for about 20% and in heart 45% of tissue protein. A further application of this factor was in the determination of the DPNH/DPN and TPNH/TPN distribution in cytosol and mitochondria after the extraction methods for the reduced and oxidized pyridine nucleotides had been modified for whole tissues [14]. Thus, I concluded that in liver more than 70% of the DPNH and 90% of the TPNH is mitochondrial whereas the DPN and TPN are largely cytosolic. These somewhat bold extrapolations were later nicely substantiated by others in fluorescence studies on perfused liver and by the non-aqueous extraction methods for separating intra- and extramitochondrial compartments.

The "comparative biochemistry of mitochondria" was extended further to the dehydrogenases for glutamate, malate and isocitrate. Interesting correlations were found which gave important clues to the actual function of equivocal relations.

These relations could be elegantly visualized in logarithmic plots where equal distances correspond to equal ratios although the absolute activities reach widely different levels in various organs. First developed by Bücher and Pette for describing the "enzyme patterns" typical for various metabolic branches, they were here applied also to the molar contents of pyridine nucleotides and cytochromes. Whereas in various mitochondria the content of malate-dehydrogenase (DH) is related to DPN content, that of glutamate (Glu)-DH is related more to that of TPN [15]. Also the content of DPN system and of DPN-linked DH is related to that of cytochrome a, but that of TPN is not [16]. These results led us to conclude that in mitochondria Glu-DH uses TPN and not DPN as a coenzyme although this enzyme is known to be active quite well with both coenzymes. This was substantiated by us later when studying the function of Glu-DH in mitochondria. The exclusive occurrence of Glu-DH in mitochondria made this enzyme also a mitochondrial marker in total tissue extracts. I calculated that in liver mitochondria Glu-DH contributes to about 5% of the total and thus is the most abundant single soluble protein. This correlates well to the very high TPN + TPNH content in these mitochondria. Also in adrenal mitochondria I found a high TPN content which was related to the high amount of TPNH-linked P-450 system [17]. Later Harald Goebell in my laboratory group found that also the DPN-linked isocitrate dehydrogenase (IDH) is in constant proportion to cytochrome *a* whereas the TPN-linked IDH is not [18].

While comparing different mitochondria also the content of the various cytochromes and the stoichiometric composition of the respiratory chain were determined. Schollmeyer and I found a simple stoichiometric composition which suggested a multicomponent chain complex: 1 cytochrome *b*, 0.5 cytochrome c_1 and 1 cytochrome a, a_3 as a constant-proportion group in mitochondria from liver, heart, brain, etc. [13,19]. In retrospect we find that this early determination of the respiratory chain composition is still valid even though today we know that cytochrome *b* and the cytochrome a, a_3 complexes contain two hemes each. The observed variability of cytochrome *c* is also in line with the shuttle function between the membrane-bound complexes.

Reversed electron transfer

Our interpretation of the reversed hydrogen transfer to DPN by a reversibility of energy transfer induced me to enforce a complete reversibility upon the phosphorylation and electron transport with ATP. The energy dependency would thus become more definitive. When putting these ideas to work dramatic and exciting effects were observed. On excluding oxygen, addition of ATP caused a widespread oxidation of the fully reduced cytochromes [20]. The effect was strongest with cytochrome a,a_3 and weakest with cytochrome *b*. It could be enhanced by placing hydrogen acceptors such as oxaloacetate at the lower end of the respiratory chain. Thus with ATP reducing equivalents were driven all the way from cytochrome oxidase into the dehydrogenase region. Most important, it was highly dependent also on ADP and phosphate. Quantitative titrations with ATP/ADP and phosphate versus the redox state of cytochrome *c* made it possible for the first time to determine a phosphorylation potential in equilibrium with the anaerobic respiratory chain [21].

The reversibility of electron transfer throughout the respiratory chain was instrumental for further understanding of the energy coupling process. The stored energy, intermediate between electron transport and ATP, is thus shown to be part of the electron transport per se. At this point I introduced the steady-state *redox patterns* of the respiratory chain, illustrating the gradients of redox state and also the interplay with energy pressure going from substrate to oxygen. A strong gradient of reduction emerged from coupled forward electron transport in the aerobic energized state, a reversed gradient in reversed electron transfer in the anaerobic state, and an even degree of reduction was found in the uncoupled state [22].

The reversed electron transfer was not only an academic experiment for revealing fundamentals of energy coupling but also turned out to be of cell-physiological significance in hydrogen transfer for TPNH, such as in fatty acid synthesis, the control of respiration not only by ADP but also by ATP level, and in hydrogen (DPNH) formation in those prokaryotes which feed only on high potential substrates.

The control of respiration by ADP, first shown by Lardy and Wellmann, and then analyzed in detail by Chance and Williams, was a milestone in understanding regulation of oxygen uptake. Our finding that ATP would inhibit respiration provided evidence for a more intimate control by the ADP/ATP ratio [23]. By quantitative appraisal, the phosphorylation potential interacting with electron transport, was determined to be $\Delta G = -15.5$ kcal [24]. The question of respiratory control, either alone by ADP or by the ATP/ADP ratio, has become a matter of strong controversy. We shall return to this point below in context with the ATP/ADP transport into mitochondria.

Considerable doubts were expressed concerning the reversed succinate-to-DPN-hydrogen transfer, especially by Hans Krebs and Nathan Kaplan, who upheld that DPN should be reduced instead by the malate generated from succinate. To make this uphill reaction more convincing we demonstrated an overall hydrogen transfer driven by ATP uphill from succinate to acetoacetate [25]. We determined the complete hydrogen balance in order to eliminate other hydrogen sources. For this reason the changes of all four metabolites and the ATP consumption were quantitatively determined. Considerable effort was spent to prepare succinate-DH and β-hydroxybutyrate-DH for these assays. A fairly even hydrogen and ATP balance was determined which finally convinced the critics of reversed hydrogen transfer.

Hydrogen transfer network in mitochondria

Having in this context also used other hydrogen acceptors such as oxaloacetate and ketoglutarate plus NH_3, I embarked on an investigation of the mitochondrial hydrogen network established by the intramitochondrial dehydrogenase. Hydrogen transfer to ketoglutarate plus NH_3 from malate and succinate, the formation of glutamate and aspartate were determined in aerobic and anaerobic systems, showing an ATP dependency for hydrogen transfer not only from flavin but also from the DPNH level to ketoglutarate [26,27]. The analysis of isotope-labeled metabolites led to the conclusion that Glu-DH is primarily coupled to TPNH rather than DPNH for reduction of ketoglutarate plus NH_3 since an energy requirement obviously reflects transhydrogenation to TPN. This was also borne out by the preferential oxidation by ketoglutarate plus NH_3 of intramitochondrial TPNH rather than DPNH. All this agreed with my previous findings: first, the energy-linked transhydrogenase; second, the high content of Glu-DH; third, the constant — proportion — relation of Glu-DH and TPN in various mitochondria; fourth, the preferential oxidation of inner mitochondrial TPNH by ketoglutarate plus NH_3; fifth, the energy dependency of hydrogen transfer for glutamate formation from malate. All these results led to the conclusion that Glu-DH in liver mitochondria is primarily destined for formation rather than oxidation of glutamate. As a result I postulated that the major role of Glu-DH in liver consists of ammonia sequestration for the formation of urea, which has now been widely accepted. These papers were in a way extremely sophisticated. The kinetics of all metabolite levels were scrutinized using the double-beam enzymatic assay technique, even at the μM level of oxaloacetate. On the basis of these data the time-related change of the involved substrate redox potential was analyzed and in a redox nomogram the influence of energy on the redox potential difference established, exhaustive work which took nearly 2 years but clearly established the physiological work of energy-linked transhydrogenation and of the TPNH/TPN system in urea synthesis.

Isocitrate-dehydrogenase regulation mechanism

In this context I studied the role of isocitrate dehydrogenase (IDH) as well. Ketoglutarate plus CO_2 was found also to oxidize intramitochondrial TPNH, although only weakly, obviously via the TPN-linked IDH (T-IDH). Under anaerobic conditions ATP addition was required for the formation of citrate from malate, again demonstrating the involvement of the energy-linked transhydrogenation.

In this context and also in connection with our studies on the dehydrogenase composition of mitochondria we embarked on the subject of DPN-linked (D-IDH) versus T-IDH isocitrate-DH. A controversy persisted in the literature about the occurrence of D-IDH in mitochondria. Whereas the group of Nathan Kaplan [28] and others maintained that there is no D-IDH, Lars Ernster [29] insisted on the presence of D-IDH in mitochondria. In all these studies, however, D-IDH was evasive because of its rapid inactivation in contrast to T-IDH and because it was not yet known that ADP is required as activator. By pure chance we discovered parallel to W. Plaut that ADP addition did not only preserve the activity in the mitochondrial extract but was in fact required for full activity of D-IDH.

Using these findings together with SH-protection we were able to retain full activity and could thus determine the distribution of D-IDH in mitochondria isolated from various organs [30]. We found that D-IDH occurs exclusively in mitochondria and is part of a constant proportion group of dehydrogenases which, in various types of mitochondria includes MDH, cytochrome *a* and DPN [31,32]. This clearly placed D-IDH into the oxidative pathway through the respiratory chain whereas T-IDH activity was found to be related to the TPN content in various mitochondria. These results settled a long-standing controversy about the occurrence of D-IDH in various types of mitochondria.

The highest content of D-IDH was found in insect flight muscle mitochondria (*Locusta migratoria*). Here we embarked on a systematic investigation of the regulatory properties of extracted D-IDH and discovered a sigmoid dependence on isocitrate [33]. These dependencies became linear on activating D-IDH by ADP, whereas without ADP full activity was reached at high concentrations of isocitrate and DPN. Most important was the finding of a very strong pH dependency of these K_m's. The data could be fitted by a rate equation with quadratic concentrations of DPN and isocitrate as well as of H^+ [34]. This was interpreted by the following assumption: the enzyme consists of two subunits, each of which bind isocitrate, DPN and H^+ on the catalytic site and ADP on the regulatory site. The sequence of binding is H^+-DPN-isocitrate. The subunits exist in an inactive and an active state. The interaction between the two subunits is strongly dependent on their protonization.

We arrived at these results in 1962 and 1963, independent of the "allosteric" models developed by Monod et al. In fact, allostery in its original sense was ruled out because full activity could be induced only from the substrate site, without using the regulatory or "allosteric" site. Unfortunately, these issues were later blurred, since the name "allosteric" was used in a sense different from the original concept. The work on the H^+ effect and the interaction of the two subunits, as well as the activation of IDH by ADP was the first investigation with such a far-reaching scope of regulation principles. The H^+ involvement in subunit interaction was also something new. An extensive paper was largely disregarded because it was published in *Biochemische Zeitschrift*, although it anticipated fundamental features of regulatory enzymes.

Fatty acid oxidation

In our comparative studies of mitochondria from various organs, the capacity of fatty acid oxidation was a particularly vexing aspect. Locust flight muscle mitochondria were expected to have a high oxidative capacity for fatty acids; however, a couple of years of efforts failed to demonstrate this capacity. Only after we noted from the work of J. Fritz that carnitine in muscle homogenates can stimulate fatty acid oxidation, we added carnitine also to isolated mitochondria. The activation of fatty acid oxidation in the insect mitochondria was so tremendous that at first we feared that by accident we had added glycerolphosphate. The ensuing studies by Christian Bode with mitochondria from a variety of organs showed that only insect flight muscle mitochondria have a complete dependence on carnitine for activation, whereas in heart mitochondria only the long-chain fatty acids depend on carnitine addition [35,36]. The dependence is still less expressed in parenchymatic organs, like liver and kidney, where carnitine requirement is low. Bode synthesized the carnitine esters of C_{10}, C_{12} which were found to be still more active than fatty acids plus carnitine. This permitted us to determine the fatty acid oxidation "capacity" in various types of mitochondria which was correlated to the cytochrome *c* or cytochrome *a* content.

In this context we worked out an assay for the carnitine transacetylase in order to measure its distribution in various types of mitochondria [37]. A good correlation of the carnitine transacetylase content with the dependence of the fatty acid oxidation on carnitine addition was found which was highest, for example, in flight and skeletal muscles and nearly absent in liver.

Comparative respiratory capacity of mitochondria and organs

In trying to relate the various oxidation values obtained from mitochondria to the tissue performance, I compared them with the working capacity and the maximum oxygen uptake of the organs as measured by the physiologists. The basis for this comparison was the "cytochrome factor", used to determine the mitochondrial content on a protein basis in various organs [13]. The ratio of respiration rate to cytochrome content was a "cytochrome" turnover compared to the oxidative capacity of the organs. For example, as a cytochrome turnover, fatty acid oxidation turned out to be evenly distributed in various organs with a somewhat lower rate in skeletal muscle or liver.

These data were evaluated in terms of the oxidative capacity and compared to the muscle performance in work and the physiological respiratory capacity. The relative inefficiency of the oxidative capacity in mammalian heart as compared to that of insect flight muscle was a striking result. These comparative studies were comprised in a review for *Ergebnisse der Physiologie* [19] (now *Reviews of Physiology*) but had been originally written in 1961 as *Habilitationsschrift* for the University of Marburg.

Mitochondrial creatine kinase

A sideline of considerable consequence was the discovery in our group by Hans Jacobs together with Hans-Walter Heldt of the existence of a mitochondrial creatine kinase [38]. Jacobs found this enzyme in mitochondria from organs metabolizing creatine phosphate, such as muscles and brain. He noted that creatine kinase is attached to the mitochondria in an easily dissociable state and concluded that the enzyme is located in the intramembrane space. Furthermore, Jacobs showed by elaborate electrophoresis studies that the mitochondrial creatine kinase is an isoenzyme to the cytosolic enzyme, both in muscle and nerve tissues. The existence of these two isoenzymes was quite intriguing and of course suggested to us that creatine phosphate instead of ADP may actually transport energized phosphate from the mitochondria to the myofibrils [19]. As a smaller molecule it should diffuse faster than ATP. This idea was then followed up by several other laboratories.

The mitochondrial phosphate pathway in substrate phosphorylation

Anion exchange chromatography with the highest possible resolution and sensitivity was a major developmental effort in the laboratory of T. Bücher, sustained by the developments of precise pressure pumps and thin columns by Heinrich Schnittker. Walter Heldt applied this nanomole scale anion chromatography to the analysis of phosphate metabolites in mitochondria. One early side track resulted in a new method for the determination of the reduced pyridine nucleotide, which was free from the fallacies of the alkaline extraction [39]. In the acid extracts of biological material the ADP ribose and ADP ribose phosphate were found in anion chromatography and shown to be correlated to the DPNH and TPNH contents. By comparing the acid extract analysis with the alkaline one, also the claim of a mysterious acid-labile TPN by Burch and Lowry was shown to be due to experimental error in my only publication in *J. Biol. Chem.*[40]. As mentioned above, the fallacious variability of the sum DPNH + DPN or TPNH + TPN was often misinterpreted to reflect the existence of an important third mystery form.

One major application of the high-resolution chromatography was the elucidation of the phosphate pathway in oxidative and substrate level phosphorylation. The starting kinetics of ^{32}P incorporation under various conditions were scrutinized with an automatic sampling device and then during chromatography the efflux of ^{32}P activity and UV absorption were monitored, followed by a half-automated phosphate analysis in the nearly 250 fractions from one extract. Thus we determined the path of ^{32}P incorporation into all mitochondrial phosphate metabolites. We found that in substrate level phosphorylation, ^{32}P was incorporated first into histidine and, after a lag, into GTP [41]. Subsequent transfer occurred into ADP and finally ATP. Of great interest was the GTP–AMP phosphate transfer which showed to us that intramitochondrial AMP can be returned only by the substrate level phosphate transfer into the ADP/ATP system [42]. This complemented our finding that the adenylate kinase is located outside the matrix space.

Also a parallel direct transfer from GTP to ADP could be discerned. This intramitochondrial "NuDiKi" had to be a regulatory enzyme since the ratio GTP/GDP was found to be higher than that of ATP/ADP and thus a "NuDiKi"-catalyzed equilibration between the two nucleotide systems did not seem to occur. The mitochondrial "G"-system obviously operates as a separate pipeline with a higher phosphate pressure than the "A"-system. This explains the separation of substrate level from oxidative phosphorylation.

Permeability of mitochondrial membranes

Up to 1960 the permeability of the mitochondrial membranes was terra obscura. Implications abounded in the literature but no real data were available, obviously because of experimental difficulties. Rapid separation was required for maintaining the solute contents. Millipore filtration was not the answer since mitochondria became anaerobic and they also leak under pressure. The best approach seemed to us the silicon-layer technique of Werkheiser and Amoor which, however, had been aborted by these originators because of experimental problems. Together with Erich Pfaff I tried to revive this "centrifugal silicon filtration", looking into the details of the filtering process. The postulate was to separate and to quench the mitochondria simultaneously to stop substrate interconversions; thus filtration into an acid layer was introduced [46]. Technical details had to be solved before the silicon method became routine. At first performed in siliconized glass vessels, it was later simplified by the advent of plastic centrifuge cups.

In the beginning we tried to differentiate between the permeability of the inner and outer mitochondrial membrane. The outer membrane was shown to be freely permeable to molecules of M_r up to 4000, but the inner membrane was impermeable even to small solutes [43,44].

Here the previous differential extraction for enzymes from mitochondria could be rationalized. In hypotonic medium the outer membrane was broken by swelling of the matrix and by increased ionic strength intramembrane proteins were released. Thus in liver mitochondria adenylate kinase was assigned by us to this space. Cytochrome *c* and creatine kinase were released only by higher ionic strength from this space [45]. The dependence of the now measurable matrix volume on the osmotic pressure was elaborated and thus the matrix was defined as the osmotic space. This was also correlated by electron microscopy [44]. Up to half of the matrix water was found to be osmotically inactive and probably bound to protein.

Substrate permeability

In context with our hydrogen transfer studies in mitochondria we now measured the distribution of the solutes linked to the reductive amination of ketoglutarate with malate using the silicon separation method. The distribution was determined for the substrates of three major interlinked reactions, the transaminase reaction, the glutamate-dehydrogenase and the malate-dehydrogenase. Clear differences in the distribution were made evident under deenergized and energized conditions by blocking respiration and adding ATP [47]. Notably for aspartate an energy-driven export from the mitochondria was shown which was later rationalized as the energy-driven aspartate–glutamate exchange. I conceived redox nomograms which permit one to discern the deviation of the reaction complex from equilibrium. In the ATP-driven reaction sequence the disequilibrium between the malate and the glutamate dehydrogenase became strikingly evident whereas the transamination equilibrated well [27]. The dual effects of energy on the transhydrogenase and on the aspartate export were thus demonstrated to be the key to the energy-driven NH_3 incorporation into glutamate which had been previously defined.

Whereas with the silicon filtration method we measured directly transport of metabolites through the mitochondrial membrane, its limitations in speed and kinetic resolution were obvious. Therefore, together with Ferdinando Palmieri in our laboratory, we used the swelling induced by valinomycin plus K^+ and its dependence on anions for following permeation into the matrix. The permeation of malate and of tricarboxylate was established and correlated to their effects on the oxidation and reduction of mitochondrial DPNH and TPNH. These studies were not completed because of the lack of valinomycin at that time. This first collaboration in Marburg resulted in further publications with Palmieri, then in Bari, on the permeation of dicarboxylates, tricarboxylates and phosphate into mitochondria (e.g. [48]). The dependence on the pH of the distribution was established and then the kinetics of the di- and tricarboxylate carriers were measured by the "inhibitor-stop" method. With John McGivan in Munich the problem was attacked to what extent H^+-movement accompanies the anion exchange system [49]. The malate against citrate exchange was shown to be electroneutral and it was concluded that the carrier picks up one H^+ specifically together with the citrate. We found a similar electroneutrality for the phosphate–malate exchanger [50].

During these studies we became interested again in phosphate transport and with the collaboration of Bernard and Martine Guérin, the "membrane sidedness" of inhibition by permeant and impermeant SH-reagents was studied [51]. The Guérins found a differential sensitivity of the carrier to SH-reagents depending on whether it performed in efflux or influx. This was an early indication for a redistribution of the carrier site, reflected in the SH-group, actually inferred from similar concepts developed for the ADP/ATP carrier shortly before (see below). In collaboration with Roger Durand the influence of the external and internal P_i concentration, and of the membrane potential and ΔpH on the SH-reactivity was analyzed, which led us to conclude that the carrier sites change their membrane orientation under the influence of H^+ and P_i concentrations [52]. In the proposed mechanism an H^+-binding group regulates the "switching" so that the deprotonized form of the carrier site is active but the protonized form is inactive. Only the H^+-carrier binds P_i forming the translocating P_i–H^+-carrier complex.

In later years the identification and isolation of the P_i carrier was pursued in other laboratories mostly by former collaborators or visitors, such as Ferdinando Palmieri, Bernhard Kadenbach, Hartmut Wohlrab, Roger Durand and Bernard Guérin. It appears that the problem in identifying and isolating the important P_i carrier is more formidable than with the ADP/ATP carrier.

The role of quinones in electron transport

After the discovery of ubiquinone (UQ) in mitochondria attempts in various laboratories to establish its role in electron transport led to quite ambiguous results. Spectrophotometric recordings in mitochondria and in extracts had led to wrong conclusions due to experimental errors. With our tradition of analyzing the redox state of pyridine nucleotides and as part of our work on the function of the respiratory chain it was a challenge to us to establish accurately the redox reactions of UQ in mitochondria. It was our goal to determine the redox change in mitochondria of UQ in extracts and to follow the redox kinetics spectrophotometrically. First together with Ludmilla Szarkovska [53] and then with Achim Kröger [54] the methods for fixing the redox state of UQ in mitochondria, for extraction, and for determination of oxidized and reduced UQ in the extracts were worked out. Considerable difficulties had to be overcome. A methanol–petrolether mixture was found to be a rapid quench and extraction medium which instantaneously isolated UQ from further redox reaction. Special equipment was constructed permitting precise withdrawal of aliquots and addition of quenching medium during spectrophotometric recording [46]. A high-intensity and ultrastable mercury arc was installed into our home-built dual-beam instrument which fortunately had high UV intensity quartz prism monochromators. Interference of cytochrome changes and mitochondrial DPNH and TPNH, which in other laboratories had obviously masked the UQ signal, had to be circumvented. The absorption changes of UQ could then quantitatively be correlated to the redox changes found in the extract on transition between the controlled, active, uncoupled and anaerobic states [52].

The redox state of UQ reflected faithfully the steady-state electron flow and perfectly fitted into our redox patterns of the respiratory chain. In fact, the redox state of UQ proved to be the most reliable component in reflecting the redox pressure and its control by oxidative phosphorylation with a wide variety of substrates linked to DPN, TPN or flavin protein. These results elevated UQ not only to a fully competent member of the respiratory chain but also established a "pool function" of UQ for the hydrogen from all substrates [55–58].

The redox kinetics were followed with the newly developed moving-mixing chamber [59] and the kinetic constants were in full agreement with the overall electron transfer, once we reckoned with the large size of the UQ pool. This pool aspect had hitherto not been accounted for by other groups which from seemingly slow kinetics concluded that UQ is not on the main electron pathway. In further experimental efforts, Achim Kröger constructed a quenched-flow apparatus for following with high resolution the redox kinetics of UQ by quench-extraction [50]. Here a quantitative correlation down to a 20 ms resolution of the reduction- and oxidation-kinetics to the electron flow rate was determined. In all these efforts we could clearly establish UQ as a direct member of the "electron transport". In particular, by showing that the redox state of UQ is well proportioned to the hydrogen donor activity of many different substrates, we conceived the "UQ pool" [55]. UQ collects and redistributes hydrogen from and to the dehydrogenases and to the cytochromes, by means of its large stoichiometric excess and the two-dimensional diffusibility on the membrane.

The redox potential of UQ was not yet known except at acid conditions. With our methods we could determine the UQH_2/UQ ratio in the mitochondria in dependence on the succinate/fumarate ratio and found a standard redox potential of 50 mV [61]. In the same studies the redox potential of cytochrome *b* was determined, based on the idea that the well known aberrations of cytochrome *b* kinetics are due to an equilibrium of cytochrome *b* with the UQ pool. In fact, for cytochrome *b* $E'_0 = 72$ mV was determined which was close to that of UQ and confirmed our concept. At this time one did not yet differentiate between two *b*-type heme groups. Also a pH dependence of the redox potential in cytochrome *b* was found, which indicated an ionizing group of pK 6.8 close to the heme. This group seemed to be a good candidate for a H^+-pumping function of cytochrome *b*.

The role of UQ was also exploited with a view to resolve problems on "branched" electron transport. Together with Tomoko Ohnishi we studied in *Saccharomyces* the entrance point of substrates which are not oxidized in animal mitochondria, D-lactate and external DPNH, and found that both fed directly into the UQ pool [62]. This made understandable their lower P/O. Another branching point, later investigated with Gebhard von Jagow, was the cyanide-insensitive respiration in *Neurospora* mutant mitochondria, which is also representative for plant mitochondria [63]. The cyanide-insensitive oxidase pathway was shown to branch from the UQ-pool and not from cytochrome *b* as hitherto postulated.

A further spin-off from the UQ pool concept was the explanation of the nonlinear titration of respiration with inhibitors such as antimycin A by Kröger and me [64]. The blockage of individual cytochrome chains by antimycin A could be overcome by UQ redistributing hydrogen to residual not yet blocked chains. This model was quantitatively developed, and with the rates of hydrogen entry and exit to the UQ pool a good fit with the sigmoidicity of antimycin A titration was calculated. Thus the UQ pool concept gave an elegant explanation for what has been "explained" by other groups as an "allosteric" effect.

In 1969, working with W. Mannheim in Marburg, I had turned my interest to respiratory chains in bacteria. We recorded cytochrome difference spectra from a wide variety of bacteria with a newly constructed high intensity "split beam" instrument. Striking at that time was the enormous cytochrome *c* content of *Paracoccus denitrificans* and its similarity to the mitochondrial respiratory chain. Our interest now turned to the function of the naphthoquinone derivative menaquinone (MQ; Vit. K_2) occurring in Gram-positive bacteria. First with Vladimir Dadak and then with Achim Kröger, the extraction and spectrophotometric methods were applied to determine the redox behavior of MQ in *Bacillus megaterium*. A favored object was *Proteus rettgeri* which is closely related to *E. coli* and contains both MQ and UQ [64a]. The two quinones were expected to play a separate role in electron transport and this was rational considering the fact that the redox potential of MQ is > 100 mV more negative than that of UQ. We suggested that analogous to the TPN and DPN system, MQ may be employed more in reductive processes whereas UQ is busy in the oxidative electron transport. We succeeded in separately recording the redox reactions of UQ and MQ in the *P. rettgeri* membranes and thus to assign to MQ the role of acceptor for the low-potential substrate formate from where it mediates hydrogen to fumarate between two dehydrogenases. UQ reoxidizes the produced succinate by transferring hydrogen to the cytochrome chain and to oxygen. These pathways function not only in sequence but also independently, e.g. in anaerobic bacteria. The further role of MQ in bacterial electron transport was then very successfully investigated by Achim Kröger.

The distribution of UQ and MQ in various bacteria was also of taxonomic interest. On the basis of the MQ and UQ contents in various bacteria, we devised an evolutionary history of quinone function according to which the obligate anaerobes, before the existence of atmospheric oxygen, did not contain quinones [58]. The advent of quinones was considered a criterion for membrane-bound electron transport and energy conservation. The first primitive electron transport systems utilized MQ with electron acceptors such as sulfate or fumarate. At a later stage, with higher atmospheric oxygen pressure, UQ evolved with acceptors of higher redox potential, such as nitrate or oxygen itself. Today some anaerobic bacteria retain transport exclusively by MQ whereas the obligate aerobes use only UQ.

Spatial organization of the respiratory chain

The studies on the composition of the respiratory chain including the cytochromes, quinones and pyridine nucleotides, as well as the content of dehydrogenases led me to calculate the occupation density of these components on the inner mitochondrial membrane. The average surface available for one cytochrome a, a_3, as reference, was calculated as 250 × 250 Å. The other components, including the coenzymes, were projected into this pattern showing that considerable distances have to be bridged between the more thinly distributed dehydrogenases and cytochromes in accordance with our concept of a diffusional UQ pool. We also suggested that electron transfer between the cytochromes might require lateral mobility in the membrane. This scheme was depicted on the jacket of the book *Biological Oxidations*, edited by Th. Singer [56], and was later extended to include also the ATPase and ADP/ATP carrier [58].

In this context we turned our attention to the transmembrane distribution of respiratory components [65,66]. The problem of the asymmetric localization of the respiratory chain had not yet been attacked. From differential extractions we had realized early that cytochrome *c* is localized on the outer face of the inner membrane [45]. Another approach we used was to test the accessibility to a respiratory component of a reactant or ligand. With Ferdinando Palmieri we noted that azide, before reacting with cytochrome oxidase, has to enter the matrix and concluded that the oxygen-binding site faces the matrix side [67].

Ferricyanide proved to be a very useful probe for sidedness. Having prepared ^{53}Fe-labeled ferricyanide with the silicon method, we showed that it does not permeate to the matrix [68]. In intact mitochondria ferricyanide reacted readily with cytochrome *c*. The ferricyanide reduction by substrates such as succinate was inhibited by antimycin A, since electrons passing from the dehydrogenase on the inside to cytochrome *c* on the outside go via cytochrome *b*. In contrast, in broken membranes the reduction was antimycin-insensitive due to direct interaction with the inner dehydrogenase. Surprisingly, my old favorite glycerolphosphate reduced ferricyanide in an antimycin-insensitive manner in all intact mitochondria tested. This implied that glycerolphosphate, generated in the cytosol, does not have to penetrate the mitochondria, which would have required an additional glycerolphosphate carrier. Whereas in animal mitochondria glycerolphosphate dehydrogenase is the only one localized on the cytosolic face, Gebhard von Jagow realized that in mitochondria from yeast and *Neurospora* dehydrogenases for NADH and NADPH are also localized on the outer face [63,99]. In these cases ferricyanide was reduced by NADH and NADPH in an antimycin insensitive manner. Also an intramitochondrial alcohol-DH, amounting to 6% of the total yeast cell activity was identified and we conceived the idea that the couple ethanol/acetaldehyde forms a hydrogen shuttle across the inner mitochondrial membrane for the transfer of the glycolytic hydrogen equivalents into mitochondria, in addition to the oxidation of external NADH found in mitochondria from fungi and plants. The common acceptor, also for the outer face dehydrogenases, is assumed to be the UQ pool which thus does not only collect hydrogen equivalents from lateral but also from transmembrane-distributed dehydrogenases [70].

Epilogue on my research on the respiratory chain

With these studies I finished my 15 year long extensive research on the respiratory chain in mitochondria. I may summarize a few major findings here: the energy-linked transhydrogenation in mitochondria, the reversed electron transport from cytochrome a, a_3 to NAD, the association of the redox state of a respiratory component with the ATP phosphorylation, the hydrogen transfer network in mitochondria, in particular the glutamate formation, the comparison of electron and hydrogen transport systems in various organs, the selective proportion constancy and functional association of dehydrogenases to DPN or TPN and cytochrome, the function of ubiquinone in the main electron pathway and as the hydrogen collecting pool, the localization of respiratory components across the inner mitochondrial membrane, including that of the dehydrogenases.

It is a highly diverse work laying foundations for many presently accepted functional relations in electron and hydrogen transport of mitochondria and its coupling to oxidative phosphorylation. Although nowadays the one sided "chemiosmotic" view dogmatically pervades the field, I believe that these early studies which required pioneering technical and experimental advances and efforts, will retain their impact.

ADP/ATP transport in mitochondria

The work on the ADP/ATP transport was to occupy increasingly my activities in the past 20 years. There were three methodological developments in our laboratory which gave us the know-how and potential to tackle the unsolved problem of how the adenine nucleotides interact with mitochondria. These were the quench filtration by the silicon layer method for the determination and segregation of intra- and extramitochondrial metabolites [46,71], the nanomole high-resolution anion exchange chromatography for nucleotides, and the sensitive enzymatic determination of metabolites and nucleotides in our dual wavelength spectrophotometer. In this unique position and with a great concerted experimental effort, Erich Pfaff, Hans-Walter Heldt and I were able to untangle the most important transport system of the eukaryotic cell, the ADP/ATP transport across the inner mitochondrial membrane.

Before we embarked on the adenine nucleotides as part of our efforts to determine the transport of metabolites into mitochondria as mentioned above, we turned to the nicotine amide nucleotides. Erich Pfaff demonstrated that NADH, NADPH, NAD and NADP only penetrate into the intramembrane but not into the matrix space of the mitochondria. This was not surprising but experimentally it was a challenge to perfect the silicon layer filtration technique since the redox state of intramitochondrial NAD is a most critical test for the filtration-quenching: a reduction of NAD would be a sign that mitochondria had become anaerobic during the filtration and that the original steady state, which was to be quenched and assayed, had not been maintained. In his diploma thesis, Pfaff described the appropriate conditions and showed that the oxidized and reduced pyridine nucleotides did not penetrate to the matrix space.

When we turned to the adenine nucleotides and found that in contrast to NAD, etc., ADP and ATP were taken up, we were sure that we were attacking something highly significant. At the 6th IUB Congress in New York in 1964, and the subsequent unforgettable "Compostium" on Britton and Lil Chance's farm in Paoli, we announced that mitochondria specifically take up external ADP and ATP in exchange against their intramitochondrial counterparts. Eventually the ADP/ATP transport system was to turn out to be the best defined and most powerful of the mitochondrial translocating systems.

Our key paper which opened up this field with numerous features of the ADP/ATP transport was presented at the first "Bari meeting" in 1964 [45]. In this paper also the differentiation between the matrix and the intramembrane space and the localization of inner membrane space proteins such as adenylate kinase, cytochrome *c*, etc. were presented. Also the inner membrane was defined as an impermeable barrier and therefore as the locale for the metabolite translocators.

In the first years we were busy solidifying the evidence for the exchange system and establishing its role in the main pathway of oxidative phosphorylation, a work which consisted often in refuting prevailing contrary opinion [43,71,73,74]. The transmembrane phosphorylation concept could be excluded by using mixtures of [^{14}C]ATP and [^{32}P]ATP in the silicon filtration. Of greater importance was the relation of the exchange to the endogenous nucleotide pool which had already been well characterized by H.-W. Heldt. This posed an important experimental challenge since three methods had to be combined, semi-automatic sampling, filtration of these samples and the chromatographic separation and determination of the ^{14}C-labeled ADP or ATP and of the ^{32}P incorporation. The endogenous ADP was shown to be phosphorylated prior to the exogenous one and the endogenous ATP was dephosphorylated prior to exogenous ATP [75]. These results proved that the endogenous ADP/ATP pool is an obligatory intermediate in oxidative phosphorylation where the exogenous nucleotides enter or exit by the counter-exchange. This question was most vital for the significance of the translocation data since these were derived from exchange with the endogenous pool.

With these findings we were now also able to demonstrate clearly that atractylate (ATR) is an inhibitor of the ADP/ATP exchange. ATR had been known for several years as an inhibitor of oxidative phosphorylation and was studied by the groups of Bruni in Padova and Vignais in Grenoble. There were peculiar and unexplained features of this inhibition. Our results showed directly that ATR does not interact with the ATPase but only with the hitherto unknown exchange [43,72]. ATR and the homologue carboxyatractylate (CAT) proved to be valuable tools in further studies of the ADP/ATP exchange system.

Measuring accurately the kinetics of the ADP/ATP exchange has been and still is quite a challenge. The equilibration between the external and internal, differently labeled nucleotides has only a short half-life, since the inner pool is often small. Some mitochondria easily lose the nucleotides; as a result, the existence of transport-deficient or inhibitor-insensitive mutants of yeast mitochondria has been erroneously claimed. Even today, reliable data on ADP/ATP transport rates of mitochondria are most difficult to obtain. In the literature data are reported which often do not correctly account for initial rates or pool sizes involved.

A great progress was our introduction of the "inhibitor-stop-method" where the transport is interrupted by addition of an inhibitor, here ATR [74,76]. This facilitated enormously manual or automated exchange studies. With great developmental efforts sampling methods became more refined up to the point of a complete preprogrammed rapid mixing and inhibitor-stop-sampling apparatus which permitted a sampling of discrete steps at 80-ms intervals. Development of this apparatus which also employs the "moving-mixing chamber" and of a more sophisticated derivative, a rapid sampling and pressure filtration machine, took over 5 years [77–79]. It kept not only the mechanical workshop occupied but also the electronic one for programming the intricate valve systems. RAMQUESA and RAMPRESA are crowned by Aztek clay figures which somehow fit those acronyms and should feel obliged to be the good ghosts of these sensible apparatus.

Energy control of ADP/ATP exchange

At an early stage we became intrigued by an influence on the ADP/ATP exchange by the energization of the mitochondria [48,78]. ATP was shown to be rejected, in contrast to ATP, by energized mitochondria whereas in the uncoupled state ATP and ADP were equally accepted. Already early, we conceived the idea that this control is due to the membrane potential, which utilizes the difference between the ATP^{4-} and ADP^{3-} charge [74,80]. Only a few years later it was possible to investigate also the control of the efflux with the "RAMPRESA" which allowed continuous sampling and separation of the internal and external space during exchange [81,82]. Labeling the internal space with ^{14}C and the external space with ^{3}H, the complete balance of in- and efflux of external vs. internal ADP and ATP was determined. In the energized state ATP was clearly found to be exported much faster than ADP.

Thus the energization of the membrane obviously directs the limited capacity of the ADP/ATP carrier primarily into the "productive mode" in line with the oxidative phosphorylation [83]. Mostly, regulation is associated with changes of K_m. However, in our assays the changes of K_m were found to be small, in contrast to the results of others. Our analysis revealed that primarily the translocation velocity is changed, in line with the concept of an electrophoretic control by the membrane potential $\Delta\psi$ [80,83]. The carrier-binding center (C^{3+}) was visualized to form an electroneutral complex with ADP^{3-} (C·ADP) and a negatively charged one with ATP^{4-} (C·ATP), by offering three positive compensating charges.

Evidence for the electrical nature of the exchange was obtained by following cation movement together with ATP transport [84]. For some time it seemed that the exchange was partially electroneutral which seemed to conform with the large energy requirement [80]. However, after accounting for all corrections when probing the electrical transport with valinomycin or uncoupler, a 1:1 K^+ or H^+-movement per ATP transport was determined [85,86].

A most important, at first profoundly surprising consequence was a higher ATP/ADP ratio outside than inside the mitochondria [80,84,87,88]. When blocking ATP synthesis the ATP/ADP ratio outside may be up to 30-fold higher than inside. When the $\Delta\psi$ is abolished with uncoupler this difference collapses [88]. The difference in ADP/ATP ratio across the mitochondrial membrane was determined together with Hagai Rottenberg to be proportional to

$\Delta\psi$ with one negative charge slope, in agreement with the other data [89].

We concluded that the energy-dependent regulation not only utilizes more fully the ADP/ATP transport capacity but also raises the free energy of external ATP compared to internal ATP. The difference in energy imposed by the ATP export as derived from four different types of data converged to about 2–4 kcal [87]. Consequently, about 25% of the free energy of external ATP originates from the export. We realized that this causes a considerable competition for energy with the ATP synthesis. As a result the picture of the P/O ratio or of H^+/e^- stoichiometry as postulated in Mitchell's chemiosmotic theory would be revolutionized. Either the P/O ratios are lower than the conventional 3 or 2, or the postulated two H^+ per synthesized ATP are erroneous. We exposed this problem already in 1969 at the FEBS Prague meeting, but of course met with great resistance by P. Mitchell, who believed that he had demonstrated an electroneutral ADP/ATP exchange.

We concluded that mitochondria in the eukaryotic cell appear to prefer producing less ATP at a higher potential than anticipated [80]. It was now clear that the ADP/ATP translocator connects the ATP/ADP systems between the cytosol and the mitochondria which operate at widely different potential, similar to the situation shown by us for the NADH/NAD system a decade earlier. In eukaryotic cells the cytosol and glycolysis have evolved to an extent where they operate at a higher ATP potential than the prokaryotic type intramitochondrial ATP system (see also review in [89a]). This fact is crucially linked to the number of H^+ per ATP or e^- in energy coupling. With 2 H^+ per coupling site, extramitochondrial ATP would require a total of 3 H^+ and an increase of 50% in energy requirement compared to the original one which seemed rather high and challenged the chemiosmotic scheme. Later by the discovery in other laboratories that a ratio H^+/e^+ of 3 to 4 H^+ was generated per coupling site, the proportion became more reasonable. With 4 H^+/site only 25% of energy would be used for the ATP export. However, in this case the hydrogen/electron loops of chemiosmosis were to be abandoned.

Definition of carrier sites

Already early I became convinced that only the molecular approach, i.e. defining carrier sites and eventually isolating the carrier protein, would advance our understanding of the transport mechanism. We started by determining the binding of ADP and ATP to mitochondrial membranes which were depleted of endogenous ADP and ATP [90]. These studies had to be performed with extreme care in order to exclude interference by ATPase and adenylate kinase. The evasiveness of the nucleotide binding was quite troublesome although we now understand that it was a feature typical for the carrier-type nucleotide binding. Only due to the heroic efforts of Maurice Weidemann and Herwig Erdelt these studies led to reproducible and highly significant results. For differentiating the binding of ATP or ADP to the carrier from those of other sites and from uptake into the mitochondrial nucleotide pool, we used ATR in an intricate double-check system. The total number of binding sites for ADP or ATP were found to be astonishingly high. This encouraged us to pursue the molecular approach to define the carrier in the membrane.

Of great consequence was a novel interpretation of the observed heterogeneous high- and low-affinity binding based on the concept that a carrier site can reorient between the two membrane surfaces. The low-affinity sites $K_d = 10^{-5}$ M were thought to look outward, whereas the smaller portion of the "high"-affinity sites ($K_d = 10^{-6}$ M) was reoriented inward[90]. Together with the internal nucleotides these "suck up" external nucleotides and this gives an apparent high affinity, although the actual affinity may be as low as that of the sites looking towards the external surface. In line with this interpretation the ratio of low- to high-affinity sites can vary and the low-affinity sites disappear when the inside is made accessible by extraction of phospholipids.

These studies were based on the assumption that ATR displaces the nucleotide from the carrier site. At this time, a quite capricious inhibitor of oxidative phosphorylation surfaced, bongkrekate (BKA), a truly exotic compound from Indonesia, where it ravaged the Javanese population. Being a 3-carboxylic isoprenoid branched fatty acid, it has an unusual structure quite different from ADP or ATR. We realized early that BKA behaves quite strangely because its inhibition of exchange and its influence on ADP binding were dependent on temperature, pH and time of incubation [91]. BKA was mysteriously more effective when ADP or ATP were present [92]. Instead of displacing ADP, it even increased ADP binding. Using complex combinations of the ligands BKA, ATR and ADP, we scrutinized a portion of binding increase which corresponded to carrier sites and not to nucleotide uptake. In these triple controls, it was found that BKA apparently increased ADP binding exactly to the same extent as atractylate displaced ADP [93]. The titration looked like an apparent affinity increase for ADP. Consequently a ternary complex between ADP and the carrier and BKA was thought to exist, in which the regulating site influences the affinity for ADP either positively by binding BKA or negatively by ATR. Whereas this dual site concept is still maintained by other groups, we soon realized that the BKA influence might be interpreted in an entirely different way.

The molecular basis of carrier action, the reorientation mechanism

At this point, with all these intricate results at hand, I had one of those illuminating moments of the type one reads about when scientific discoveries are made. It is interesting but not atypical that I had this enlightenment while listening at a small meeting in the Black Forest to a lecture on an entirely different subject and thinking at the same time about my own upcoming contribution. It suddenly struck me that the maze of BKA effects could be elegantly rationalized by assuming that BKA binds to the carrier only from the inside of the membrane whereas ATR binds only from the outside. With one stroke this assumption could explain all the data. The prerequisites were that ADP or ATP can bind from both sides and that the carrier site can switch from the inside to the outside only when loaded with ADP or ATP. Also we postulated that there is one common center in the carrier for binding ADP and ATP, as well as the inhibitors ATR and BKA [94]. Only the modulation of the carrier site when going from the outside to the inside changes its affinity from ATR to BKA. If we had interpreted these binding data in the classical manner, we would have had to accept high- and low-affinity binding equilibria with the ligands and superimposed allosteric effects changing the affinity of these ligands by the inhibitors. Instead, by focusing on the essential function of the carrier-binding sites, namely the switching from one side of the membrane to the other, we were able to give these binding phenomena an entirely new meaning. Thus we were not only collecting static data about affinities but instead gaining insight into the dynamics of the translocation mechanism.

The idea that BKA first had to penetrate the mitochondrial membrane before reacting with the carrier, came from our studies on the striking pH dependence of BKA action. Not only respiration, but also the time dependence of ADP binding were strongly increased by lowering the pH [94,95]. A quantitative insight was obtained by observing a strange, not yet fully explained, swelling-shrinkage of heart mitochondria associated with the nucleotide carrier [96]. BKA caused a shrinkage, i.e., a turbidity increase and its rate was shown to follow with approx. 2.8 × pH. This means that the rate would increase nearly 100-fold when lowering the pH by one unit. We explained this by assuming that only the acid form BKA-H_3 is permeant to the membrane. The relatively low concentration of BKA-H_3 in equilibrium with BKA^{3-} is limiting the rates of the BKA effect. BKA^{3-} remains the actual ligand. In fact, in inverted membranes the BKA inhibition is pH independent and there is no time lag of BKA-binding.

The apparent increase by BKA of ADP-binding affinity could now be explained to reflect the reorientation of binding sites which take up ADP from outside and then turn to the m-side, where the ADP is replaced by the high-affinity ligand BKA. Thus the bound ADP represents ADP which is released to the inside of the mitochondria where it is trapped because the carrier is blocked by BKA. Exactly so much ADP is trapped inside as there were originally carrier sites on the outside. This demonstration of the crucial translocation step on the molecular level was made possible through the fortunate supply by nature of side-specific inhibitors which permitted us to place all carrier sites either into the m- or the c-state by transition into inhibitor-trapped states.

All these studies relied on labeled ADP or ATP. It took us a long time to prepare labeled ATR and BKA. For a number of years, with some difficulties, we grew seeds from atractylis on [^{35}S]sulfate, but obtained only low yields of a useful, however, short-lived [^{35}S]ATR [97,98]. Later by other procedures, we succeeded in obtaining [^{3}H]ATR, [^{3}H]CAT and finally also [^{3}H]BKA [99]. In extensive binding studies on the interaction of all these ligands it became clear that the number of binding sites for CAT and BKA is about the same as for ADP and ATP. Whereas the ADP and ATP binding is evasive and easily decreases on aging of the membranes, that of CAT is more persistent and has proved to be the most reliable means for determining the number and whereabouts of carrier binding sites. Never two different ligands were found to bind to the carrier at the same time [100]. With the help of all these data we substantiated our earlier contention that all ligands occupy the same site at the carrier. Moreover, these results substantiated the reorientation mechanism of the carrier site in all its various implications. If we had invoked "allosteric" positive and negative effects of the inhibitors on ADP/ATP binding site, as others did, we would have missed this major finding.

A striking visualization of the transition between c- and m-state of the carrier, mentioned above, is the structural change observed in heart mitochondria on bringing the carrier sites either to the c or to the m state [96]. Thus, when adding ATR, the c face of the membrane appeared to be enlarged as compared to the m face and vice versa with BKA. This enabled us to study the kinetics of the transition between the c and m states and also the catalytic influence of ADP and ATP. Also the partial distribution of the carrier between c and m states could be observed from these data. Whether the configuration change of the membrane is due to changes in the surface potential or to a slight physical transposition of the carrier molecule from the c to the m side, remains an open question. The high density of carrier molecules in the membrane might cause this change of the membrane curvature. The kinetics of the redistribution of the carrier sites under the influence of BKA and ATR are then fitted with an enzyme computer program through the kind support of Benno Hess in Dortmund, based on the reorientation mechanism [101]. A quite good fit was reached with the actually observed kinetics of the configuration change.

Many of these results and concepts were published only in conference monographs, in particular in the series of the Bari meetings. Much of this work did not gain the attention of normal publications. A nearly unforgivable delay in publishing these important data permitted parallel literature to flourish which often due to experimental problems resulted in different interpretations. Some of these were, however, converged on our models. A first review was published in 1970 in *Essays of Biochemistry* [50]. A more detailed account of the ADP/ATP translocation work appeared in 1976 in *The Enzymes of Biological Membranes* [102].

The isolated ADP/ATP carrier

It took a couple of years — starting in 1970 — of trying and failing, but learning, before we succeeded in isolating the first intact carrier protein, which was the ADP/ATP carrier of mitochondria [103,104]. The learning process involved three measures, each of them employing novel approaches to the purification of membrane proteins. First, how to keep the protein intact and identifiable. Second, how to select the appropriate solubilization procedure. Third, how to purify the protein without denaturation.

The criterion for isolating the intact protein was the binding of [^{35}S]CAT in a specific saturable manner. Deceptive binding with [^{35}S]CAT was observed in the cholate and/or ammonium sulfate extracts commonly used [105]. Other "solubilizers" en vogue, such as organic solvents, phospholipases, etc., also failed. Only when we scrutinized the actual steps where the binding was lost, we had a handle to establish the proper conditions and detergents for solubilization. An important measure for success was the prior loading of the carrier with [^{35}S]CAT while still in the membrane before solubilization, to protect the carrier against denaturation or proteolysis. Later also other forms, in particular the BKA-carrier complex, were isolated, although they were much more labile [106].

Detergents

Instead of using the widely popular cholate, we finally arrived at using the — at that time — disfavored nonionic polyoxyethylene detergents, in particular Triton X-100. Proper handling and understanding of the physical properties of detergents in their application to membrane proteins were just emerging. Our philosophy was first to isolate the protein in an intact state, disregarding the disadvantages of the nonionic detergent which deterred so many from its wider application, and then, at the second state, to develop ways and means to overcome the other difficulties. This approach eventually fully paid off.

The experience with the ADP/ATP carrier gave us a handle for classifying detergents according to their usefulness in protein extraction. This protein provides a more subtle scale than most other proteins. We found a graded stability of the carrier in the CAT-, BKA- and unloaded forms. The most useful detergents in our hands have been Triton X-100 and, to some extent, lauroyl-propyl-aminoxide [107–108a].

The isolated protein

When, one day in March 1973, Paolo Riccio told me that he saw [^{35}S]CAT binding in the supernatant by the dialysis method, there was at first disbelief after 2 years of frustration until we established specificity of binding by chase with "cold" CAT, etc. From that day on we had a confident measure for following the protein during further gel chromatography. A great progress was the introduction of the hydroxylapatite step which together with Triton, enormously facilitated the purification procedure.

The ADP/ATP carrier protein turned out to be rather small, with an M_r = 30 000, but there seemed to be only one molecule

CAT bound per 2×30 kDa, indicating a dimer structure [103]. The dimeric structure of the isolated protein was then demonstrated by hydrodynamic studies [109], as will be told below.

In a search for the carrier in other sources by enriching the [^{35}S]CAT carrier complex with hydroxylapatite, we found that the M_r in liver and kidney is the same as in beef heart [110]. A higher M_r $\simeq 34\,000$ was found in *Neurospora* mitochondria isolated by Heinz Hackenberg. Our main interest in *Neurospora* focused, however, on biogenesis of the ADP/ATP carrier. With the well-proven criteria developed in Bücher's laboratory by W. Sebald and W. Neupert, it was shown to be synthesized in the cytosol [111]. This of course is understandable in view of the obligatory eukaryotic origin of the ADP/ATP carrier. In subsequent years, in the able hands of Walter Neupert, the ADP/ATP carrier proved to be a treasure house for understanding the insertion of membrane proteins into mitochondria because of its abundance and its particular properties.

The ADP/ATP carrier was estimated to amount to 15% of the inner mitochondrial membrane protein. It is the most abundant single peptide in beef heart mitochondria. Actually it amounts to about 3% of the total protein of heart. We concluded that the ADP/ATP carrier is the most abundant membrane protein in the total organism of a mammal, which would be in line with its function as a catalyst for the most important transport process in the aerobic cell; most important not only in functional but also in quantitative terms. We also realized that the high amount is there to compensate for the relatively low turnover of this carrier, and thus to meet the high ATP export rate into the cytosol. The low turnover is understandable if one considers that the ADP/ATP carrier squeezes through the membrane, not only a large but also a highly charged solute. When I inaugurated the first EBEC Meeting at Urbino in 1976, I presented a slide comparing the ADP/ATP carrier activity with carriers transporting smaller solutes such as lactose, glucose and chloride, showing a good correlation of the turnover to the M_r of the solutes. The activation energy required for translocation can be expected to increase with the size of the transported solute.

Antibodies

With Bob Buchanan we embarked on raising antibodies against the ADP/ATP carrier. Under certain conditions the protein proved to be a strong antigen and, being both novices in this technique, we were quite proud about the strong positive immune response [112]. The antibodies have an interesting conformation specificity only for the CAT-protein complex. The antibodies did not interact with the binding of CAT or BKA itself and did not inhibit transport in the mitochondria [113]. Our engagement with antibodies was to continue in the following years with Wolfgang Eiermann. We raised antibodies also against the BKA-protein complex, which was a much more difficult task [110]. With this antibody we could even record the transformation into the BKA-protein complex due to the turbidity.

Several years later Heinz Schultheiss developed more quantitative assays for the antibodies, employing two-dimensional immunoelectrophoresis and radioimmunoassays with protein A. The unusual electrophoretic properties of the ADP/ATP carrier required the development of a charge shift electrophoresis in the immunoprecipitation [114]. The organ specificity of the antibodies, first noted by Eiermann, was thus further quantitated [115].

Due to my long acquaintance with Peter Berg from Tübingen, who collected sera without antibodies against mitochondria from patients with primary biliary cirrhosis, pseudolupus, etc., I cultivated the idea that the ADP/ATP carrier may be a primary antigen in these sera. This reasoning was based on the abundance of the carrier and on indications that it is the major protein protruding to the outer surface of the inner membrane.

With Heinz Schultheiss strong antigenic activity was in fact observed in these patient sera [116]. Moreover, Schultheiss found strong carrier antigenic activity in sera from dilated cardiomyopathies. Most interesting was an organ specificity of these sera such that the liver disease sera reacted with the liver carrier and the cardiomyopathy sera reacted only with the carrier from heart. Thus a major defined antigen was found for mitochondrial autoimmune antibodies. These findings raise considerable interest with respect to clinical diagnostic application, in particular, for a differential diagnosis of cardiomyopathy. For us it was a long anticipated result which showed how fundamental research can turn out to have unexpected clinical significance.

The c/m-state transition

In view of the concept that the binding center can face either the c side when blocked with CAT or the m side when blocked with BKA, we were intrigued to isolate the carrier also as BKA complex. Then we would have an isolated nucleotide carrier in two different translocation states. Although the BKA-protein complex proved to be labile, we were rewarded when BKA could be specifically replaced by CAT in the solubilized protein and found that, most importantly, this replacement was dependent on catalytic amounts of ADP or ATP [106,117]. Exactly this dependence on ADP was the clue that we were reproducing the specific translocation step in solution, because only the ADP-carrier complex should be able to undergo the transition from the m to the c state. Particularly useful in assembling a functional ADP/ATP carrier in solution was the isolated ATR complex. It is nearly as stable as the CAT complex; however, ATR can fairly easily be replaced. Thus the transition from the c to the m state can be performed by replacing [^{35}S]ATR with [^{3}H]BKA. In this way, ATR-protein proved to be the best source for an m state carrier since it can be easily stored. We are still using this source for most of our functional studies.

SH groups are often relevant in biomembrane carrier action. The ADP/ATP carrier at first appeared to be resistant to SH reagents until it was discovered that the functional SH group appeared only on addition of ADP or ATP to mitochondria. This cooperation could be prevented by prior addition of CAT. Based on these findings in other laboratories we soon discovered that the SH group appearance has all the characteristics of the m state [102]. Not only did NEM induce the shrinkage of mitochondria typical for the m state, but also it did not disturb BKA binding, indicating that the SH group is not directly at the binding site. Under all circumstances the unmasking of the SH group was dependent on the addition of ADP, a condition also characteristic for the transition to an m state. Actually Heinrich Aquila had used the ADP-induced NEM incorporation to first identify the 30 kDa band in mitochondria as the ADP/ATP carrier [105]. Most importantly, the appearance of the SH group could be used to monitor directly the transition to the m state by incorporating DTNB. The dependence on the catalytic amount of ATP was thus impressively visualized [118,119]. Also, with other fluorescent SH reagents the transition to the m state could be recorded in this way. Although the SH group emergence was the best defined indicator of a conformation change, also lysine and arginine groups were found to appear to be more exposed in the m state. For example the incorporation of the arginine reagent phenylglyoxal is facilitated and the sensitivity to trypsin is increased [120].

Reconstitution of the ADP/ATP carrier

One of the most important goals when isolating a protein is to investigate its proper function. In the case of a carrier this requires reinsertion into artificial phospholipid vesicles. Reinhard Krämer embarked on this problem first by isolating the carrier in the unloaded, i.e. uninhibited state using a short-cut hydroxylapatite batch procedure [108]. The 70% enriched carrier warrants reconstitution assays where with confidence the activities can be assigned to the ADP/ATP carrier. By means of CAT and BKA binding, the number of carrier molecules incorporated in the lipid vesicles could be determined [121]. Moreover, the sidedness of the incorporation could be established in this manner and thus accurate accounting of the incorporation was achieved, in contrast to most of the unqualified reconstitution claims in this and other systems en vogue at that time. Only about 6% of the carrier nucleotides were right-side out and fully active. Inverted molecules as defined by BKA inhibition were only weakly active, and the rest was inactive [122].

In these systems we carefully screened, step by step, many

parameters influencing the transport activity. At first only the binding activity with CAT and BKA was reconstituted and only by trying several incorporation and vesicle preparation procedures did Reinhard Krämer succeed, with a combined sonication step after the freeze/thawing, in getting reconstitutional activity. The activity was strongly dependent on the phospholipids, favoring phosphatidylethanolamine and cardiolipin [123]. The inactivity in phosphatidylcholine impressively demonstrated the structural parameters of the phospholipid composition in facilitating the carrier translocational conformation change.

The reconstituted system was particularly useful for investigating the influence of membrane potential on the ADP/ATP transport. By imposing a potassium diffusion potential on the vesicles, the modulation of the exchange rate could be studied in detail in dependence on many parameters, such as the size and the direction of the membrane potential, as well as concentration of ADP/ATP inside and outside [124]. A linear response to the membrane potential was found, indicative of a simple electrophoretic effect on the transport. The transport would even respond linearly to the inversion of the membrane potential although the carrier molecules are asymmetrically inserted in the membrane. Reinhard Krämer's scrutinizing analysis of the complex kinetics showed that the K_d's which are inherent in the K_m's were not changed under the influence of the membrane potential. Instead the V_{max} was modulated and responded to the electrical field [125]. Moreover, he showed that only the ATP-carrier complex senses the field and is either accelerated or decelerated by the membrane potential. This agrees with the proposal that one extra negative charge, which results from the binding of ATP^{4-} to three positive charges at the binding center, is instrumental in the response to the membrane potential. Apart from this center there appears to be no other machinery built into the carrier to "regulate" the response to the membrane potential. This was the first case where a membrane potential influence on a reconstituted carrier system was verified and quantitatively analyzed.

Structural characteristics

After purification, the ADP/ATP carrier exists as a mixed protein detergent micelle. For any structural studies by hydrodynamic methods it was most important to measure the Triton bound to the protein, since it will largely determine the hydrodynamic properties. An unusually high detergent binding of 1.5 weight ratio to protein indicated a large hydrophobic surface in accordance with the embedding of the protein in the membrane [109]. Heinz Hackenberg surmounted the obstacles in the ultracentrifugal sedimentation analysis due to high Triton UV absorption which obscured the protein. It was an ingenious solution to use the Triton absorption itself to follow the sedimentation of the protein micelle. This was possible at a relatively low Triton concentration and in the UV scanner the Triton micelles could be clearly separated from the slower sedimentation of the mixed Triton/protein micelles. Thus the M_r of the total micelle was found to be larger (160 000) and after subtraction of Triton and phospholipids an M_r of 64 000 for the protein was determined, which would be double that found in the sodium dodecyl sulphate (SDS) gels.

This proved to us that the carrier is a dimer which was suggested first on the basis of a functional M_r per binding site that was twice that found for the peptide. With these various hydrodynamic data a carrier-Triton micelle was "constructed" as a large oblate ellipsoid with a short axis corresponding to the twofold axis of the protein which is surrounded by a large annulus of 160 Triton molecules. These cover with their hydrophobic heads the hydrophobic surface of the protein and extend to the aqueous surroundings their polyoxyethylene chains producing the large Stokes' radius of about 65 Å.

Establishing the amino acid sequence of the ADP/ATP carrier was quite a difficult task for our group and an enterprise hesitantly embarked upon, as there was no experience in sequencing and moreover, the facilities were lacking. At any rate, Heinrich Aquila, at first assisted by several colleagues, finally completed the amino acid sequencing after 5 years in 1982, after surmounting many difficulties with the blocked N-terminus, the hydrophobic cleavage products, etc. [122,126]. It was the first conventional sequence of a translocator or carrier protein. Striking were the many hydrophilic or polar amino acids distributed over the sequence, making the protein relatively hydrophilic, which contrasts to its large hydrophobic surface covered by the detergents or phospholipids. There are numerous ion pairs and a large excess of basic amino acids, in particular of lysine.

Secondary structure predictions, based on the various methods developed for soluble proteins gave quite contradictory results and are obviously not useful for membrane proteins. On the other hand, by circular dichroism 42% α-content was determined, similar as with other nucleotides binding soluble proteins. Only three hydrophilic stretches of 18–22 residues were found which are candidates for the popular hydrophobic transmembrane helices. Integrating hydrophobicity plots detected two further candidates for transmembrane α-helices. More information about the folding of the sequence across the membrane was obtained by a new application of the membrane-impermeant lysine reagent pyridoxal phosphate. Specifically in the ADP/ATP carrier, the high frequency of lysine residues made this lysine probe very fruitful. For example, by incorporating this reagent from either the "c" or the "m" side in the original membrane and then isolating the carrier and determining the location of reacted lysine in the sequence, one detects their membrane sidedness. "Neckline" lysines are covered by phospholipid headgroups and uncovered to pyridoxal phosphate in the nonionic detergent micelles. Also some lysines were attributed to line the hydrophilic channel through the membranes. Only one lysine at position 22 could be associated with the ligand binding. Thus a tentative folding picture of the ADP/ATP carrier could be derived [128,128a].

Carrier mechanism

Since our studies on the interaction of ligands with the ADP/ATP carrier in the membrane, we had several unexpected opportunities to gain new insights into the carrier mechanism. Herefrom we developed models of carrier action which were not only based on deductions from kinetic measurements, as found widespread in literature, but on ligand interaction at the molecular level [83,113,117,129]. As mentioned above, the reorientation mechanism is based on the interplay between binding of ADP and ATP with CAT and BKA. Although there were ramifications and more precise attributions, the original concept of one reorienting binding center has basically been unchanged, because it could accommodate all the further data with the isolated protein. This mechanism was also designated as the "gated pore" mechanism [113]. Although early I had called it a "mobile carrier", it was always meant and stated to be a mobile-carrier binding center in a stationary membrane protein. The classical mobile-carrier cycles of Wilbrandt and others could in my view just as well reflect a mobile site instead of mobility of the total protein.

My model differed from the assumption of two or more binding sites along a diffusional channel across the carrier, as it was for example discussed for the glucose carrier and is still for the ADP/ATP carrier by others. Whether the carrier is in the original membrane, in the isolated state or in the reconstituted vesicles, the seemingly complicated responses to various probes and ligands can be explained by a single flexible binding center. There is, however, a fundamental difference between inhibitor ligands which stick to one state of the binding center, either the "c" or the "m" state and the substrates ADP and ATP with their binding to the variable state. This basically simple model explains a wide variety of binding phenomena observed with the membrane bound and the soluble carrier. To others these results seemed to indicate a more complicated model. Sometimes the diffuse concept of allostery was used without any explanatory value to the translocation mechanism. Regulatory sites have been invoked by others with allosteric effects on the binding and thus the crucial switching of the binding site between the two states was missed which is at the heart of biomembrane transport. Also with amino acid reagents two different binding sites were derived although a more carrier-typical explanation is that some residues are active only in the "m" state, thus preventing transition to the "c" state and binding to CAT. In this way, amino acid reagents serve to determine the transition and distribution between "c" and "m" state.

The consequences and implications of the single site gated pore to the oligomeric structure of membrane proteins were discussed by me in a review article in *Nature* [129]. The effect of half-site reactivity and the implied oligomeric structure suggest the central twofold axis to be the locale of the gated pore. Half-site reactivity would simply result from a sterical obstruction by one ligand of the opposing subunit site. Whether this aspect is a universal principle also for other transport systems, seems to be doubtful. However, the single-site gated pore principle, being independent of dimeric structure, seems not to become accepted also for other solute carrier and even for ion pumps, although in no case the evidence is so clearcut as in the ADP/ATP carrier because of the two-state inhibitors.

For any catalytic process the energetics involved are essential. In particular ligand–protein interactions are key elements in understanding enzyme catalysis. Already in 1973 I pondered about the various forces involved in lipid-carrier-ligand interactions [101], but recently I more precisely elaborated the specifics of carrier analysis versus enzyme catalysis [130]. Energy for the activation process in carriers is required mainly for the enormous conformation changes of the protein, since the substrate does not have to be deformed for the reaction, as in enzymes. In both cases a strong intrinsic binding energy of the ligand to the protein is utilized for the activation and as a result the effective affinity for the substrate is decreased. For carriers, inhibitors are side-specific or "c" or "m" state-specific, whereas in enzymes the best inhibitors are often transition state analogs. In carriers, inhibitors complement more the ground state, forming an "abortive-ground state", whereas in enzymes they complement the transition state of the substrate which does not exist as such in carriers. This concept is in line with my early proposal that chemically different inhibitors complement either the "c" or the "m" state of the same site. I visualized that by its very flexibility the conformation of the carrier site follows a very specific trajectory in an energy space which defines the translocation process, but which also can complement the structure of the antibiotic-type inhibitors as they are developed by nature exactly against these carrier site conformations.

The key element in the carrier catalysis is the postulated flexibility of the topochemistry of the binding center. The accompanying conformation changes should be large for more bulky substrate and therefore be strongly pronounced in the ADP/ATP carrier. The study of these changes is an important part of our efforts toward understanding the translocation process. There are several kinds of evidence for these changes in the ADP/ATP-carrier conformation, such as by physical methods, e.g. fluorescence, circular dichroism by immunological methods, and by some amino acid reagents. In general the “c” state appears to be more tight than the “m” state. ADP or ATP binding appears to “labilize” or “fluidize” the structure as a result of the translocation activation, whereas inhibitor binding “rigidizes” the protein. With ADP and ATP there is a continuous oscillation between the “c” and “m” states, whereas with inhibition the carrier is fixed in extreme “c” and “m” states.

The “uncoupling protein” from brown fat mitochondria

In brown fat mitochondria with azido-ATP a specific component was labeled by Nicholls and Kemp, which had an M_r similar to that of the ADP/ATP carrier [131]. This finding triggered my interest in isolating also this membrane protein. The similarity in M_r and in nucleotide binding suggested to me that structural and functional analogues might exist between the ATP carrier and the GTP-binding protein. However, only several years later could I embark upon the isolation of this protein with the collaboration of Dr. Chi-Shui Lin from Shanghai. First we had to establish a continuous source for brown fat in our laboratory such as cold-adapted hamsters. The idea of a similarity to the ADP/ATP carrier worked well for the isolation of the “uncoupling protein”, setting another example for the highly successful principle of membrane protein isolation, using nonionic detergents and hydroxylapatite [132,133]. Fortunately, this protein still retained nucleotide binding during the isolation and this was an important assay during the purification and for its intactness. Most interesting was the similarity of the protein-detergent micelle to that of the ADP/ATP carrier as shown by hydrodynamic studies [134]. The protein also proved to be a dimer of 2 × 32-kDa subunits. Also the number of nucleotide-binding sites indicated half-site reactivity and a functional dimer similar to that for the ADP/ATP carrier.

An easily accessible functional property of the protein was the binding of nucleotides which was extensively studied by Chi-Shui Lin [133]. This binding has very strong pH dependency, indicating the involvement of the ionisation of the carboxyl group for the binding. The pH dependency certainly is a vital function for the regulation of the protein by the nucleotide binding. Also the competition of anions with the nucleotide binding was unusual. The kinetics of the binding proved to be surprisingly slow, taking 1 to 2 min at pH 6. The parallel inhibition is also slow and H^+-dependent [135].

Fortunately, also the "DAN"-nucleotides fluoresce on binding to the uncoupling protein. However, differently from "free" ATP, this binding is undelayed. Interestingly, binding of DAN-ATP does not inhibit, which indicates that a slow, conformationally conditioned "deep binding" of ATP leads to the inhibited state, whereas this transition cannot be triggered by the ATP analogue. In general, the comparison of the nucleotide interaction with the uncoupling protein and the ADP/ATP carrier turned out to be very rewarding, more so than I originally expected [130]. For example, in UCP, ATP inhibits and binds tightly, whereas in AAC, ATP activates and binds only loosely. Obviously in UCP less conformational energy and consequently debinding are involved.

Reconstitution of the UCP in vesicles encountered many difficulties until we drastically changed the methods. Since the UCP was sensitive during isolation against the high "cmc" detergents which had to be used to form vesicles by detergent removal, it was protected by addition of the phospholipids, required for the vesicles, prior to detergent exposure [136]. A rapid protein isolation and detergent removal procedure gave phospholipid vesicles with high H^+ transport activity, which was attributed to UCP by its sensitivity to GTP, etc. The UCP-containing proteoliposomes could be well defined in terms of number and direction of molecules incorporated. The H^+-transport activity was found to be linearly dependent on a membrane potential induced by a potassium gradient. However, there was little pH dependency suggesting that the carboxyl group involved in nucleotide binding is not involved in the H^+-translocation process. There are many controversies regarding the regulation of the protein which could be answered in the reconstituted system such as control by fatty acids, chloride, CoA, etc. The linear response to the membrane potential, similar as to that in the ADP/ATP carrier, again showed that an electrophoretic effect on the maximum rate controls the transport location velocity. This may be a general principle of membrane potential control on translocation and quite different from an unlinear regulatory effect, for example as proposed for the ATP-synthesis.

The amino acid sequence of UCP was established by Heinrich Aquila with a newly constructed automatic sequencer of considerably higher sensitivity [137]. Sequencing was relatively straightforward and easier than for the ADP/ATP carrier due to the free N-terminal and a quite favorable distribution of the peptide-splitting sites. Whereas at first the quite different amino acid composition was a disappointment to our view that UCP and AAC may be similar, the sequence comparison turned out to give quite striking similarity. On three different levels this suggestive homology was seen, the tripartite structure of three "repeats" about 100 residues long was again found in UCP, similar as in AAC. This suggests that a common ancestral structural gene for about 100 residues underlies both AAC and UCP. Second, a quantitative comparison using homology indices for the amino acid residue gives a good similarity between the two proteins. Third, in hydrophobicity plots also a similar distribution of hydrophobic and hydrophilic regions in the sequence of the two proteins was found, suggesting a similar folding through the membrane.

Proton transport through membranes has been associated mostly with H^+-channels such as in the F_0 section of the ATPase. However, in the H^+-cotransport carriers such as phosphate, glucose or lactose, a single turnover cycle is taken for granted in parallel to the substrate translocation. It may very well be that the UCP is analogous also to other H^+-cotransport carrier proteins, in particular to the phosphate carrier, and that therefore the UCP is a rudimental H^+ carrier just by omitting a substrate-binding site. By developing this idea, several difficulties on the role and function of UCP are resolved such as its low turnover, it being a H^+ carrier, not an OH^- carrier. Although UCP is a more simple H^+ carrier, it may actually be derived from H^+-substrate cotransport systems rather than the other way around. UCP may be the simplest H^+-transport system yet known. Thus the conditions for understanding H^+ translocation in UCP are now very favorable because of so many structural and functional data at hand.

Magnetic resonance studies

In a laboratory of physical biochemistry last not least magnetic resonance methods are also applied to biochemical systems. First with Dieter Brunner, but then for the last 8 years with Klaus Beyer we established a nuclear magnetic resonance facility with continuous frequency and multinuclear capability. Based on our work of solubilizing membrane proteins, Klaus Beyer studied the mechanism of detergent phospholipid interaction using H and ^{31}P resonance. Most suitable was a detergent, "LAPAO" which has similar distinct H-methyl signals as lecithin. With a paramagnetic shift reagent Pr^{3+} the accessibility and the sidedness of the headgroups on the membrane can be probed [141]. Results show that the detergent is incorporated into the bilayer as a whole micelle and with increasing concentration forms larger nonvesicular structures until mixed detergent phospholipid micelles are formed.

In further studies the incorporation of the protein-detergent micelle into bilayers was analyzed using the phospholipid headgroup ^{31}P signal [142]. Phosphorous resonance permits the differentiation of multilayer larger liposome structures from small vesicles. Furthermore, protein–phospholipid interaction permits the differentiation of those vesicles which contain protein. These show that the protein addition strongly decreases the vesicle size when added to larger monolayer or multilayer liposomes. The results indicate that under the influence of the protein, monolayers are pealed off from the liposomes when the detergent is removed by polystyrene beads. The resulting vesicles contain the protein often in dense arrays as seen by freeze-etching electronmicroscopy.

Also electron spin resonance equipment was installed. Spin-label probes were applied, for example by synthesizing the carboxyatractylate spin label (CAT-SL) for the ADP/ATP carrier [143]. This is one of the rare cases where spin label has been attached to a specific ligand of a protein. The extreme hyperfine splitting beyond the "rigid limit" indicates binding of the label in an ionic environment probably fixed by hydrogen bonds. On solubilization and by high temperature the splitting is decreased. Additional "less immobilized" signal component also strongly increases with temperature and is attributed to a librational anisotropic motion which becomes isotropic at higher temperature. Also saturation transfer data agree with the concept of high immobilization of CAT at the AAC binding site.

Prospects of membrane carrier research

Although our knowledge of the mitochondrial ADP/ATP transport system is relatively advanced, we are far from understanding what I consider the most central issue of biomembranes, the mechanism of carrier catalysis. The major obstacle is the lack of atomic structure of the ADP/ATP carrier, since no X-ray-amenable crystals are yet available for any carrier, despite considerable efforts. But there are other approaches to the molecular analysis based on comparison of primary structures. Increasingly, sequences are projected into secondary structures, particularly for membrane proteins, based on hydrophobicity of the residues. With more sequences becoming known, their comparison will permit more solid structure and folding predictions. Again, the comparative principle, so vital for biology, will be an important guide since it reflects the history and direction of evolutionary development towards the function-determined structure. For example, the surprising similarities in the primary structure between the ADP/ATP carrier and the uncoupling protein gave us strong guidance to significant structural features which are obviously common to both proteins and which possibly are also structural principles for other carriers. However, primary structures of the family of mitochondrial carriers may become available only when a systematic search for their genes has become feasible. All this may help to understand the atomic structure of the carriers and these predictive methods may become more important the longer we have to wait for a crystal structure analysis. Also the other procedures used for structure–function analysis such as site-directed chemical modifications and the application of monoclonal antibodies, etc. may help us in probing the primary structure and protein surface for functional residues and relationships.

A conclusion

Nearly all my scientific activities had to deal with membranous systems. Starting with "microsomes", I entered the field of mitochondria which I never left until today. Here at first I dealt with the components, organization and function of the electron and hydrogen transfer and then of the phosphate transfer on a broad scale. A particular effort was directed at the analysis of the nucleotide metabolite systems in hydrogen and phosphate transport. From here I went into membrane transport by first discerning the specific transport systems and then defining the carriers, followed by research into the molecular basis of transport.

My research has more often than not been coupled to technical developments. When attacking a new problem, I first tried to concentrate on new experimental solutions. These were aimed at obtaining data which would directly answer the questions pertinent to the research problem. For example, instead of relying on swelling or contraction of mitochondria, I developed methods to directly measure entry and release of substrates of mitochondria

TABLE I

	Subject — Methods — Development	Discoveries — Results	Ref.
1954	Electron transport in microsomes.	CO-binding pigment (450 nm).	1a
	Kinetics stopped flow.	DPNH-reductase-cytochrome *b*-oxidase	4
1957/8	Insect flight muscle mitochondria.	Respiratory control.	5
	High UV in dual beam construction.	Redox pattern of cytochromes. μmol enzymatic UV-assays.	
1958/61	Role of glycerolphosphate in respiration.	Glycerolphosphate-hydrogen transfer shuttle.	8, 11
	Mitochondrial pyridine nucleotide systems in liver.	Redox pattern of DPNH/TPNH systems. Energy-linked transhydrogenation.	7, 9, 14
	Distribution and function of hydrogen transfer systems in mitochondria from various tissues.	Energy-linked DPN-reduction by flavin-linked substrates — reversed hydrogen transfer.	
1961/7	Distribution of cytochromes, enzymes and coenzymes in mitochondria from various organs.	Proportion constancy between DPN, cytochromes, DPN-linked DH. Variability of TPN-linked systems. Constant relation of TPN to Glu–DH.	15, 16
1961/2	Cytochromes *c* and *a* composition and content in mitochondria various organs.	Cytochrome factor for mitochondrial tissue content. Stoichiometric composition of respiratory chain.	13, 19
1959/61	Reversed electron transport in cytochrome chain.	ATP/ADP ratio dependence. Concept of phosphorylation potential — redox equilibration.	20, 21
1962/6	Hydrogen transfer between DPN-, TPN-linked systems in mitochondria.	Energy-linked H-transfer from succinate to acetoacetate.	25
	Role of Glu–DH.	Energy-linked transhydrogenation to TPN-linked Glu–DH.	26, 27
		Function of Glu–DH as TPN-linked reductase and in NH_3-sequestration for urea synthesis.	

(Table I, continued)

1963/5	DPN-linked isocitrate dehydrogenase (D-IDH), assay and distribution in various mitochondria.	Proportion constancy D-IDH to cytochromes in various organs.	18, 30
		Activation mechanism of D-IDH by ADP and H^+. Concept of a regulatory enzyme dimer. Differentiation of regulatory and catalytic site.	34, 32
1962/5	Moving-mixing chamber for rapid kinetics in absorption cuvette.	Kinetics of electron transfer in respiratory chain. Ca^{2+}-, K^+-, inhibitor-effects on cytochromes, UQ and NADH.	59
1963/5	Mitochondrial creatine kinase.	Creatine kinase associated with mitochondria from muscle and brain. Isoenzyme. Location in intramembrane space.	38
1961/5	Nanomole assays for intramitochondrial nucleotide pool and P_i-transfer pathways. Kinetics, sampling methods.	Pathway of substrate phosphorylation via histidine phosphate and GTP-AMP/ADP transferases. Intramitochondrial ADP as primary acceptor for oxidation phosphorylation.	39, 42, 73
1961/4	Methods for determining intramembrane and matrix space of mitochondria, for distribution of substrates, for permeability of outer and inner membrane, for determining transport rate. The silicon layer filtration.	Outer membrane permeant to molecules of $M_r <$ 4000. Inner membrane impermeant to DPN, DPNH, etc. Energy-linked export of aspartate, transmembrane gradient of DPN- and TPN-linked redox system. Specific permeability of ADP and ATP, impermeability for AMP.	43, 45, 46, 71
1963/70	Inhibitor stop methods for kinetics of ADP/ATP translocation, sampling, multiple stops, etc.	Specificity, kinetics, site of atractylate action as inhibitor of exchange. Exchange imposes properties on ox. phosph. in mitochondria, i.e. specificity, high temperature dependence, Mg^{2+} independency, etc.	71, 72, 74, 76
1966/72	Sidedness of respiratory-chain components. The ferricyanide method.	Cytochrome *c* external of inner membrane cytochrome oxidase site. Glycerolphosphate–DH external, all other DH internal, NADH–DH yeast, *Neurospora* external.	65, 66, 67, 68

(Table I, continued)

1966/71	Respiratory chain, in *Neurospora* and yeast.	External NADH–DH, NADPH–DH. UQ as branching point for CN insensitive respiration.	63, 69
1966/70	Methods for quenching and determining the steady-state redox level of UQ in mitochondria. Determination by chemical and absorption recording.	UQ functions in electron main path. Pool function of UQ as hydrogen collector. UQ as a most faithful follower of steady-state redox pattern of respiratory chain.	53, 54, 55, 58, 60
1961	**Molecular distribution of respiratory components on the membrane surface, UQ diffusable coenzyme**		56, 58
1969/72	Transport of anions in mitochondria, cotransport of H^+.	Citrate–malate exchange as an electroneutral H^+-cotransport. Phosphate-H^+ cotransport requires H^+ at P-carrier.	48, 49, 51, 52
1968/71	Electron transport in bacteria, role of menaquinone.	Menaquinone functions more as a reducing hydrogen coenzyme.	64a, 58
1969/77	Simultaneous recording of K^+, H^+ movement and measurement of ADP/ATP exchange kinetics.	Electrical nature of ADP/ATP exchange H^+ or K^+ to ATP 1:1 stoichiometry. ATP export requires 1/5 to 1/3 of total phosphorylation energy.	84, 85, 86, 80
1969/79	Differentiation ATP/ADP ratio inside and outside.	$(ATP/ADP)_e/(ATP/ADP)_i = < 30$ in energized mitochondria. Linear dependent on $\simeq 0.85 \times \Delta\psi$.	88, 84, 87, 89, 80
1969/72	Carrier as control of ox.phos. Hexokinase-trap titration.	$\Delta G_{ATP\ e\text{-}i} \leqq -3.5$ kcal	80, 87
from 1968	Definition of ADP/ATP carrier (AAC) sites in mitochondria by differentiation of ADP/ATP binding with atractylate. Interaction of BKA with ADP/ATP binding.	High density of AAC-binding sites, transmembrane flexible binding sites. Bongkrekate (BKA) contrary to ATR seems to enhance affinity for ADP/ATP. Reorientation mechanism of binding site. BKA binds from inside, ATR from outside.	90, 91, 92, 93, 94, 95

(Table I, continued)

1972/4	Macroscopic changes of inner membrane induced by ligands to AAC. Light absorption recording.	Kinetics of c to m state transition. BKA penetrates only as $BKAH_3$ c m state distribution.	96
1972/6	Synthesis of [^{35}S]ATR, [^{35}S]CAT, [^{3}H]BKA/[^{3}H]CAT. Binding to AAC.	Determination of affinity and binding to AAC. 1 to 1 displacement of ATR/ATP/ADP/BKA.	97, 98, 100
1972/4	Development of methods to isolate intact carrier (membrane) proteins with Triton X-100 and hydroxylapatite.	Purification of AAC as stable CAT–protein complex. M_r 30 000, CAT binding 1 mol per 2 × 30 kDa.	103, 104, 107
1975/8		Isolation of the BKA–AAC-complex. Transition between m to c state (translocation step) in solution.	106
1975/8	Ultracentrifugal analysis based on Triton UV absorption.	High Triton binding to protein, M_r of micelle = 165 000, M_r of protein = 67 000. Dimer structure of AAC. Oblate ellipsoid Triton protein micelle.	109
1975/6	Antibodies against AAC.	Conformation specificity to CAT–protein complex. Organ specificity.	112
1975/9	Rapid mixing-quenching and filtration equipment with discrete sampling ("RAMPRESA and RAMQUESA") and "programmed" time course.	Differentiation between ADP and ATP during in- and out-transport. Energy influence on ADP versus ATP export. Internal ADP/ATP pool as intermediary of oxidative phosphorylation and transport.	77, 78, 79, 81, 82
1977/81	Reconstitution of ADP/ATP transport into liposomes. Accounting of active AAC molecules. Orientation of AAC molecules. Membrane potential on liposomes by K^+ gradient. Segregation of $\Delta\psi$ influence on K_d and V_{max}.	Stability of AAC in phospholipids. Increase of BKA, CAT binding. Transport activity dependent of negative phospholipids. Linear dependency of transport on $\Delta\psi$, dependency on sidedness. $\Delta\psi$ changes V_{max} not K_d. Only ATP–AAC complex is influenced by $\Delta\psi$.	108, 121, 122, 124
1977/8	c/m state transition in isolated protein. Conformational changes.	Essential SH-groups only with m-state. Kinetics of transition.	118, 119

(Table I, continued)

1980/4	Fluorescent ADP analogues as probes of AAC carrier.	Binding of fluorescent DAN-ADP only to m state, nonfluorescent in c state. Visualization of asymmetry of binding center in c versus m state.	138–140
1977/83	Sequencing of AAC.	Complete amino acid sequence. First of a carrier protein.	
1981/84	The pyridoxal method for lysine localization.	Membrane sidedness, P-L headgroup-collar lysines, channel lysines, tentative transmembrane folding.	128, 128a
1979/82	Isolation of uncoupling protein (UCP) from brown adipose mitochondria. Purification also in breakthrough of hydroxylapatite. Nucleotide-binding properties.	Binding of nucleotide (GTP) for solubilized UCP. From binding effective M_r 65 000. Hydrodynamic results M_r 170 000 for UCP-Triton micelle, M_r 66 000 for protein → dimer. High (H^+) dependency, anion competition.	132–134
1982/4	New procedure for reconstitution of UCP by octyl-E_{2-8} + phospholipid.	H^+ transport by active UCP in liposomes linear dependent on $\Delta\psi$.	136
1980/2	Immunochemical differentiation of AAC from heart, liver, kidney. Charged shift immunoelectrophoresis.	Quantitative organ specific differences. Autoimmune antibodies against AAC in sera from patient of primary biliary cirrhosis and viral myocarditis. Organ specificity of these sera.	114, 115, 116

when I embarked on the problem of membrane permeability and transport. Not only radioactivity distribution was used but also the biochemical assay on the nanomolar scale of metabolites was applied before I drew conclusions on the intra- and extramitochondrial solute distribution.

The survey in Table I on my research activities stresses the relation of the methodological developments to the research findings. I felt that progress is mostly linked to development and application of new methods, particularly in biology. New concepts, models and theories which can sustain for some time the critical onslaught will develop only from new experimental results. This attitude is not a very comfortable one, since the technical developments may consume the greater part of one's energy and time. Unforeseen technical difficulties and the resultant frustration have to be overcome before eventually the new methods can be applied. Often only after several years can the first experiments be performed and the first fruits of long labor be harvested. But it may happen that in the meantime others have attacked the problem in a different manner so that priorities are missed. Also these methodological efforts are less appreciated by the public in proportion to the invested efforts than a more routine accumulation of biochemical results. On the other hand, there are highly satisfactory and rewarding moments, when, after long waiting, the new methods are applied with great apprehension and the first new data are acquired. Rarely can any other scientific experience match this exciting event.

In stressing the methodological approach, I would not regard theories and concepts as secondary. However, experience shows that in biology, much more so than in physics, theories mostly have a short survival time. Nature just does not make use of the biophysicist's concepts. Humans tend to anticipate the construction and function of biological systems. Nature by its very principle of evolution, has found only one solution out of nearly infinite possibilities which of course we are unable to anticipate. The laws of physics are obeyed, but we never know the myriads of conditions to which the particular system is optimally adapted. Our task is to unravel the existing living world and thus to overcome our blindness. It is not to force our anthropomorphic preoccupations upon the scientific research process.

From this viewpoint, scientists have to be "uncreative", always critical to their own preoccupations and ready to change their views once new contrary data are obtained. This is a profoundly unhuman postulate and therefore nearly impossible to adhere to. Human nature likes to be ahead of the process of discovery, since it gives a feeling of superiority. Therefore, theories are greatly admired and believed in, even if they might be contradicted by the facts shortly after their creation. This greatly retards the scientific progress and no lesser scientist than Max Planck once bitterly concluded that theories only die together with their proponents. There is a tendency to overestimate strongly the merits of theories and of the theoretician as pointed out by analysts of the scientific discovering process. For example, two theories or concepts I have encountered have long ceased to be scientifically fruitful. Allostery may originally have had some merits for helping enzymologists to analyze their data. However, it has no scientific value, when by invoking allostery, the underlying mechanism may be dangerously camouflaged. Chemiosmosis certainly was a great stimulus in bioenergetics. However, although it may have helped to understand systems functioning in bioenergetics, it failed and cannot help to explain mechanism. At the present stage of research, the dogmatic chemiosmotic interpretations may retard progress. My attitude has been that of an experimentalist in biology who of course operates with working hypotheses but who will always critically appraise these with new emerging data. Ideas should be proposed with modesty and only with greatest respect towards the continuous unfolding of new experimental data.

As we advance further in solving a particular problem, the climb tends to become steeper. Sometimes the walls seem so steep that despite enormous efforts little progress is achieved. New paths will eventually be found by new methods which take us up another step towards the never-ending goal of full understanding. Also as we advance further, the vistas become larger, new interrelations appear. For example, a specific protein may not only be a catalyst for a particular reaction, but being embedded in the incomprehensible highly organized biological network, may interrelate to other proteins in its micro-environment and may have functions in the so far unknown topography of metabolite traffic within the cell. All these and other prospective functions are assumed to be programmed into the protein primary structure.

During my scientific work I had the excellent collaboration and support of relatively few scientists whom I mentioned in my overview. Therefore, to a large part I supervised and performed the experiments myself. In these daily chores I could rely on the unrelented and unselfish help of a number of technicians who also figure as coauthors on my publications. Here I would like to name and praise the assistance in Marburg of Ello Ritt, Helga von Häfen, Gertrud Wenske and Karin Grebe, and in Munich Marlies Buchholz, Barbara Schmiedt, Maria Appel, Ingeborg Mayer and Edith Winkler. Since my presence in the laboratory was required most of the time, I had never the chance to take a leave of absence in a sabbatical or similar arrangement.

In addition to the experimental work and evaluating the data, I had to write and edit publications and unfortunately never kept up with the pace of the results obtained. In particular I could not fulfill my commitments in writing reviews or reviewing my work. On the other hand I can maintain that in most of my papers I am not only the coauthor but an author who has actively taken part in all stages of the work until the final publication. Furthermore, I therefore feel fully responsible for the published scientific results.

REFERENCES

1 Kl. Schäfer and M. Klingenberg, Z. Elektrochem., 58 (1954) 828–836.
1a M. Klingenberg, Arch. Biochem. Biophys., 75 (1958) 376–386.
2 B. Chance, Adv. Enzymol., 12 (1951) 153–185.
3 B. Chance, M. Klingenberg and E. Boeri, Fed. Proc., 15 (1956) Abstract 751.
4 M. Klingenberg, Proc. Am. Soc. Meeting Minneapolis (1955) Abstract 43.
5 M. Klingenberg and Th. Bücher, Biochem. Z. 331, (1959) 312–333.
6 M. Klingenberg, Proc. IVth Int. Cong. Biochem., Vienna (1958) Abstract 67.
7 M. Klingenberg, W. Slenczka and E. Ritt, Biochem. Z., 332 (1959) 47–66.
8 Th. Bücher and M. Klingenberg, Angew. Chemie, 70 (1958) 552–570.
8a M. Klingenberg, Int. Ed., Angew. Chemie, 3 (1964) 54–61.
9 M. Klingenberg and W. Slenczka, Biochem. Z., 331 (1959) 486–517.
10 B. Chance and G.P. Williams, Adv. Enzymol., 17 (1956) 65–73.
11 M. Klingenberg and Th. Bücher, Biochem. Z., 34 (1961) 1–17.
12 E. Zebe, A. Delbrück and Th. Bücher, Biochem. Z., 331 (1959) 254–264.
13 P. Schollmeyer and M. Klingenberg, Biochem. Z., 335 (1962) 426–439.
14 M. Klingenberg, 11. Mosbacher Kolloquium, Springer Verlag, Heidelberg, 1960, pp. 82–114.
15 M. Klingenberg and D. Pette, Biochem. Biophys. Res. Commun., 7 (1962) 430–432.
16 D. Pette, M. Klingenberg and Th. Bücher, Biochem. Biophys. Res. Commun., 7 (1962) 425–429.
17 M. Klingenberg, Proc. GDNA Symp. Rottach-Egern, Springer Verlag, Heidelberg (1962) pp. 69–85.
18 H. Goebell and M. Klingenberg, Biochem. Biophys. Res. Commun., 13 (1963) 212–216.
19 M. Klingenberg, Ergebn. Physiol., Bd. 55 (1964) 129–189.
20 M. Klingenberg and P. Schollmeyer, Biochem. Z., 335 (1961) 243–262.
21 M. Klingenberg, Biochem. Z., 335 (1961) 263–272.
22 M. Klingenberg and P. Schollmeyer, V. Int. Cong. Biochem., Moscow, 1961, Pergamon, Oxford, 1963, pp. 46–65.
23 M. Klingenberg and P. Schollmeyer, Biochem. Z., 335 (1961) 231–242.
24 M. Klingenberg, in S. Papa et al. (Eds.), The Energy Level and Metabolic Control in Mitochondria, Adriatica, Bari, 1969, pp. 185–188.
25 M. Klingenberg and H. von Häfen, Biochem. Z., 337 (1963) 120–145.
26 M. Klingenberg, H. von Häfen and G. Wenske, Biochem. Z., 343 (1965) 452–478.
27 M. Klingenberg, Biochem. Z., 343 (1965) 479–503.
28 A.M. Stein, N.O. Kaplan and M.M. Ciotti, J. Biol. Chem., 234 (1960) 979–986.

29 L. Ernster and F. Navazio, Exp. Cell Res., 11 (1956) 483-494.
30 H. Goebell and M. Klingenberg, Biochem. Biophys. Res. Commun., 13 (1963) 209-212.
31 H. Goebell and M. Klingenberg, Biochem. Biophys. Res. Commun., 13 (1963) 212-216.
32 B. Kadenbach, H. Goebell and M. Klingenberg, Biochem. Biophys. Res. Commun., 14 (1964) 335-339.
33 H. Goebell and M. Klingenberg, Biochem. Z., 340 (1964) 441-464.
34 M. Klingenberg, H. Goebell and G. Wenske, Biochem. Z., 341 (1965) 199-223.
35 C. Bode and M. Klingenberg, Biochim. Biophys. Acta, 84 (1964) 93-95.
36 C. Bode and M. Klingenberg, Biochim. Z., 341 (1965) 271-299.
37 A.M.Th. Beenakkers and M. Klingenberg, Biochim. Biophys. Acta, 84 (1964) 205-207.
38 H. Jacobs, H.W. Heldt and M. Klingenberg, Biochem. Biophys. Res. Commun., 16 (1964) 516-521.
39 H.W. Heldt, M. Klingenberg and E. Papenberg, Biochem. Z., 342 (1965) 508-517.
40 H.W. Heldt, N. Greif, M. Klingenberg, R. Scholz, U. Panten, J. Grunst and Th. Bücher, J. Biol. Chem., 240 (1965) 4659-4661.
41 H.W. Heldt, H. Jacobs and M. Klingenberg, Biochem. Biophys. Res. Commun., 17 (1964) 130-135.
42 H.W. Heldt and M. Klingenberg, Biochem. Z., 343 (1965) 433-451.
43 E. Pfaff, M. Klingenberg and H.W. Heldt, Biochim. Biophys. Acta, 104 (1965) 312-315.
44 E. Pfaff, M. Klingenberg, E. Ritt and W. Vogell, Eur. J. Biochem., 5 (1968) 222-232.
45 M. Klingenberg and E. Pfaff, in J.M. Tager et al. (Eds.), Regulation of Metabolic Processes in Mitochondria, Elsevier, Amsterdam, 1966, pp. 180-201.
46 M. Klingenberg, E. Pfaff and A. Kröger, in Rapid Mixing and Sampling Techniques in Biochemistry, Academic Press, New York, 1964, pp. 333-337.
47 M. Klingenberg, Symp. Redoxfunktionen Cytoplasmatischer Strukturen, Wien, 1962, pp. 163-187.
48 F. Palmieri, G. Prezioso, E. Quagliariello and M. Klingenberg, Eur. J. Biochem., 22 (1971) 66-74.
49 J.D. McGivan and M. Klingenberg, Eur. J. Biochem., 20 (1971) 392-399.
50 M. Klingenberg, in P.N. Campbell and F. Dickens (Eds.), Essays in Biochemistry, Vol. 6, Academic Press, London, 1970, pp. 119-159.
51 B. Guérin, M. Guérin and M. Klingenberg, FEBS Lett., 10 (1970) 265-268.
52 M. Klingenberg, R. Durand and B. Guérin, Eur. J. Biochem., 42 (1974) 135-150.
53 L. Szarkowska and M. Klingenberg, Biochem. Z., 338 (1963) 674-697.
54 A. Kröger and M. Klingenberg, Biochem. Z., 344 (1966) 317-336.
55 A. Kröger and M. Klingenberg, in D.R. Sanadi (Ed.), Current Topics in Bioenergetics, Vol. 2, Academic Press, New York, 1967, pp. 152-190.

56 M. Klingenberg, in T.P. Singer (Ed.), Biological Oxidations, Interscience, New York, 1968, pp. 3-54.
57 M. Klingenberg and A. Kröger, in J.M. Tager (Ed.), Electron Transport and Energy Conservation, Adriatica, Bari, 1970, pp. 135-143.
58 A. Kröger and M. Klingenberg, in Vitamins and Hormones, Vol. 28, Academic Press, New York, 1970, pp. 533-674.
59 M. Klingenberg, in B. Chance et al. (Eds.), Rapid Mixing and Sampling Techniques in Biochemistry, Academic Press, New York, 1964, pp. 61-65.
60 A. Kröger and M. Klingenberg, Eur. J. Biochem., 34 (1973) 358-368.
61 P.F. Urban and M. Klingenberg, Eur. J. Biochem., 9 (1969) 519-525.
62 T. Ohnishi, A. Kröger, H.W. Heldt, E. Pfaff and M. Klingenberg, Eur. J. Biochem., 1 (1967) 301-311.
63 G. von Jagow, H. Weiss and M. Klingenberg, Eur. J. Biochem., 33 (1973) 140-157.
64 A. Kröger and M. Klingenberg, Eur. J. Biochem., 39 (1973) 313-323.
64a A. Kröger, V. Dadak, M. Klingenberg and F. Diemer, Eur. J. Biochem., 21 (1971) 322-333.
65 M. Klingenberg, in B. Hess and H. Staudinger (Eds.), Biochemie des Sauerstoffs, Springer Verlag, Berlin, 1968, pp. 131-135.
66 M. Klingenberg and G. von Jagow, in J.M. Tager et al. (Eds.), Electron Transport and Energy Conservation, Adriatica, Bari, 1970, pp. 281-290.
67 F. Palmieri and M. Klingenberg, Eur. J. Biochem., 1 (1967) 439-446.
68 M. Klingenberg and M. Buchholz, Eur. J. Biochem., 13 (1970) 247-252.
69 G. von Jagow and M. Klingenberg, Eur. J. Biochem., 12 (1970) 583-592.
70 M. Klingenberg, in E. Quagliariello et al. (Eds.), Energy Transduction in Respiration and Photosynthesis, Adriatica, Bari, 1970, pp. 23-34.
71 M. Klingenberg and E. Pfaff, Methods Enzymol., X (1967) 680-684.
72 H.W. Heldt, H. Jacobs and M. Klingenberg, Biochem. Biophys. Res. Commun., 18 (1965) 174-179.
73 H.W. Heldt and M. Klingenberg, Biochem. Z., 343 (1965) 433-451.
74 E. Pfaff and M. Klingenberg, Eur. J. Biochem., 6 (1968) 66-79.
75 H.W. Heldt and E. Pfaff, Eur. J. Biochem., 10 (1969) 494-500.
76 E. Pfaff, H.W. Heldt and M. Klingenberg, Eur. J. Biochem., 10 (1969) 484-493.
77 M. Klingenberg, Eur. J. Biochem., 76, (1977) 553-565.
78 R. Palmieri and M. Klingenberg, Methods Enzymol., LVI (1979) 279-301.
79 M. Klingenberg, K. Grebe and M. Appel, Eur. J. Biochem., 126 (1982) 263-269.
80 M. Klingenberg, in Mitochondria: Biomembranes, Elsevier, Amsterdam, 1972, pp. 147-162.
81 M. Klingenberg, in Energy Transformation in Biological Systems, Ciba Foundation Symposium, 31 in 1974, Elsevier, Amsterdam, 1975, pp. 105-124.
82 M. Klingenberg, in K. van Dam and B.F. van Gelder (Eds.), Structure and Function of Energy-Transducing Membranes, Elsevier, Amsterdam, 1977, pp. 275-282.

83 M. Klingenberg, J. Membrane Biol., 56 (1980) 97-105.
84 M. Klingenberg, R. Wulf, H.W. Heldt and E. Pfaff, in L. Ernster and Z. Drahota (Eds.), Mitochondria: Structure and Function, Academic Press, London, 1969, pp. 59-77.
85 K. LaNoue, S.M. Mizani and M. Klingenberg, J. Biol. Chem., 253 (1978) 191-198.
86 R. Wulf, A. Kaltstein and M. Klingenberg, Eur. J. Biochem., 82, (1978) 585-592.
87 M. Klingenberg, H.W. Heldt and E. Pfaff, in S. Papa et al., (Eds.), The Energy Level and Metabolic Control in Mitochondria, Adriatica, Bari, 1969, pp. 237-253.
88 H.W. Heldt, M. Klingenberg and M. Milovancev, Eur. J. Biochem., 30 (1972) 434-440.
89 M. Klingenberg and H. Rottenberg, Eur. J. Biochem., 73 (1977) 125-130.
89a M. Klingenberg and H.W. Heldt, in H. Sied (Ed.), Metabolic Compartmentation, Academic Press, London, 1982, pp. 101-122.
90 M.J. Weidemann, H. Erdelt and M. Klingenberg, Eur. J. Biochem., 16 (1970) 313-335.
91 M. Klingenberg, K. Grebe and H.W. Heldt, Biochem. Biophys. Res. Commun., 39 (1970) 344-351.
92 M.J. Weidemann, H. Erdelt and M. Klingenberg, Biochem. Biophys. Res. Commun., 39 (1970) 363-370.
93 M. Klingenberg, H. Buchholz, H. Erdelt, G. Falkner, K. Grebe, H. Kadner, B. Scherer, L. Stengel-Rutkowski and M.J. Weidemann, in G.F. Azzone et al. (Eds.), Biochemistry and Biophysics of Mitochondrial Membranes, Academic Press, New York, 1972, pp. 465-486.
94 H. Erdelt, M.J. Weidemann, M. Buchholz and M. Klingenberg, Eur. J. Biochem., 30 (1972) 107-122.
95 M. Klingenberg and M. Buchholz, Eur. J. Biochem., 38 (1973) 346-358.
96 B. Scherer and M. Klingenberg, Biochemistry, 13 (1974) 161-170.
97 M. Klingenberg, G. Falkner, H. Erdelt and K. Grebe, FEBS Lett., 16 (1971) 296-300.
98 M. Klingenberg, K. Grebe and B. Scherer, Eur. J. Biochem., 52 (1975) 351-363.
99 W. Babel, H. Aquila, K. Beyer and M. Klingenberg, FEBS Lett., 61 (1976) 124-127.
100 M. Klingenberg, B. Scherer, L. Stengel-Rutkowski, M. Buchholz and K. Grebe, in G.F. Azzone et al. (Eds.), Mechanisms in Bioenergetics, Academic Press, New York, 1973, pp. 257-284.
101 M. Klingenberg, in L. Ernster et al. (Eds.), Dynamics of Energy-Transducing Membranes, Elsevier, Amsterdam, 1974, pp. 511-528.
102 M. Klingenberg, in A.N. Martonosi (Ed.), The Enzymes of Biological Membranes: Membrane Transport, Vol. 3, Plenum, New York, 1976, pp. 383-438.

103 P. Riccio, H. Aquila and M. Klingenberg, FEBS Lett., 56 (1975) 129–132.
104 P. Riccio, H. Aquila and M. Klingenberg, FEBS Lett., 56 (1975) 133–138.
105 M. Klingenberg, P. Riccio, H. Aquila, B. Schmiedt, K. Grebe and P. Topitsch, in G.F. Azzone et al. (Eds.), Membrane Proteins in Transport and Phosphorylation, Elsevier, Amsterdam, 1974, pp. 229–243.
106 H. Aquila, W. Eiermann and M. Klingenberg, Abstracts 10th Int. Congress of Biochemistry, Hamburg, 1976, p. 345.
107 M. Klingenberg, P. Riccio and H. Aquila, Biochim. Biophys. Acta, 503 (1978) 193–210.
108 R. Krämer, H. Aquila and M. Klingenberg, Biochemistry, 16 (1977) 4949–4953.
108a M. Klingenberg, in A.N. Martonosi (Ed.), Membranes and Transport, Vol. 1, Plenum, New York, 1982, pp. 203–209.
109 H. Hackenberg and M. Klingenberg, Biochemistry, 19 (1980) 548–555.
110 W. Eiermann, H. Aquila and M. Klingenberg, FEBS Lett., 74 (1977) 209–214.
111 H. Hackenberg, P. Riccio and M. Klingenberg, Eur. J. Biochem., 88 (1978) 373–378.
112 B.B. Buchanan, W. Eiermann, P. Riccio, H. Aquila and M. Klingenberg, Proc. Natl. Acad. Sci. USA, 73 (1976) 2280–2284.
113 M. Klingenberg, P. Riccio, H. Aquila, B.B. Buchanan and K. Grebe, in Y. Hatefi and L. Djavadi-Ohaniance (Eds.), The Structural Basis of Membrane Function, Academic Press, New York, 1976, pp. 293–311.
114 H.-P. Schultheiss, O.J. Bjerrum and M. Klingenberg, Biochim. Biophys. Acta, 771 (1984) 235–240.
115 H.-P. Schultheiss and M. Klingenberg, Eur. J. Biochem., 143 (1984) 599–605.
116 H.-P. Schultheiss, P. Berg and M. Klingenberg, Clin. Exp. Immunol., 54 (1983) 648–654.
117 M. Klingenberg, H. Aquila, R. Krämer, W. Babel and J. Feckl, in G. Semenza and E. Carafoli (Eds.), Biochemistry of Membrane Transport, Springer Verlag, Berlin, 1977, pp. 567–579.
118 H. Aquila, W. Eiermann and M. Klingenberg, Eur. J. Biochem., 122 (1982) 133–139.
119 H. Aquila and M. Klingenberg, Eur. J. Biochem., 122 (1982) 141–145.
120 M. Klingenberg and M. Appel, FEBS Lett., 119 (1980) 195–200.
121 R. Krämer and M. Klingenberg, Biochemistry, 16 (1977) 4954–4961.
122 R. Krämer and M. Klingenberg, Biochemistry, 18 (1979) 4209–4215.
123 R. Krämer and M. Klingenberg, FEBS Lett., 119 (1980) 257–260.
124 R. Krämer and M. Klingenberg, Biochemistry, 19 (1980) 445–560.
125 R. Krämer and M. Klingenberg, Biochemistry, 21 (1982) 1082–1089.
126 W. Babel, E. Wachter, H. Aquila and M. Klingenberg, Biochim. Biophys. Acta, 670 (1981) 176–180.
127 H. Aquila, D. Misra, M. Eulitz and M. Klingenberg, Z. Physiol. Chem., 363 (1982) 345–349.

128 W. Bogner, H. Aquila and M. Klingenberg, in E. Quagliariello and F. Palmieri (Eds.), Structure and Function of Membrane Proteins, Elsevier, Amsterdam, 1983, pp. 145-156.
128a W. Bogner, H. Aquila and M. Klingenberg, FEBS Lett., 146 (1982) 259-261.
129 M. Klingenberg, Nature, 290 (1981) 449-454.
130 M. Klingenberg, Ann. N.Y. Acad. Sci., (1985) in press.
131 G.M. Heaton, R.J. Wagenvoord, A. Kemp Jr. and D.G. Nicholls, Eur. J. Biochem., 82 (1978) 515-521.
132 C.S. Lin and M. Klingenberg, FEBS Lett., 113 (1980) 299-303.
133 C.S. Lin and M. Klingenberg, Biochemistry, 21 (1982) 2950-2956.
134 C.S. Lin, H. Hackenberg and M. Klingenberg, FEBS Lett., 113 (1980) 304-306.
135 M. Klingenberg, Biochem. Soc. Trans., 12 (1984) 390-393.
136 M. Klingenberg and E. Winkler, EMBO J., (1985) in press.
137 H. Aquila, T. Link and M. Klingenberg, EMBO J., (1985) in press.
138 M. Klingenberg, in H. Eggerer and R. Huber (Eds.), Structural and Functional Aspects of Enzyme Catalysis, Springer Verlag, Berlin, 1981, pp. 202-212.
139 I. Mayer, A.S. Dahms, W. Riezler and M. Klingenberg, Biochemistry, 23 (1984) 2436-2442.
140 M. Klingenberg, I. Mayer and A.S. Dahms, Biochemistry, 23 (1984) 2442-2449.
141 K. Beyer and M. Klingenberg, Biochemistry, 17 (1978) 1424-1431.
142 K. Beyer and M. Klingenberg, Biochemistry, 22 (1983) 639-645.
143 A. Munding, K. Beyer and M. Klingenberg, Biochemistry, 22 (1983) 1951-1947.

G. Semenza (Ed.) Selected Topoics in the History of Biochemistry: Personal Recollections (Comprehensive Biochemistry Vol. 36) © 1986 Elsevier Science Publishers

Chapter 8

An Eventful Life Around Flavins

PETER HEMMERICH

Fachbereich Biologie, Universität Konstanz, D-7750 Konstanz (F.R.G.)

Memoirs Dictated in the Last Weeks of his Life, August 1981

With an Introductory note by

HELMUT BEINERT

Institute for Enzyme Research and Department of Biochemistry, College of Agricultural and Life Sciences, University of Wisconsin, Madison, WI 53706 (U.S.A.)

On a Saturday afternoon in late summer 1981 Peter Hemmerich, in the last weeks of his life, was visiting at his home with a few friends and members of his family. In one of the ever more rare spells of relative wellbeing he began vividly recounting events from the most decisive years of his life, the early years of his professional ascent. One of the listeners suggested that he should record these recollections. It was a time when he had resigned to the fact that "all is over" — as he used to say in moments of depression — and when, deprived of much of his physical strength and abilities, he found himself condemned to idleness, an experience hard to bear for a man as active as he used to be. Somewhat reluctantly Peter Hemmerich followed the suggestion and began some recording, but soon took to this new task with obvious pleasure and satisfaction, developing some of the fervor that previously had characterized all of his endeavors. He had found here a last and unique opportunity to assert himself, an opportunity of which he had not yet been

Plate 13. Peter Hemmerich (1929–1981)

deprived. The resulting document is essentially what will be found in the pages that follow.

The original transcript of the recordings was not suitable for publication. It is an interesting sign of our times that even Peter Hemmerich — a man who loved his German homeland — preferred to dictate all professional matter in English, while the more personal part was dictated in German. To make his remembrances accessible to a wider audience, the German part was translated* and the English part corrected as far as grammar, wording and sentence structure was concerned**, in either case with great care to preserve the meaning and to convey the flavor of the original text. In a few instances the recording as typed was unintelligible, but reconstruction was possible from knowledge of Peter Hemmerich's scientific field and of previous remarks and preferred expressions of his***. The explanatory footnotes were provided by V. Massey and S. Ghisla.

Peter Hemmerich's brief memoirs will remain a touching document of the farewell message to the world of his endeavors. Despite his hopeless situation his message at the end is reconciling and positive. Facing death at a time that could have meant the zenith of his life, he calls himself a "Glückspilz" (translated on p. 424 as "I have been uncommonly lucky").

What might Peter Hemmerich's words have to tell future generations? Much of what he says — in retrospect — sounds so straightforward and simple, unexpected from a man who led a multifaceted and seemingly rather stormy life. It is also obvious that human fates and careers many a time depend on subtle chance-events. But it is exactly here that there is room for personal contribution. Peter Hemmerich has keenly perceived his opportunities and has seized them with dedication, often passionately. So he has not only left behind friends in his life.

May his message — irrespective of this — be received with indulgence as the last assertion of a man of unusual vitality and of original thoughts and vision, who had the determination and strength to follow them up.

---

* By Margot Massey.

** By Vincent Massey and Helmut Beinert.

*** By Helmut Beinert, Vincent Massey and Sandro Ghisla.

Peter Hemmerich (1929–1981): memoirs

It was in 1954 that I took up thesis work at the Inorganic Institute of the University of Basel under the directorship of Prof. Hans Erlenmeyer. Prof. Erlenmeyer was not an inorganic chemist in the true sense of the word. He was a typical representative of what was later on called inorganic biochemistry. He was fascinated by Warburg's work on heavy metals in living matter and was in a state to be convinced that practically all important biochemical reactions were in some way metal-determined or metal-catalyzed. At first he offered me a thesis on the metal-binding properties of α-hydroxy acids, which appeared a little bit boring to me. But after three weeks he came up with the latest issue of the *Journal of the American Chemical Society*; there was a contribution from Wisconsin by two people named Foye and Lange who were studying the metal affinity of vitamin B_2[1]. Erlenmeyer was elated and he told me that I should continue with this topic. The Wisconsin people claimed that they had found a high heavy-metal affinity for riboflavin, which they attributed to its similarity in structure to 8-hydroxyquinoline. At that time in the early fifties metal affinity of organic compounds was mainly estimated and determined according to acidification of organic compounds upon addition of heavy metals, and from the observation of a shift of the compound's pK during titration in presence of heavy metal. When I tried to reproduce the data of Foye and Lange, I became aware they were wrong and that no acidification of riboflavin — especially not by acidification of the N(3)H-position — was obtained in the presence of heavy-metal ions. On the contrary, what happened was hydrolysis of the metal ions before any complexation of the ligand could be observed.

From this I learned that the enol tautomer of riboflavin is a high-energy compound and that all the formulae in the textbooks describing riboflavin with hydroxy functions in either position 2 or 4 must be wrong[2]. This was in agreement with the fact, which later on helped Watson and Crick establish their code, that nitrogen functions in heteroaromatic compounds always have the amino-tautomer structure while oxygen functions have the keto and not the enol or iminol structure. At this time nobody could learn this from a textbook; the reason for this was very simple: it is much easier to call a compound a hydroxy derivative according to the wrong tautomer than a keto dihydro- or diketotetrahydro-derivative which requires a lot more description. So it happened that I had to explain to my beloved teacher, Erlenmeyer, that the literature he gave me was wrong. He was not elated about this. Apart from Foye and Lange his old friend Adrian Albert had been working on the metal affinity of flavin and had been making the same mistake, as I found out, isolating largely metal hydroxide instead of flavin metal chelates. Furthermore, I saw that the only site in the flavin which could possibly liberate protons upon addition of heavy metal in solution is the ribityl side chain which is, of course, a trivial case since any type of sugar side chain, especially with vicinal hydroxy groups, would lend itself to such an unspecific metal complex formation. This could not be brought into any connection with the biological activity of the vitamin, however.

Now I was in a bad situation, and I was close to dropping the whole subject. But what I tried first, since it appeared to me that flavin was unique in resisting acidification of the N(3)H-group, and I wanted to know for sure, was to look into the whole heterocyclic literature for exceptional cases of hydroxy- vs. keto-tautomer stability, which was, of course, thermodynamically unfeasible. For quite some time I was trying, nevertheless, together with the first assistant of Prof. Erlenmeyer, Dr. Fallab, to find such exceptions, but without success. As a next step, I ran into the early work of Warburg who isolated riboflavin by its red silver salt, whereas with all other kinds of heavy metals no coloration upon complex formation could be observed. Silver apparently made an exception in forming the desired type of chelate, but why silver? The reason was that silver is a soft d^{10} ion which is strongly polarizable. This contributes to the well-known biological toxicity of silver ions. But here I got caught again since it turned out that in spite of being a so-called water-soluble vitamin, riboflavin was everything else but easily soluble in water. The experimental work with metals and riboflavin always yielded uncontrollable amounts of free ligand in the preparations. Hence, I decided that, in order to continue my study, I had to synthesize (which I liked since I was a chemist) really water-soluble flavin derivatives which at the same time would be expected to have a higher heavy-metal affinity. Going through the literature I learned that practically no modifications of the basic flavin molecule were known at that time, especially no functionally substituted derivatives. Furthermore, I learned that the ribityl side chain would only make my intended work more difficult, especially due to its light sensitivity, which was confined to the side chain. This photosensitivity had been studied quite a bit by photochemists and biologists, but the photoderivatives were absolutely unbiological; lumichrome and lumiflavin have no biological activity whatsoever. Here I saw that in order to stick to heteroaromatic flavin chemistry one had to forget about the sugar side chain and to move into Richard Kuhn's early flavin chemistry dealing with the most simple flavin chromophore having a methyl group in position 10, namely lumiflavin. The original Kuhn-synthesis [3], however, was hampered by the fact that the aromatic starting materials were not commercially available. Kuhn had obtained them from friends in the German industry and their synthesis was tedious. This was the reason why nobody cared for the much more comfortable lumiflavin model instead of the meanwhile commercially available vitamin and its undesired photochemistry and sugar side chain chemistry. Hence, I developed a handy lumiflavin synthesis based on the earlier Tishler-synthesis of the vitamin [4]. This synthesis was the first breakthrough since it allowed me to deal with large amounts of flavin chromophore, although the solubility problem was by no means solved. I learned to master this problem by recognizing the fact that I had to block the N(3)H-group in lumiflavin since this group, though it has an acid function, was thought to be unbiological because of its high pK of approx. 10. I knew that methylation of acidic NH in heteroaromatics would increase water solubility since it would remove or decrease the formation of stable hydrogen bonds in the crystals, e.g. *N*-methyl barbituric acid is more soluble than barbituric acid itself. And thus I decided to expand [5] the lumiflavin synthesis into a 3-methyl lumiflavin synthesis which, however, was not too easy for the following reasons: first, the methylation had to be done at a pH higher than the pK, this means more than 10. This led to a competitive reaction by hydrolytic opening of the pyrimidine subnucleus of the flavin chromophore, which was long known.

In order to cope with this hydrolytic side reaction, I decided to move out of the water and to introduce the newly developed solvent dimethyl formamide. Still I had to make the choice of a base catalyst which ought to be devoid of all nucleophilic properties. At that time overcrowded trialkylamines were not available commercially, and thus I had to be afraid that direct alkylation of the base catalyst might take place. Finally, I landed on my feet with good old potassium carbonate as base catalyst which would render the whole reaction inhomogeneous. This turned out to be no disadvantage, since the solubility of the carbonate in dimethylformamide proved to be unexpectedly high. But the next obstacle was already on the way, since it turned out that I could not apply heat to the relatively slow alkylation reaction or else the whole mixture turned into a green marmalade. Hell knew why! Thus I had to stick to temperatures only as high as 30° or 40°C and let the reaction run for days or weeks. And this finally yielded nearly quantitative amounts of 3-monoalkylated flavin. These flavins in the oxidized state had a sufficient solubility in chloroform as well as in water, while most other properties of the natural chromophore remained unchanged. Thus we had a model compound which was very stable and with which we could do lots of new chemical reactions. There remained the explanation for the green color upon heating. Upon treatment of this green snake pit with aqueous acid, I finally was able to isolate red and green colored new flavins which turned out to be covalent flavin dimers[6]. When looking for the connection point of the two flavin halves which were covalently bound, I saw right away that the methyl group in position 8 was no longer intact. And this taught me immediately that, in view of the electron deficiency of the flavin chromophore, I was confronted with a reaction very well known for electron deficient aromatic compounds (for example, nitrotoluenes exhibiting a CH-active methyl group under aprotic basic conditions). This was the discovery of the activity of the 8-methyl group which later led to the structural elucidation of lots of new flavins, i.e. covalently bound flavins, such as the flavocoenzymes from succinate dehydrogenase and monoamine oxidase. This work developed later after I had gone to California to cooperate with the group of Tom Singer. Once I knew about it, it was easy enough (with a pencil) to remove a proton from the methyl group in position 8 and write what we later on were calling the benzoquinoid flavin tautomer. This was a quinone methide and, therefore, yielded all the characteristic dimerization reactions known for such methides. This was the long story of a seemingly trivial alkylation reaction which turned out to be a milestone for all the further developments. Up to that time I had only been working with the oxidized state of the flavin redox system which I began to call flavoquinone because of its quinoid character. Still I couldn't satisfy my master Erlenmeyer in whose eyes I was doing silly chemistry without concern for either metals or for biology. He was not elated and he was even less so when I decided to turn to the chemistry of the reduced flavin, since I became aware that nobody knew anything about the chemistry of this system. The reason for this was a psychological one: especially because it was apparently colorless and nonfluorescent, and because one would not even know whether it had a biological importance, besides the long known semiquinone redox state, which had already been described, by Leonor Michaelis, in solution and, by Richard Kuhn, in the solid state. Just at this time I had an illumination: when there was neither metal affinity in the oxidized nor in the reduced flavin, as far as I could see, then the metal affinity must be in the semiquinone, for the very simple reason that the semiquinone must be more acidic than the oxidized flavin but would have the same hydroxyquinolinate profile, and should, therefore, liberate protons upon addition of heavy metals[7]. But how to prove it? It was known since Michaelis that the semiquinone had a distinct but low thermodynamic stability, which meant that upon mixing of oxidized and fully reduced flavin one would obtain an equilibrium concentration of semiquinone up to the amount of a few percent per total flavin present, but this dismutation equilibrium would be expected to have a very complicated pH-dependence[8]. I now had three redox states and since the flavin was an amphoteric chromophore in each state at least two if not three different chromophores[9], the metal affinity of each of which had to be checked. And even worse, all the partially or fully reduced chromophores were obviously sensitive to molecular oxygen, while excess of reducing agent present in solution would strongly disturb all investigations. Here began my search for suitable reducing agents which would allow me to work comfortably under anaerobic conditions. The reducing agents should, of course, not contribute to the spectral properties of the system and excess of reducing agent should be inert, while it was desirable to isolate reduced forms of flavin which would not immediately return into the oxidized state by autoxidation. Then checking the dihydroflavin formula I became convinced that the newly formed NH-group in position 5 should easily undergo acylation, and thus I tried reduction of flavoquinone with zinc in acetic anhydride. To my big surprise, I obtained, together with the excess of zinc metal, a nearly colorless flavin residue which precipitated upon acidification and could be redissolved in dilute ammonia without immediate reoxidation and reappearance of fluorescence[10]. This was the second milestone, especially since a slightly more acid treatment removed the introduced acetyl substituent easily and quantitatively. This was particularly so when nitrous acid was used as additional oxidizing agent. Thus I finally had a stable dihydroflavin preparation with which I could undertake chemistry, although everybody warned me saying that dihydroflavin is an absolutely unbiological redox state of the vitamin, and is uninteresting, because it is colorless. But this colorless compound opened new paths into B_2-chemistry by the simple fact that all the alkylation reactions now run in quite different ways as compared to the alkylation with the oxidized chromophore. I thus started with the N(5)-blocked acetyldihydroflavin from which I obtained, upon alkylation under nearly neutral conditions and in a smooth reaction, a number of products. From the structure of these it became evident that the energy difference between the keto- and enol-forms of the chromophore had to be much less than in the oxidized state; leading to the result that alkylation would not only occur at the nitrogen functions of the reduced flavin, but also at the tautomeric iminol positions O(2), O(4) and also at position 4a[10].

The N-alkyl and the O-alkyl derivatives could be easily differentiated by infrared spectroscopy as well as by hydrolysis, since the O-alkyl derivatives were easily hydrolyzed while, of course, the N-alkyl derivatives were not. Surprisingly the complicated mixture of alkyl derivatives lent itself to fractional crystallization. Nitrous acid oxidation followed by 5-deacetylation gave a new revelation, since upon addition of perchlorate as stabilizing anion I could for the first time crystallize alkyl flavoquinonium salts, substituted with methyl in position 1 and in position 2α. The 2α derivatives lent themselves to further substitution of the O-alkyl group by amines, leading into the long-desired iminoflavin region [11]. Furthermore, it turned out that a protonated flavin had the same chromophore as the N(1)-alkylated flavoquinonium cation, which proved that protonation of the flavoquinone was occurring at N(1) and not at N(5) [12], as many people had thought before, and that this protonation was accompanied by a very characteristic hypsochromic shift. The final achievement, however, in the whole alkylation business was omission of the protecting 5-acetyl group which led to a preferential alkylation of the soft position N(5) in spite of the fact that N(5)-H is not acidic [13]. The 5-alkylated dihydroflavins, however, led upon autoxidation to stable blue-green radicals which upon further oxidation lost the 5-alkyl substituent and reverted back to flavoquinone. Thus we now suddenly had stable quinonium salts, stable radicals and stable alkyl derivatives: O-alkyl and N-alkyl of all kinds, and this was the foundation of the later formulated four magic questions concerning flavin reaction mechanisms, namely the question of redox stoichiometry being $1e^-$ or $2e^-$, and the question of catalytic intermediate adducts being π- or σ-adducts. The final discovery in this context was alkyl derivatives fixed to the flavin chromophore in the angular position 4a, from which they could again easily be removed by nitrous acid oxidation or photooxidation in the presence of molecular oxygen, leading back to the starting flavin. By the way, the easiest path to such derivatives was opened up accidentally by Gregorio Weber when he substituted phenylacetic acid for benzoic acid in fluorescence quenching studies and suddenly saw that he got a decolorization of the flavin which was not reversed by oxigen [14]. When we looked into this photoreaction in Massey's lab in Michigan, we found that the colorless flavin solution contained two alkyl derivatives, namely a 4a-benzyl and a 5-benzyl dihydroflavin. In these compounds the benzyl groups arose from decarboxylation of phenylacetate and were attached onto the flavin nucleus covalently — from which, however, they could be easily removed: the 4a-derivative by photooxidation and the 5-derivative by oxidation in the dark, e.g. with O_2, a process which went through a radical intermediate. Again, upon addition of nitrous acid the 5-alkylated derivatives could be intercepted as 5-alkyl quinonium flavins which, as they were absorbing at around 560 nm, turned out to be a new and very important chromophore in the flavin series [15]. Thus it was very important to learn that quaternisation of the oxidized flavin by alkylation in the 1,2α-region would give a hypsochromic effect while the much more difficult substitution, which could only be obtained via the reduced state in position 5, would give a strong bathochromic effect. On the other hand, alkyl groups in position 5 would be even more labile than those in position 1,2α and would act as potential transalkylating agents. In this context one must also mention the fact that in contrast to the ground state of flavoquinone, the excited triplet state has its most basic position not in position 1 but in position 5, which leads to the action pK for all triplet photochemical reactions [16]. This was long disputed and led to the outcry of mine, that "a pK is a pK is a pK" [17]. This was indeed a great surprise in the present case since the pK of position 5 in the oxidized triplet was against Förster's theory. But Förster's theory, of course, applies only if the protonation point is the same in the singlet and the triplet chromophore, which turned out not to be the case for the flavin system. But in any case one should always remain aware that the only easy way into 5-substitution of flavins in the chemical system is through the neutral and anionic reduced flavoquinone. This species will react rapidly with, say a soft alkylating agent at the positions C(4a) and N(5), which are frontier orbital-controlled positions in contrast to the other positions, O(2) and N(1) of the flavin, which will react in charge-controlled ways [18]. This is a very important fact.

But now I must add some remarks about my coworkers. The first man who helped me considerably was Dr. Peter Bamberg [19], with his evaluation of the flavoquinone silver complex. And the second man was my first American postdoctoral fellow, Kenneth Dudley, who first isolated the flavoquinonium salts. The third man was Franz Müller, later a professor at Wageningen, who did all the functional substitutions in positions 2 and 4, which broadened the whole field tremendously. One must always have in mind that at this time nobody had an idea about the working mechanism of vitamin B_2, although this was the first vitamin to be structurally elucidated, and handled commercially. But the apparent commercial uselessness of this vitamin was also the reason for the very slow further development and the reason that practically nobody would go into the underdeveloped B_2-chemistry while at the same time B_2-biochemistry began to be developed many years after the first landmarks set by Leonor Michaelis and by Richard Kuhn, Otto Warburg and Hugo Theorell. One knew meanwhile that flavin was an indispensable redox carrier, but one did not know what was carried and why and in which context! After the fundamental work of the early thirties it was 20 years until Helmut Beinert came up with his fundamental spectral investigation into the flavosemiquinone [20]. I should also mention the work of Henry Mahler and the erroneous conclusion that butyryl-CoA dehydrogenase is a copper flavoprotein [21], which in my early time with Erlenmeyer had great impact and was very important for me in getting further support from my research director. An additional stimulus was then given by the fact that in the CIBA laboratories, where all Erlenmeyer products were checked for biological activity, my first new compound, namely 2-thioflavin, showed some bacteriostatic properties and was even expected to possibly have some anticancer activity. This, of course, in the clinical approach turned out to be wrong. Still it gave me much support and provided conditions under which my investigations could be nicely continued.

Meanwhile time had progressed into the early sixties, and I developed more and more international connections, the first in the territory of Bill Slater in Amsterdam, who was introduced to me by Daniel DerVartanian. Here I found the first general interest, especially from the side of Cornelis Veeger, into the whole and complicated absorption spectral properties of flavin chromophores;

and here I succeeded to show in one of the first years when a suitable apparatus was available, that upon addition of heavy metal ions, solutions of half-reduced flavin would change their dismutation behavior in favor of the radical, that means in the presence of the metal-radical chelate [22]. This was a breakthrough, which led me into cooperation with Anders Ehrenberg in Hugo Theorell's lab at the Karolinska Institute in Stockholm, where we were able to elucidate the tautomeric structure of the radical. This was important because it was wrong as reported in the early Kuhn literature. Richard Kuhn anticipated without discussion that the position of the hydrogen in the neutral radical would be the N(1)-position [23]. We showed, however, by our ESR work in Stockholm that the hydrogen atom is fixed in position 5, leading to the 5-protonated radical. This was one of the most important findings of my career since a 1-protonated radical ought to be very acidic, in contrast to a 5-protonated radical, which showed a pK in the range of 8. Furthermore, it turned out that a 5-protonated radical is blue. This was consistent with the later observed biologically active flavosemiquinone, which first became established, as mentioned before, by Helmut Beinert and then became known through my work with Vincent Massey. Vince, at the same time, demonstrated that not all flavin redox reactions would run through radicals [24], in contrast to what Michaelis had been postulating after his first discovery. On the contrary, it had to be checked carefully with every biological flavin reaction whether it would make use of the radical state or not. This was actually the beginning of the development of the flavoprotein classification by Massey and Hemmerich. This classification is based on the recognition that we have three mechanistically differing principal activities of the flavin, namely (de)hydrogenation, electron transfer and oxygen activation, activities which must be differentiated from each other. Furthermore, it must be distinguished whether one is talking about an input or output reaction because flavin action may change two 1e$^-$-redox equivalents to one 2e$^-$-redox equivalent and vice versa; and when feeding in 1e$^-$ equivalents one may obtain 2e$^-$ equivalents given off by the system [25].

In the meantime, we were writing of the early sixties; and my promotion had been long executed with the support of our Basel Nobel hero, Thadäus Reichstein, who was fascinated by my achievements and gave me a *summa cum laude* mark. My research director Erlenmeyer then provided for me a safe position as Swiss National Research Fellow which I kept for about 10 years. This gave me the opportunity of having in Basel a kind of research empire including the Enzyme Institute in Wisconsin, Bill Slater's lab in Amsterdam, Anders Ehrenberg's place in the Karolinska Institute in Stockholm and finally Thomas P. Singer's new installations at the Golden Gate of San Francisco. It was a fortunate development when I wrote the first letter about semiquinone stabilization to Helmut Beinert that he insisted I should immediately show up in Madison[26]. It was fortunate also, that I could go there without abandoning my Basel position, for any time I wanted, with the support of my Basel directors, and, subsequently, to have the same arrangement with Stockholm, leaving my teaching obligations to other people.

When I first came to Wisconsin, Helmut Beinert urged me to participate in the 1st ISOX-Symposium which was very important to me. The next thing was when I could participate in the copper symposium at Arden House, which will yield a separate chapter of these memoirs. In the U.S.A. I got tremendous encouragement from everybody, and it was the happiest time of my life. In the meantime I had completed my habilitation thesis, and I had obtained a lectureship at the University of Basel, though not much loss of time was connected with this business. But since I was very fond of Basel I never neglected my obligations at this place; and all the swirling around in Wisconsin, Holland, Karolinska and San Francisco could not keep me away from sticking to my old and first sources of support, up to the time when I would get the final call to a chair at the newly founded University of Konstanz[27].

Two things must now be dealt with, and this is my engagement in copper and my beginning political activities.

At this time copper coordination chemistry was nearly exclusively confined to Cu(II) because, again like flavoquinone, bivalent copper is nicely colored, nicely stable towards oxygen, nicely paramagnetic, and is ubiquitous in living matter, though it acts in living matter apparently as a redox catalyst, and as a redox catalyst it could not stay in one redox state. Coordination chemistry of univalent copper was hampered above all by the instability of the univalent hydrate which gave rise to awful precipitations and inhomogeneities of reduced copper systems. Based on Meerwein's discovery[28], I introduced acetonitrile as an inert auxiliary ligand which would stabilize the univalent copper redox state to the extent that now, in the presence of low acetonitrile concentrations, univalent copper could be handled as if it were a stable metal hydrate. This then allowed me to go through the whole spectrum of biologically active prosthetic groups and to check their copper (I) affinity to determine stability constants of copper (I) complexes with biologically active ligands, especially protein prosthetic ligands. I rapidly arrived at the hypothesis that people were thinking wrongly in the biology of univalent copper because they were applying ideas which would only fit into bivalent copper chemistry. And I predicted at the Arden House Copper Symposium[29] that the well-known copper-dependent formation of peptide complexes by liberation of the peptide proton was not the biologically important one, but that one had to go into the interaction of two kinds of prosthetic groups, namely all kinds of biologically active sulfur (and this meant methionine, sulfide sulfur as well as cysteine mercaptide sulfur) and second histidine imidazole as a ligand of high specific univalent copper affinity. Quite in general one could subdivide the biologically active prosthetic groups into copper(I)- and copper(II)-specific groups, as soon as copper(I) could be handled in a homogeneous system containing nothing but the auxiliary ligand acetonitrile, which would, at the required concentration, not disturb any biological activity. The corresponding acetonitrile complexes of univalent copper with suitable inert anions, such as perchlorate or tosylate, could be isolated in the crystalline state and were easy to handle, especially since they turned out to be not particularly O_2-sensitive. This even allowed us in many reactions to work without precautions against atmospheric oxygen. This univalent copper chemistry[30], which I developed, was later on pursued in my Konstanz research but is by no means exhausted even now, especially in its biological aspects. In fact, one has learned that one can approach even sensitive biological preparations such as proteins by inert polar solvents such as dimethyl sulfoxide, and the same is true for acetonitrile. In my opinion, the possibility of eliminating copper reversibly from active copper enzymes with the aid of acetonitrile is not at all exhausted. I think that many interesting results can be obtained in the same way as they have been obtained by treatment of iron sulfur clusters with dimethyl sulfoxide. Especially copper mercaptide chemistry is not exhausted at all, nor is univalent copper imidazole chemistry especially since imidazole must in this case be taken as a ligand which can bind two coppers with liberation of two protons, and can give rise to copper–copper interactions in interesting model compounds.

This development in univalent copper reminded me of a very similar neglect concerning flavoquinone. Thus I only learned on going to San Francisco to Edna Kearney's and T.P. Singer's group, that apart from the long established flavocoenzymes there were derivatives, for example, in succinate dehydrogenase where the flavin is bound covalently to the apoprotein. I predicted immediately, according to what was said above, that the connections between flavin and protein had to be through the active methyl position 8α. Here I have to mention my most potent disciple who joined me in Konstanz and stayed there for the rest of my life, Dr. Sandro Ghisla. With his aid and Tom Singer's preparations of succinate dehydrogenase I was able to disclose with our new ESR that indeed the connection must be via position 8α. We succeeded in reacting an 8α-brominated flavin with histidine imidazole and the ESR of the product was indeed identical with that of the succinate dehydrogenase flavosemiquinone [31]. The next step by Singer and us was to do the same thing with monoamine oxidase flavin preparations. The new flavin of this enzyme turned out to be 8α-cysteinyl flavin, and was again synthesized by Ghisla [32]. Monoamine oxidase remained a challenge to me since it does not fit into my system of flavin substrates. This is because the monoamine hydrogen to be removed in the first step of catalysis is not activated at all and would not form a carbanion such as postulated by the Massey–Ghisla–Hemmerich theory [33].

Finally, I cannot avoid formulating yet another blame for my fellow researchers in this field and this concerns photochemistry. Many fellow scientists have always been laughing at me because I did photochemistry as a kind of biological model chemistry simply because among the three main flavin activities which I have mentioned, the most important one, dehydrogenation, could not be imitated chemically but only photochemically; and many people asked me whether I had a light burning in my stomach which would then catalyze biological dehydrogenation. It was Vince Massey's merit that I overcame this handicap (which was apt to provide some inferiority complexes to me as a bad biochemist), and it was for me a tremendous boost and encouragement when I saw during Massey's sabbatical year in Konstanz that we were able to develop an entirely new photoreduction method with the aid of 5-deazaflavin as sensitizer[34]. By this means we could put single electrons into multicenter redox enzymes and follow the sequence of events to the end; we got rid of the use of nasty reducing agents such as dithionite; we rather had the opportunity now to fill redox catalysts with single electrons (one by one) doing nothing but press a button for illumination. The question is still open which types of multicenter redox clusters may respond to this technique. Not all will do it as was mainly shown by Cees Veeger in his most important work on nitrogenase[35]. But this method will remain a potent means of handling biological oxidoreductions stepwise and getting more insight into multielectron oxidoreduction, such as encountered in flavin-dependent sulfur metabolism, for example, and nitrate metabolism, and especially in the handling of Helmut Beinert's iron-sulfur clusters. I have made clear in my latest work[36] that 5-deazaflavin is a modified flavin that allows one to differentiate between the two possible types of radicals. These are the red radical which is "blocked", as we say now, i.e. protonated, in position 1 and the "blue" radical, which in the natural flavin system is blocked in position 5. While the modified deazaflavin system is colorless, it is still a natural system which stabilizes the blue radical and destabilizes the red radical according to the position of the various one-electron potentials[37]. This natural system is a simple tautomeric equilibrium which we might write out:

$$\text{1-HFl(red)} \rightleftharpoons \text{5-HFl(blue)}.$$

But this equilibrium is a prototropic equilibrium by which protons are moved from position 1 to position 5, and the question is how fast this can occur, especially under biological conditions. One must keep in mind here that protons cannot just walk around but must make their way through preformed hydrogen bridges if they should move rapidly. And according to our theory every single switch from one activity to another in a flavoprotein must be accompanied by a change in conformation between an interaction blocking lone pairs in position 1,2α on the one hand, or 5 on the other hand. Thus we must have a traffic of protons between the upper right hand and the lower middle side of flavocoenzymes; and we should have, furthermore, specific arrangements of water molecules which catalyze this traffic through directed hydrogen bonds.

This is the very beginning of what we call anisotropic flavin chemistry, since we have to assume that the active flavin is embedded in a water layer which is anisotropic, and this being the secret contained in membrane-bound flavins and flavoproteins [38]. In the 5-deazaflavin case, however, 5-deprotonation implies removal of a proton from a carbon center, which by all means is a slow reaction. It is a very interesting question, which is unresolved, as to how fast the equilibrium between the two tautomeric radicals is in the case of 5-deazaflavin. In any case one must be aware that the red deazaflavin is a powerful reductant whereas the 5-blocked tautomer is a mild oxidant [39]. This work was done in my last outstanding resource, namely the Weizmann Institute, under the auspices of my friend Israel Pecht and our beloved young fellow Michel Goldberg, who at the same time developed about the same malignant tumor that I have. But now I have to leave modified flavins though they constitute the largest part of recent progress in B_2 chemistry and biochemistry, mainly thanks to the efforts of Vincent Massey. All I want to emphasize is the fact that this work is not exhausted at all, especially not in the case of 5-thiaflavin which absolutely should be continued by the young people in the Berlin group of Fenner, above all by Grauert [40]. In this context, it appears a silly postulate that a young man doing his habilitation thesis should switch from one biologically active heteroaromatic to another one, in order to prove his abilities, instead of sticking to the most promising case. In the present case this appears to be 5-thiaflavin, since 5-deazaflavin has been worked out much better while the corresponding case of 5-thiaflavin has not yet been exhausted. However, we do not know what deazaflavin still will yield in the hands of the Wolfe-group in Illinois, who find 8-hydroxy-5-deazaflavin active in biological methane metabolism[41]. Another very neglected area of life-long interest to me is pteridine. It is very difficult to understand why pteridines, quite in contrast to flavins, have been considered since their discovery in the thirties as a matter of chemistry and not as a matter of biochemistry, which now turns out to be more urgent than anything else. I can only recommend every young man in the field to come back to pteridines, but not as a chemist, no, as a biochemist, and to deal with xanthine oxidase and nitrogenase as pteridine containing systems[42] and to develop a biopterine vitaminology as indicated by Viscontini in Zürich and which might have most interesting medicinal consequences as is being shown by my friend Curtius[43]. Pteridines are cofactors at least as fascinating as flavins and they are far from completely exhausted. Under the auspices of my friend Pfleiderer I have been doing pteridine chemistry from time to time in my life. For example, I was the first to determine the true structure of the primary hydrogenation product of pteridines, namely 7,8-dihydropteridine and not 5,8-dihydropteridine as it was thought by others [44]. I came to this with the help of my Scottish friend Hamish Wood in the early sixties. And I am still dissatisfied that I will not be able to develop more pteridine chemistry for the rest of my life. It still appears inconceivable to me that academic research as well as industry have not been investing more, especially into biopterin, and that still nobody knows what biopterin is doing in the brain, since one has many illusions about the blood-brain barrier and what is permeable and what is not.

I shall now return to my initial endeavors in the B_2-field. There are some additions I would like to make. For someone starting his career as a chemist, it was a distinct handicap in those years, viz. the 50's, that B_2, in contrast to vitamins that were discovered later, is a "breadless" vitamin, i.e., it has no lucrative pharmacological aspects to it. It was found ubiquitously, it is not expensive to produce, so that the producing company (Hofmann-LaRoche), which held a quasi-monopoly on this vitamin, could not realize much more out of it than chicken feed and an, in every respect, harmless food coloring. There was neither a marked hypo- nor a marked hypervitaminosis of B_2, and to generate a B_2 deficiency in any biological system was a rather expensive and uncertain undertaking. Pharmacologically active derivatives of B_2 were not known and have barely been developed until now, so that interest in B_2 was at a low point, particularly in medicine. This situation is only changing now and we owe this change to a large extent to the flavin research group at the Max Planck Institute, Heidelberg, headed by Georg Schulz and Heiner Schirmer, who have been working on the enzyme glutathione reductase from human blood [45]. Here we have an involvement of B_2, even in human medicine, namely in the oxygen metabolism of red cells.

This is related to the third main function of flavins, which so far in this writing has received short shrift. What stands out here is the fact, surprising in the eyes of a chemist, that a purely organic molecule should be able to activate oxygen. It was only in the most recent past and under the influence of Vincent Massey that it dawned on biochemists' brains how extraordinary a process this is. Here again, concerning the underlying reaction of dihydroflavin with triplet O_2, three questions must be asked: (1) how many electrons does O_2 accept in the rate-limiting step? (2) is there formation of an intermediary adduct, i.e. a flavin-O_2 complex? and (3) how are the two components of this complex linked together? Again, in this case, I had proposed a σ bond instead of previously assumed π-interactions; and history has shown that I was right [46]. I hasten to add though that we owe much of this knowledge to my friend Vincent Massey, whose studies on flavin-dependent oxygenases clearly showed that the active intermediate has the structure of a 4a-HFlOOH [47].

This structure was then also found in the flavin-dependent luciferase by our friend Hastings. This finding revealed the molecular basis of yet another surprising property of flavins, namely the ability, in the bacterial system, to reduce oxygen with emission of light. Nevertheless, the question as to how much oxygen is reduced in one-electron steps, to superoxide — with superoxide itself then being susceptible to dismutation — and how much is reduced directly to hydrogen peroxide in two-electron steps in a straight oxidase reaction, has remained unanswered up to the present, despite intensive research. One can expect that from this work decisive new knowledge on the toxicity of oxygen may be forthcoming, which in turn is related to the chemistry of flavin. This is to be particularly welcomed after the triplet molecule O_2 — particularly its biological implications — has been neglected in German chemistry for decades; actually to the point that one may run into chemists holding a diploma (about the equivalent of an M.S. degree) who never in their life have been taught the concept of singlet oxygen. We sense here once more the aversion of the German chemist toward excited states, photochemistry and short-lived intermediates, which is rooted in the belief that anything that cannot be crystallized and locked away in a cabinet has nothing in common with honest-to-god chemistry. The chemistry of B_2 will see considerable changes and will, in the long run, be anything but a "breadless" undertaking. I venture to predict that this is going to happen after this branch of chemistry will have outlived me. Flavin chemistry came to life in Richard Kuhn's division of the Kaiser-Wilhelm Institute in Heidelberg and will, in the same Institute (presently called Max Planck Institute), now witness a new step toward perfection in Schulz's and Schirmer's work.

But coming back to what I might call my memoirs, I must now mention a second topic of my life which has influenced me quite a bit, namely at first academic education and academic politics and then political ecology. My interest in the reformation of academic education began towards the end of the fifties by an accidental development. When my second son was born we got an advertisement by the German weekly *Die Zeit* and, incidentally, I found there an article on *How to study medicine in Germany* which confirmed all my doubts about recent German education. Thus I accepted an offer by the editor to write an analogous article of *How to study chemistry in Germany*, where I became very nasty towards German academic chemistry as well as chemical industry

[48]. At that time I had to write anonymously or else those who did not like it would have tried to take revenge with the result of a "Berufsverbot". The pity is that even 20 years later, namely now, my sons suffer from the same indifference and the same lack of involvement in German chemical education which still appears to be what we called at home "Justus Liebig selige Witwe Nachfolger Sohn". The outlook of the German chemist — and this is true to some extent even for chemists internationally — has been, notoriously, encumbered by and directed toward the past in contrast to that of the German biologist. This can be seen penetrating down into the academic faculties of chemistry and biology, where no human communication is encountered among chemists, although among biologists cooperation, teamwork, and mutual acknowledgement is high. This applies in particular to the Konstanz faculties of chemistry and biology, although in this so-called "reformed" institution it is somewhat better and easier, and one may open one's mouth without immediately being killed. But one has to imagine that it is a great achievement if a faculty has a common lunch once a week as we have had it through all the years in biology, but not in chemistry. Apparently in chemistry there is too much money, which precludes human communication; and when I wrote my first article about *How to study chemistry in Germany*, the very honorable society of German chemists decided that its president had to run to Hamburg into the editorial conference of the weekly *Die Zeit* and try to force the editors to deliver up to them the man who had written this article [49]. And they were indeed so arrogant as to believe that the editorial conference would obey this demand and would give up every little bit of journalistic ethics just for chemistry, that means for industrial money. And in the next week the top people of the chemical society claimed that this weekly must have hired an idiot and a man who failed all examinations, and so I had to produce my diplomas to have them copied in the weekly in order to show that I was not an imbecile [49]. Again one week later the article became known in the universities and especially in the freshman labs, and people began to speak up against professors and against the industry since this was the first time that anybody dared to say something against Liebig's successors.

Thanks to the fact, however, that I was at Basel in Switzerland, and that people in Basel did not ever care what German industry, or what German academics were doing or thinking, I had nothing to be afraid of!

I was allowed to publish whatever I wanted in Basel, so long as I was content to restrict the subject to Germany and not, perchance, attack Swiss institutions. Thereafter, from time to time, for approximately 10 years in the sixties I wrote for *Die Zeit* a series of articles on university politics and at the same time tried to work in new didactic opinions [50].

In time these articles gave me a nationwide reputation. It was through *Die Zeit* also that I made the acquaintance of Max Delbrück and of his new Genetic Institute at the University of Cologne, where I was sent as press representative to the dedication ceremonies. On this occasion Max took a liking to me and the result of this liking was, much to my surprise, that he considered me for a chair at the University of Konstanz, his next enterprise. Meanwhile, during my stay in San Francisco, I had visited him in Pasadena and was much impressed by his life's work and the way he had of tackling problems. On his side, he was impressed by the bluntness with which I was wont to tell people my opinion without regard for possible repercussions.

And so came the year 1967, the year that saw us assembled for our first preliminary meeting in Konstanz. From sheer love for Wisconsin, Helmut Beinert, of course, declined the chair but was at least prepared to contribute his services to the new university as Permanent Guest Professor. And so it came about that I was suddenly confronted not only with my own full professorship, but also the necessity of finding people prepared to share their fate with my Constantian future.

After Max Delbrück had blown life into molecular genetics in Cologne, he took a decisive step in the direction of the latest fashionable trend, namely membranology, a step I admired especially since, already then, this was moving into the next as well as next-most-difficult trend, molecular physiology of the senses. In those days Max must have thought of me as a membranophile, which I was far from being. However, I did have a membranophile at hand, of whom I was convinced that he would cause a scientific sensation, and that was Läuger from Hans Kuhn's laboratory in Basel. On the other hand, my old friend Sund was waiting, as we had promised each other that the one who received the first academic call would invite the other to follow him to the place of his employment. In Sund's case, the call turned out to be Kiel and I was extremely grateful not to have to keep my promise but to be able instead to steer Sund to Konstanz. And so after 6 months, Konstanz was able to take off with Sund at the helm as protein chemist, Läuger as membranologist, and a few years later, my old friend Brintzinger, on whose didactic methods I was particularly keen, as inorganic chemist, and also with my other old friend Wolfgang Pfleiderer as true-blue German organic, and especially heterocyclic chemist. Thus even the organization of German chemists was unable to find fault, although from time to time they tried to brand Constantian chemistry as "madmanship".

In the meantime, it had become widely known that Hemmerich was the perpetrator of those wicked articles in *Die Zeit* and already fights had taken place in the chemical teaching laboratories of German universities because people were now unwilling to work according to the methods of Biltz and Biltz and had the arrogance to demand of their instructors that they think up some newer teaching methods. However, my oldest son is right now studying chemistry in Heidelberg and nothing, but NOTHING, has changed there in the training of beginners since I was a student. How German chemistry can survive such a degree of complacence remains a mystery to me. It seems to me as though my teachers in Basel, Cyril Grob and Max Brenner, who later became my friend, were chemically didactic demigods in comparison to what beginners in chemistry have to put up with today in German universities.

Reforms in higher education had, in the meantime, been accomplished in both their good and their bad aspects. We had become permissive, much more permissive than I found acceptable; and with a total lack of responsibility a whole generation of students had been raised into a permissiveness from which later there was no escape and which then led to an inability to achieve, and to a drug paradise. The worst was that during the course of building the new universities, everyone willing to cooperate was promised a marshall's staff in his backpack; and that spelled the beginning of reform universities becoming babysitting universities, and everyone was expected to enrich himself (if possible with the help of his union) at the cost of the State if he did not want to be branded a scoundrel.

Now I was forced to perform a significant political turn-about, a fact that brought me no small amount of aggravation in discussions with the Left of the Sixties, despite the fact that in my heart I was of the Left and proved this by ostentatiously becoming a Social Democrat after Brandt's genuflexion in the Warsaw Ghetto[51].

From the beginning I was determined to do my best to prevent the traditional party from sliding further to the left, and I saw my political duty not as many of my contemporaries did in the creation of an intellectual elite, but rather in the preservation of that which had been handed down to us by the so-called creators. And that which had been handed down to us, I soon recognized, was first of all the work of God, which it was important to protect against our destructive drive.

Meanwhile in Konstanz I had developed an absolutely inexhaustible love for Hermann Hesse's landscape, where I came to the conclusion that in the final analysis my scientific work was perhaps less necessary to humanity than the now-indispensable ecological participation of every decent human being; and so as a green Social Democrat, I took on the Constantian city council, spiting the provincial technocrats and occasionally provoking the derision of my academic brethren[52]. The Minister of Education was immediately informed that scientifically I was burnt out. However, I was able to laugh heartily at that accusation, since the scientific production during my tenure was never in question. Unfortunately, as a direct result of my generation's irresponsible permissiveness, which had started with the first steps towards social reform and led to the irresponsibility of the university teacher, the student body in the middle sixties was predictably poor.

And so today I am sitting with my tumor on the balcony, and wondering whether it is more important to investigate further iron-sulfur clusters with Helmut Beinert and better to understand the mechanism of nitrogenase with Cees Veeger, or of saving the Bodensee from the 400 million liters of leaking Swiss oil reserves that are to be pumped by diligent entrepreneurs into unsealed caverns of the Alps in order to protect the Swiss Confederation for evermore from a possible lack of oil.

I have corresponding doubts regarding the sincerity of Reagan. And when I think of oil caverns and neutron bombs, I find it more bearable to live with my tumor despite the fact that I would dearly love to show the light of day to many things, starting with molybdopteridines and finishing with an autobahn-free Bodensee, a landscape offering rest and recuperation to hikers and bicyclists and altogether a generous reorganization of all the inherited cultural building blocks with which the Bodensee country abounds and which we, in our lack of restraint and our arrogance and greed, have neglected for 20 years.

I feel that I have been uncommonly lucky and that life has fulfilled nearly all my dreams. So I take leave with the feeling of having missed nothing and of having done my best for my family, as far as I was able, and not only for my family but for all who are of good will. I die at peace with my church towards which I had harbored many doubts without being able nor wishing to deny it its claim to the monopoly of the means of grace. Thanks are due to Father Körner who will celebrate my funeral rites. The Hallelujah Ramblers will accompany the event with gospel music. So help me God the Father, the Son, and the Holy Ghost, as it was in the beginning, as it is now and in all time and in eternity, Amen.

And I thank my wife and children, who have given me a lifetime of nothing but joy which only few of my contemporaries can claim so easily.

May I hope that at some time these memoirs will form the basis of a Hemmerich Memorial Symposium at the University of Konstanz, to be held under the sign of B_2 chemistry and gather together once again all my good friends.

NOTES

1 W.O. Foye and W.E. Lange, J. Am. Chem. Soc., 76 (1954) 2199–2201.

2 In his first paper (P. Hemmerich, S. Fallab and H. Erlenmeyer, Helv. Chim. Acta, 39 (1956) 1242–1252) the tautomeric structure B was given, reflecting the assumptions common at that time. The predominant tautomeric structure A was proved by his later work. The first thorough discussion is given in K.H. Dudley, A. Ehrenberg, P. Hemmerich and F. Müller, Helv. Chim. Acta, 47 (1964) 1354–1383.

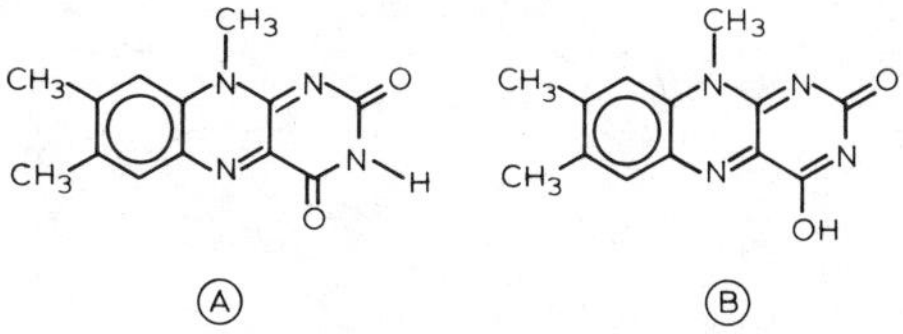

3 R. Kuhn and K. Reinemund, Berichte, 67 (1934) 1932–1936. This synthesis starts with an N-substituted *o*-nitro xylidine (I) followed by its reduction to the o-phenylenediamine (II), which is then condensed with alloxane (III) to form the isoallazine ring system (IV). The sequence shown is for lumiflavin.

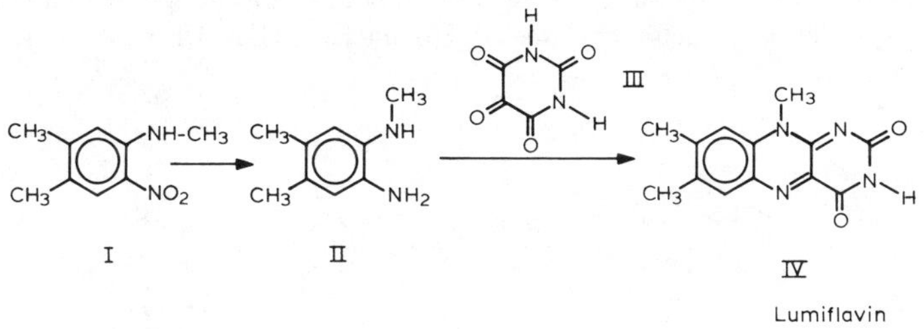

4 P. Hemmerich, S. Fallab and H. Erlenmeyer, Helv. Chim. Acta, 39 (1956) 1242–1252.
The original Tishler synthesis was for riboflavin (M. Tishler, K. Pfister, R.D. Bobson, K. Ladenburg and A.J. Fleming, J. Am. Chem. Soc., 69 (1947) 1487–1492). This was adapted for the lumiflavin series as shown below. Substituted xylidines (I) are diazotized and the aminoazo compound (II) is condensed directly with barbituric acids to yield the flavin (III). The main advantage over the Kuhn synthesis is the accessibility of flavins modified in the pyrimidine ring.

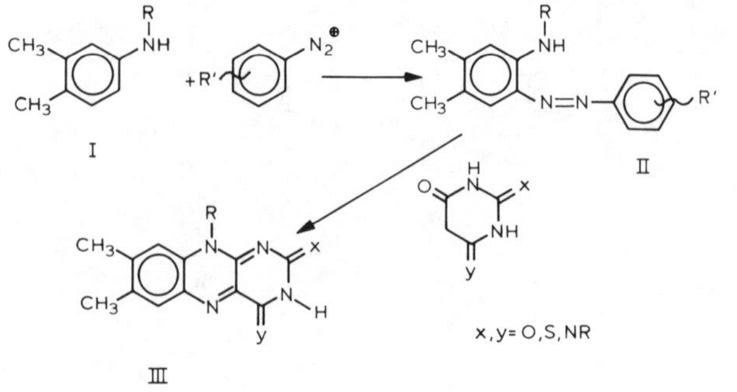

x,y = O,S,NR

5 Hemmerich refers here to the alkylation at position 3 as shown below. This method, which turned out to be of great practical synthetic value works well only at low temperature. The original description (P. Hemmerich, B. Prijs and H. Erlenmeyer, Helv. Chim. Acta, 48 (1960) 372–393 reports the synthesis at 100°C, where a maximal yield of 1% was found. The fact that alkylation proceeds in almost quantitative yield at room temperature was buried in a later paper on flavin-metal chelates (P. Hemmerich, Helv. Chim. Acta, 47 (1964) 464–475.

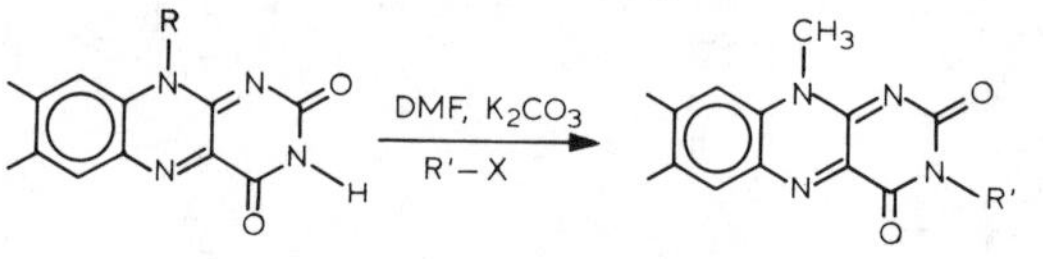

DMF, K_2CO_3 / R'–X

6 P. Hemmerich, B. Prijs and H. Erlenmeyer, Helv. Chim. Acta, 42 (1959) 2164–2177.

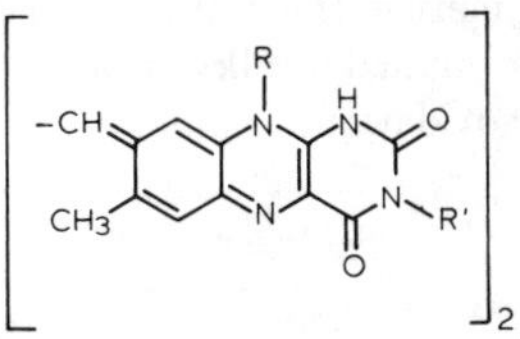

7 Complex formation of flavin neutral radical with metal ions.

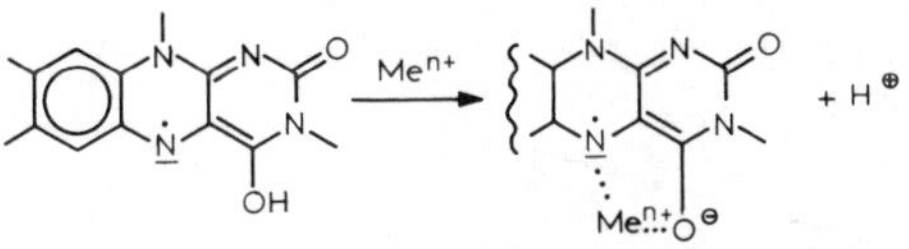

8 L. Michaelis and G. Schwarzenbach, J. Biol. Chem., 123 (1938) 527–542.

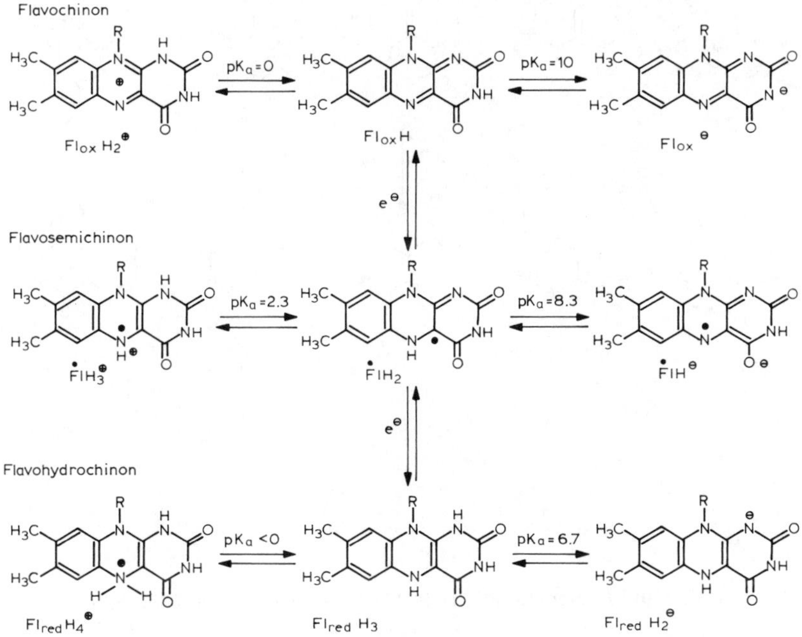

Flavochinon

$Fl_{ox}H_2^{\oplus}$ — $pK_a = 0$ — $Fl_{ox}H$ — $pK_a = 10$ — $Fl_{ox}^{\ominus}$

$e^{\ominus}$

Flavosemichinon

$\dot{F}lH_3^{\oplus}$ — $pK_a = 2.3$ — $\dot{F}lH_2$ — $pK_a = 8.3$ — $\dot{F}lH^{\ominus}$

$e^{\ominus}$

Flavohydrochinon

$Fl_{red}H_4^{\oplus}$ — $pK_a < 0$ — $Fl_{red}H_3$ — $pK_a = 6.7$ — $Fl_{red}H_2^{\ominus}$

9 Scheme showing the three flavin redox forms and the corresponding ionisation states.

10 P. Hemmerich, B. Prijs and H. Erlenmeyer, Helv. Chim. Acta, 48 (1959) 372–394.

The scheme below summarizes the pathways leading to some of the different products which can be obtained by alkylation of N(5)-acetyl-dihydro-flavins, followed by mild acidic oxidation.

Zn^{2+}, $H^{\oplus}$

Ac_2O

RX

$H^{\oplus}$ $-2e^{\ominus}$ $(NO_2^{\ominus})$

$H^{\oplus}$

11 F. Müller and P. Hemmerich, Helv. Chim. Acta, 49 (1966) 2352–2364.

$R-NH_2$

12 K.H. Dudley, A. Ehrenberg, P. Hemmerich and F. Müller, Helv. Chim. Acta, 47 (1964) 1354–1383.
By alkylation at position N(1) (right) the same chromophore is obtained as upon protonation of the flavin nucleus (left). This proves that protonation occurs at N(1).

λ_{max}390 nm λ_{max}390 nm

13 F. Müller, M. Brüstlein, P. Hemmerich, V. Massey and W.W. Walker, Eur. J. Biochem., 25 (1970) 573–580.
S. Ghisla, U. Hartmann, P. Hemmerich and F. Müller, Liebigs Ann. Chem., 1973 (1973) 1388–1415.
The scheme shows the alkylation of unprotected 1,5-dihydroflavins (I) to yield the 5-alkyl dihydro-flavin (II). One-electron oxidation of (II) leads to the blue neutral radical species (III). Further oxidation, which occurs readily only under acid conditions leads to the 5-substituted flavin quinonium salts (IV). This may either eliminate the 5-substituent to give normal oxidized flavin (V) or add H_2O to give the pseudobase (VI), depending on the conditions.

14 P. Hemmerich, V. Massey and G. Weber, Nature, 213 (1967) 728–730.
Scheme showing adduct formation occurring upon illumination of flavin in the presence of phenylacetate.

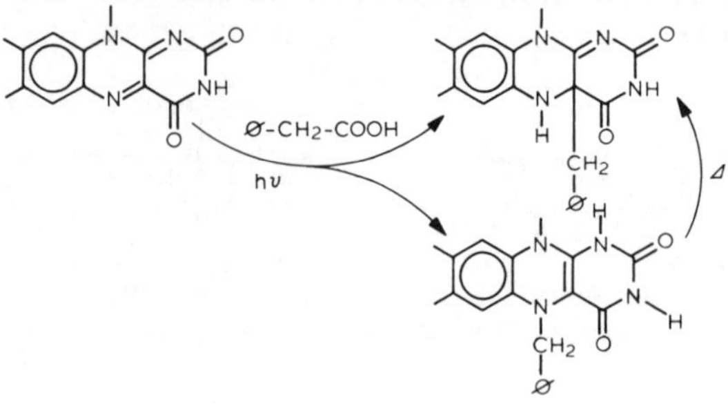

15 W.H. Walker, P. Hemmerich and V. Massey, Helv. Chim. Acta, 50 (1967) 2270–2279 (see also [13]).

16 S. Schreiner and H.E.A. Kramer in T.P. Singer (Ed.), Flavins and Flavoproteins, Elsevier, Amsterdam, 1976, pp. 793–799.
W. Haas and P. Hemmerich, Biochem. J., 181 (1979) 95–105.

17 As noted by Kamin, this passionate statement became a motto for the 3rd International Flavin Symposium.
P. Hemmerich in H. Kamin (Ed.), Flavins And Flavoproteins, University Park Press, Baltimore, 1971, p. 122.

18 C.R. Jefcoate, S. Ghisla and P. Hemmerich, J. Chem. Soc., 1971 (1971) 1689–1694.

19 P. Bamberg and P. Hemmerich, Helv. Chim. Acta, 44 (1961) 1001–1011.
Contributions of Dudley and Müller have been referred to already in [11–13].

20 H. Beinert, J. Am. Chem. Soc., 78 (1956) 5323–5328.

21 H. Mahler, J. Biol. Chem., 206 (1954) 13–26.
Subsequent work from Beinert's laboratory: (E.P. Steyn-Parvé and H. Beinert, J. Biol. Chem., 233 (1958) 853–861) showed that the green color of the enzyme which Mahler attributed to copper persisted even on complete removal of copper from the preparation.

22 F. Müller, P. Hemmerich and A. Ehrenberg, Eur. J. Biochem., 5 (1968) 158–164 (cf. [7]).

23 F. Müller, P. Hemmerich, A. Ehrenberg, G. Palmer and V. Massey, Eur. J. Biochem., 14 (1970) 185–196.
Structures I and II show the tautomeric forms of the neutral flavin radical referred to in the text. Structure I was that assumed by R. Kuhn and R. Ströbele, Berichte, 70 (1937) 753–760, while Structure II was that shown by the ESR studies.

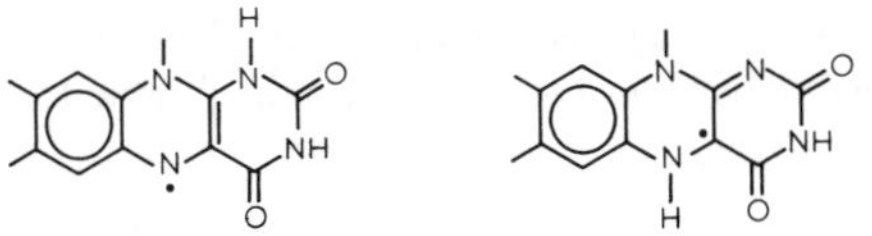

I

II

24 V. Massey and S. Ghisla, Ann. N.Y. Acad. Sci., 227 (1974) 446–465.

25 P. Hemmerich, V. Massey and H. Fenner, FEBS Lett., 84 (1977) 5–21.

26 This was the occasion of Hemmerich's first visit to the United States in 1964.

27 The move to Konstanz was made in the summer of 1967.

28 H. Meerwein, V. Hederich and V. Wunderlich, Ber. Deut. Pharm. Ges., 63 (1958) 548.
The contribution of Meerwein was to introduce acetonitrile as a Cu(I)-ligand, stabilizing this valence state of Cu against autooxidation.

29 P. Hemmerich in J. Peisach, P. Aisen and W.E. Blumberg (Eds.), The Biochemistry of Copper, Academic Press, New York, 1966, pp. 15–34.
This meeting, held at Arden House in Harriman, NY, in 1965, was the first of a series of international symposia devoted to biological aspects of copper. Many of the ideas presented by Hemmerich were quite speculative, but have largely been proven correct in later work. One of the important concepts advanced by Hemmerich was that the only biological ligands which would stabilize both Cu(I) and Cu(II), thus permitting efficient electron transfer, would be imidazole nitrogen or cysteine sulfur. This prediction has been borne out for example in the case of the blue copper protein plastocyanin (P.M. Colman, H.C. Freeman, J.M. Guss, M. Murata, V.A. Norris, J.A.M. Ramshaw and M.P. Venkatappa, Nature, 272 (1978) 319–324) and superoxide dismutase (J.S. Richardson, K.A. Thomas, B.H. Rubin and D.C. Richardson, Proc. Natl. Acad. Sci. USA, 72 (1975) 1349–1356).

30 C. Sigwart, P. Kroneck and P. Hemmerich, Helv. Chim. Acta, 53 (1970) 177–185.
V. Vortisch, P. Kroneck and P. Hemmerich, J. Am. Chem. Soc., 98 (1976) 2821–2826.
P. Kroneck, V. Vortisch and P. Hemmerich, Eur. J. Biochem., 109 (1980) 603–612.

31 S. Ghisla, U. Hartmann and P. Hemmerich, Angew. Chem. (Int. Ed.), 9 (1970) 642–643.
W.H. Walker, T.P. Singer, S. Ghisla and P. Hemmerich, Eur. J. Biochem., 26 (1972) 279–289.
This scheme shows the key step in the synthesis of succinate dehydrogenase: Flavin. 8-α-Brominated tetraacetyl riboflavin was reacted with *N*-benzoylhistidine to yield the imidazole (N)3-derivative as the major product. Acid hydrolysis removed the protecting groups to yield 8α-histidyl(3)-riboflavin, which was shown to be identical with the flavin obtained from succinate dehydrogenase.

32 S. Ghisla and P. Hemmerich, FEBS Lett., 16 (1971) 229–232.
W.H. Walker., E.B. Kearney, R. Seng and T.P. Singer, Eur. J. Biochem., 24 (1971) 328–331.

33 Hemmerich had an aversion to the involvement of radicals in substrate dehydrogenations by flavins. In the meantime it does indeed appear that monoamine oxidase functions by a radical mechanism (R.B. Silverman, S.J. Hoffman and H.B. Catus in V. Massey and C.H. Williams (Eds.), Flavins and Flavoproteins, Elsevier, New York, 1982, pp. 213–216).

34 V. Massey and P. Hemmerich, Biochemistry, 17 (1978) 9–17.
In this work which was begun during Massey's Konstanz sabbatical in 1973–1974 it was recognized that the driving force in the reduction was the generation of the very low-potential deazaflavin radical. The radical was postulated as being derived from photodissociation of the stable radical dimer, which was formed initially in a group transfer reaction involving a transient deazaflavin-EDTA adduct. The reaction was later shown to involve direct radical formation from excited deazaflavin and EDTA; the deazaflavin radical so produced readily dimerizes, as well as being a potent reductant (M. Goldberg, I. Pecht, H.E.A. Kramer, R. Traber and P. Hemmerich, Biochim. Biophys. Acta, 673 (1981) 570–593).

35 Hemmerich here refers to work where the role of flavodoxin in the *Azobacter vinelandii* introgenase complex was clarified (G. Scherings, H. Haaker and C. Veeger, Eur. J. Biochem., 77 (1977) 621–630). It was shown that flavodoxin semiquinone, produced in the photoreaction, served as reductant of the iron-sulfur protein of the complex.

36 M. Goldberg, I. Pecht, H.E.A. Kramer, R. Traber and P. Hemmerich, Biochim. Biophys. Acta, 673 (1981) 570–593.

37 This highly speculative hypothesis was one of which Hemmerich was very fond. It is discussed in some detail in a review article written shortly before his death, which was prompted by his hosting of a Symposium in Konstanz 9–13 March 1981, in honor of Helmut Beinert's 65th birthday.
(P. Hemmerich, V. Massey, H. Michel and C. Schug, Struct. Bond., 48 (1982) 93–123).

38 Here Hemmerich probably means a conducting cage of water molecules surrounding the flavin in order to accommodate the proton transfer addressed in the preceding paragraph.

39 At this point the sequence of arguments might seem confusing. The concept for the mobile proton pool arose from the work on 5-deazaflavin radicals carried out in collaboration with Israel Pecht and Michel Goldberg in Rehovoth, and Rainer Traber and Horst Kramer in Stuttgart (M. Goldberg, I. Pecht, H.E.A. Kramer, R. Traber and P. Hemmerich, Biochim. Biophys. Acta, 673 (1981) 570–593).

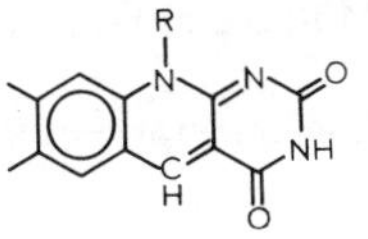

5-Deazaflavin

40 H. Fenner, R. Grauert and P. Hemmerich, Liebigs Ann. Chem. (1978) 193–213.
H. Fenner, R. Grauert, P. Hemmerich, H. Michel and V. Massey, Eur. J. Biochem., 95 (1979) 183–191.
Hemmerich held great hopes for the modified flavin, which was found to retain only the one-electron-transfer activity of native flavins. Its utility in biochemical systems has, however, been restricted because of the high redox potential of the reversible 1e$^-$ oxidation-reduction (E_m, pH 7, + 0.38 V).
The scheme shows the 1e$^-$ oxidoreduction of 5-thiaflavins.

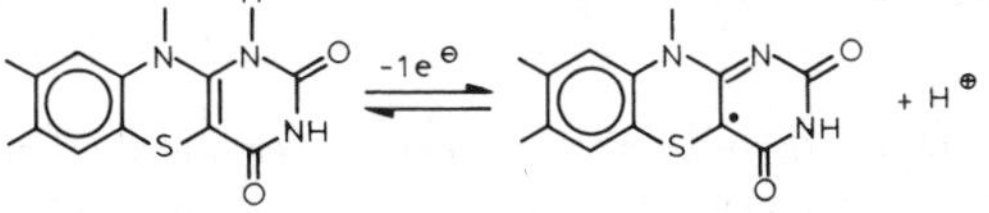

$-1e^{\ominus}$ $+ H^{\oplus}$

41 L.D. Eirich, G.D. Vogels and R.S. Wolfe, Biochemistry, 17 (1978) 4583-4593.
It should be noted that 8-hydroxy-5-deazaflavin, by virtue of its *p*-quinod structure, is less kinetically constrained during oxidoreduction, than is the parent 5-deazaflavin.

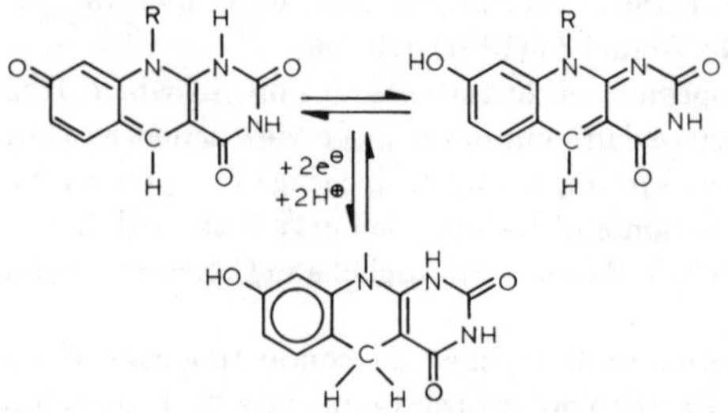

42 Hemmerich refers here to the recent work from Rajagopalan and colleagues, who present evidence for a molybdenum-pterin cofactor in sulfite oxidase and xanthine dehydrogenase (J.L. Johnson and K.V. Rajagopalan, Proc. Natl. Acad. Sci. USA, 79 (1982) 6856–6860).*

43 Prof. H.C. Curtius was a fellow student of Hemmerich in Basel, and they had maintained a close relationship over the years. See several articles in Biochem. Clin. Aspects Pter., 1 (1982) 27–49.

44 A. Ehrenberg, P. Hemmerich, F. Müller and W. Pfleiderer, Eur. J. Biochem., 16 (1970) 584–591.
Course of hydrogenation of pterines and formation of the 7,8-dihydro isomer.

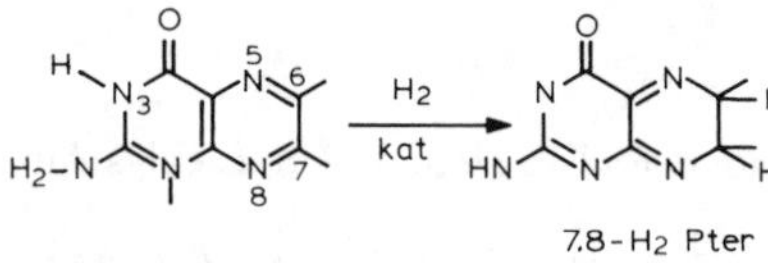

7.8-H_2 Pter

45 Here Hemmerich refers not only to the determination of the complete three-dimensional structure of glutathione reductase (G.E. Schulz, R.H. Schirmer and E.F. Pai, J. Mol. Biol., 160 (1982) 287–308) but also to the interest of this group in pharmacologically active inhibitors of the enzyme, which have considerable potential as anti-malarial drugs. He was also aware that the same workers had found that glutathione reductase was the probable target enzyme of some potent anti-tumor drugs.

---

*At the time the memoirs were written it was not certain whether nitrogenase also contained pteridin component. It is now known that it does not.

46 Historically, the first suggestion of a covalent flavin-oxygen linkage (FMNH-O-OH) appears to have been made by Q.H. Gibson and J.W. Hastings, (Biochem. J., 83 (1962) 368–377), while the first structure (that of a flavin C(10a)-hydroperoxide) see below, was proposed by Berends et al. at the first Flavin Symposium, held in Amsterdam in 1965 (W. Berends, J. Posthuma, J.S. Sussenbach and H.I.X. Mager in E.C. Slater (Ed.), Flavins and Flavoproteins, Elsevier, Amsterdam, 1966, pp. 22–36). In the 1968 Mosbach Colloquium Hemmerich predicted that the flavin C(4a)-hydroperoxide would be a more likely possibility (P. Hemmerich, Proc. 19th Mosbach Colloquium, Springer Verlag, Berlin, 1968, pp. 249–255).
Structure of the postulated flavin (10a) hydroperoxide.

47 The first demonstration of a flavin hydroperoxide in an enzymic system was with *p*-hydroxybenzoate hydroxylase (T. Spector and V. Massey, J. Biol. Chem., 247 (1972) 5632–5636). In the next year it was also found in bacterial luciferase (J.W. Hastings, C. Balny, C. LePeuch and P. Douzou, Proc. Natl. Acad. Sci. USA, 70 (1973) 3468–3472). This was commonly assumed to be the C(4a) hydroperoxide, because of similarities with the spectra of known flavin C(4a)-derivatives; the structure was proved unequivocally in the case of luciferase by means of [13C-]NMR (S. Ghisla, J.W. Hastings, V. Favaudon and J.M. Lhoste, Proc. Natl. Acad. Sci. USA, 75 (1978) 5860–5863).
Structure of flavin (4a) hydroperoxide, the active intermediate which has been shown to occur in several oxygen activating enzymes such as the phenol hydroxylases, bacterial luciferase and *N,S*-monooxygenase.

48 This was a series of four articles which appeared in the Feuilleton section of the weekly Die Zeit. 10 June–4 July, 1960.

49 In the issue of Die Zeit of 22 July 1960, there are a number of letters of reaction to the series on *How to study chemistry in Germany*. Included among these is the reply of the president of the Gesellschaft Deutscher Chemiker in which he expressed his outrage at the things Hemmerich had said. It also contains a certificate by the editor that the author had in fact completed his studies summa cum laude!

50 Many articles during this time were published in the Travel and Modern Living sections of Die Zeit under the name Peter Mörser, a particularly appropriate one which was assigned to him by the editors. For the political articles his own name was always used.

51 Chancellor Brandt's trip to Warsaw took place in 1970. During this time he made a dramatic gesture of reconciliation by kneeling at the memorial to the former Warsaw ghetto. This event had wide repercussions in Germany, and led Hemmerich to join the Social Democrats shortly after.

52 Hemmerich first joined the Konstanz City Council in 1978 and was a constant provocation to most of the council right until the time of his death.

G. Semenza (Ed.) Selected Topics in the History of Biochemistry: Personal Recollections (Comprehensive Biochemistry Vol. 36) © 1985 Elsevier Science Publishers

Chapter 9

Adventures and Research

EDGAR LEDERER

Laboratoire de Biochimie, C.N.R.S. 91190 Gif-sur-Yvette and Institut de Biochimie, Université de Paris-Sud, 91405 Orsay (France)

Vienna (1908–1930)

I was born on June 5th, 1908 in "Schloss Hacking", a 19th century mansion, in the outskirts of Vienna, which my grandparents Przibram had bought and where four families lived: my grandmother Charlotte, the two brothers of my mother, Hans and Karl, which I shall mention later, and my parents.

My mother, Friederike Przibram was born in Vienna in 1881; her mother Charlotte was née Baronin Schey (1851–1939). Friedrich Schey (1815–1881), father of my grandmother Charlotte was born in Güns, a small Hungarian town near the Austrian border and had moved to Vienna in 1832; he became one of the foremost financiers of his country, furthering industrial development, expanding the Austrian railways, building theatres, a hospital, etc. In 1869 he was knighted and became "Freiherr von Koromla"; he received decorations from the emperors of Austria, Mexico, France, Brasil, Russia, etc. [1]. A brother of my grandmother, Josef Schey (1853–1938) was professor of law at the University and the author of the Austrian civil law code; he also became member of Parliament and got the title "Hofrat".

Gustav Przibram (1844–1904), my mother's father, had been a wealthy textile manager in Prague and deputy of the "Landtag of Bohemia"; he had moved to Vienna in 1871. In his spare time he

Plate 14. Edgar Lederer (1982).

wrote poetry and theatre plays, some of which were published under the pseudonym "Hans Walter". He was a very progressive man; his house, where I later lived from 1918–1930, was one of the first in Vienna to have electric lighting.

The family of my father Alfred was of a more modest origin. His mother, Sophie, née Janovsky (1848–1916) was born in Teplitz (now Teplice) in Moravia. I still have a little book with the title *Poésies à Sophie Janovsky* where she had copied French poems with a very neat and regular handwriting*. She married Julius Lederer (1838–1919), a textile merchant living in Zwittau (now Zvitavy). My father was born there in 1872; they moved to Vienna around 1885.

My father was the eldest of seven and had to earn his living very early. From his modest origin, until he became a wealthy and very successful lawyer in Vienna, he had kept his habit of living simply and never really felt at home in the company of my mother's relatives (the Schey barons, the Lieben bankers etc.).

My eldest sister and myself had an English nanny, so that my first language was English, followed of course by German (with a slight Viennese accent). My mother had been brought up by an English lady, who stayed with us and gave me English lessons.

I can hardly say how lucky I was to have lived at home in an "English atmosphere"; in later years English was a great asset for my scientific career and for close contacts and collaborations with foreign colleagues.

I learnt some Italian with my mother, French much later, with an old aunt; then I started to learn Russian: at that time I became an ardent admirer of the Russian Revolution, after having read John Reed's *Ten days that shook the world*. The interest in foreign languages and the ease to learn them was quite usual in our Viennese society. My grandmother had a large library with German, English and French literature. I still have some of her volumes of Zola, Loti, Anatole France and others. My mother was an admirer of Byron,

*One of her ancestors was Benjamin Wolf Popper (1680–1767) who became "Primator of the Jews of Bohemia" in 1749; his son, Joachim Edler von Popper, became Primator in 1790.

H.G. Wells and G.B. Shaw; all their works are still with me and needless to say also a 32-volume edition of Goethe's *Gesammelte Werke*, the works of Schiller, Heine, Lessing, etc. I have read many of them in my youth and now they are an essential part of my environment in my study at home.

From my "Red period" I still have some Russian books of Lenin and Stalin, which were hidden in the cellar during the last war, when we were in the so-called "free zone" of France near Lyon. Stalin's works were especially appreciated by the rats in our cellar.

I grew up rather independently, my father being very occupied in his office and my mother with her social obligations and literary interests. My father loved music and played the violin, my mother and her brothers had a great talent for drawing, but I have inherited none of these talents.

A few words about Vienna in the post-war years. I was ten when the first World War ended and have of course only few memories of the years following the war; Austria had shrunk to a tiny country; 30% of the population lived in its big capital, where many marvelous buildings reminded one of the glorious past. The economic situation was very bad and food was scarce. In the Akademische Gymnasium which I attended from 1918–1926 there was once a general check up of the pupils: I was one out of four of a class of about 30 to be in a normal nutritional state! I also remember some riots when one of the big hotels was looted.

During the war more than a hundred thousand Galician Jews had found refuge in Vienna and had to earn their living by trade. This of course increased the more or less latent antisemitism of the population.

For my family, entirely assimilated for nearly two generations, being a jew was not based on any religious belief, but mainly on other people's attitude. There was a deep feeling of solidarity for our kinship and we had a strong incentive to work hard and to do better than the more or less different or hostile "others".

My interest in biology was aroused and stimulated very early, through an excellent teacher in the Akademisches Gymnasium, but still more by the elder brother of my mother, Hans Przibram, an eminent biologist, who was "ausserordentlicher" professor at the University of Vienna (at that time a jew could not become "ordentlicher"; even Freud never became "ordentlicher Professor"!). My uncle Hans had founded in 1903 and equipped with his private funds the first institute of experimental biology in the world (the "Biologische Versuchsanstalt" in the Prater); in 1914 he handed the "Vivarium" (as it was also called) over to the Academy of Sciences. When the nazis occupied Vienna in March 1938, my uncle was forbidden to enter the Institute he had founded. After the Anschluss he left Vienna with his second wife and lived for some time in Holland. After the Germans occupied the country they advised the refugees to go to Teresin (Theresienstadt, south-west of Prague) where they would be "in security". The nazis had organised a ghetto there from which most people were sooner or later deported to death camps. My uncle died miserably in 1944 in Teresin, at the age of 70; his widow committed suicide after his death*.

The "Vivarium" was destroyed by an air-raid and was not rebuilt.

My mother's younger brother, Karl Przibram (1878–1973) was "ausserordentlicher" professor of physics and worked at the Viennese Radium Institute; he introduced me very early to delicate manipulations with radioactive compounds. Having an aryan wife, he spent the war years unharmed in Belgium and then returned to Vienna after the war to become full professor, director of the Radium Institute and member of the Academy; he was also awarded the title of "Ehrenbürger" of Vienna.

Other members of my mother's family were also scientists: Adolf Lieben (1836–1914), professor of chemistry at the University**, who had married a sister of my grand-mother, their son Fritz Lieb-

*After the Anschluss two of his sons-in-law had committed suicide, his eldest daughter died in deportation . . . For (more or less correct) details on Hans Przibram as director of the Biologische Versuchsanstalt see Arthur Köstler's book *The Case of the Midwife Toad* (Random House, New York, 1971). For biographies see [2].

**He had discovered the aldol reaction; he had close scientific relations with French scientists, Jean-Baptiste Dumas in particular and was made "Commandeur de la Légion d'Honneur" in 1908.

en (1890–1966), a biochemist, and Otto von Fürth (1867–1938), a former professor of medicinal chemistry of the then German town of Strassburg, well known for his early work on adrenalin, author of a book *Vergleichende chemische Physiologie der niederen Tiere* (Verlag Gustav Fischer, Jena, 1903), which was later very valuable to me for my work on pigments of invertebrates. I visited him often at the Medizinisch-Chemisches Institut in Vienna. He died soon after the Anschluss, but his widow, his son and daughter and even his 92-year-old mother-in-law were deported by the nazis and disappeared.

During my first years of high school, I spent most of my free time with a microscope, looking at the wonders of the unicellular world. I was especially fascinated by the amoebae, by their phagocytic activities and by Metchnikoff's macrophages.

Why did I study chemistry? Simply because it was the only non-mathematical science promising a living to its students (at that time at least). I would have liked to study microbiology, but this was only possible at the Medical Faculty and I was horrified by the idea of dissecting corpses*.

My first two years of chemistry at the Chemical Institute of the University have left me few memories: several explosions and fires which I caused were mainly due to the fact that I was in a hurry (as I still am) and worked on two benches at a time.

Twice a year there were "Judenkrawalle" at the University which consisted in attacks of the "deutschnationale" students against their jewish colleagues; once my neighbour in the lab (blond and blue-eyed were we both) turned to me and said "I am glad that there are no jews in this lab". I was quick to reply "but I am a jew"; despite that incident our relations remained good.

In September 1928, after two preparatory years, I wanted to start my doctor's thesis in the "Dissertantensaal" of the boss, Prof. Ernst Späth. The day before I was scheduled to start work there, a

*Later, in Heidelberg, after having read some textbooks on anatomy and physiology I wanted to see how the "real thing" looked like, so I attended a few sessions of anatomy at the University; but when the leg I had tried to dissect kicked me on the nose, I was utterly disgusted and stopped this ghastly occupation.

petition was circulated amongst the 15 students of that lab: "we do not want the jewish student Lederer to come here". Needless to say that I found an icy atmosphere there.

Prof. Späth, well known for his brilliant work on the structure of alkaloids and coumarins was very friendly to me for the first 5 or 6 months, but all of a sudden he stopped coming to my bench. A few days later he called me in his office and showed me an anonymous letter "you have a jewish student E.L. who says wicked things about you, etc . . ."; as a result of that letter I had to finish my doctor's thesis without his help. Working very hard, the first to arrive in the morning, the last to leave in the evening I finished my thesis (*On the synthesis of indole alkaloids and isoflavones*) in two years*. My first three papers were published in 1930 [3].

From Späth I learnt to manipulate milligram quantities of compounds and the delicate art of microanalysis. I realized later how much I owe him for these fundamental and so simple techniques which became essential for my future work on isolation and degradation of natural compounds.

Späth was not an antisemite, but he informed me that he could not take a jew as an assistant, so I had to look elsewhere for my further career. I asked him to write to Richard Kuhn to recommend me, but Kuhn refused to take me; I later learnt, Späth had written that I had a very bad temper; it is curious, how this reputation followed me to France!

A friend and colleague of one of my uncles then wrote to Kuhn, who immediately accepted me.

---

*I had no time for amusements. Those who know Vienna and the life of Viennese students, at that time, will be astonished that I had never been to a "Heurigen" in Grinzing before being a guest of Viennese Colleagues, after the war.

Heidelberg (September 1930–March 1933)

Richard Kuhn (born in Vienna in 1900), a pupil of Richard Willstätter* had been appointed professor of chemistry at the Eidgenössische Technische Hochschule in Zürich, at the age of 27. In 1930 he became head of the Chemistry Department of the brand-new Kaiser Wilhelm Institut für medizinische Forschung in Heidelberg**.

He was not only a most brilliant scientist and an inspiring teacher, but also a very pleasant companion in the lab, or in the cafeteria or for chess, or ping-pong games.

My years with Kuhn were the happiest in my life. I arrived there in September 1930 at a crucial moment, when it was essential to develop a method for separating and purifying natural carotenoids (carotene had just been shown by Euler and Karrer to be a provitamin A). This led me to the first preparative applications of Tswett's chromatography and the discovery of the isomeric α- and β-carotenes and several other carotenoids***.

The Russian botanist Michael Tswett (1872–1919) had described his method of adsorption chromatography in a book published in Russian in 1906 [5]. The value of his method for preparative purposes had been criticized by Richard Willstätter, because he had, with A. Stoll, tried without success to purify chlorophyll on a column of alumina; the pigment was destroyed; they had not noticed, however, what Tswett had already stated: the chlorophylls are very unstable pigments and should be chromatographed on a

---

*R. Willstätter was born in Karlsruhe, 1872; he received the Nobel prize for chemistry in 1915; he resigned from his chair at the University of Münich in 1925 because of the antisemitism prevailing at the Faculty, and died in Locarno in 1942. See his autobiography *Aus meinem Leben*, Verlag Chemie, Weinheim, 1949.

**R. Kuhn received the Nobel prize for chemistry in 1939 "for his work on carotenoids and vitamins"; the nazi government did not let him go to Stockholm to receive the award; he died in 1967 in Heidelberg. After the war the Kaiser Wilhelm Gesellschaft was renamed Max-Planck Gesellschaft.

***See *La renaissance de la méthode chromatographique de M. Tswett en 1931* [4].

'late 15. Edgar Lederer (left) and Richard Kuhn, September 1963.

"gentle" adsorbent, i.e. powdered sucrose, or inulin. Thus, Tswett's method did not seem interesting to chemists and was nearly forgotten, except for some papers of botanists or others, who had used it to separate natural pigments on small columns and to measure the absorption spectrum of the eluates.

The day I arrived in Heidelberg, headlines in the local newspapers stated that "scientists at the Kaiser Wilhelm Institut have made egg yolk from grass". In reality, Kuhn and his assistant Alfred Winterstein (1897–1960) had been comparing the properties of leaf xanthophyll and the yellow pigment of egg yolk, both first crystallized by Willstätter and described as isomers $C_{40}H_{56}O_2$.

My first task was the precise comparison of melting points and rotations of various preparations of leaf xanthophylls with lutein from egg yolk*; Kuhn then suggested that the differences in spectra, melting points and rotations I had observed could be explained if lutein was a mixture of leaf xanthophyll with zeaxanthin (also $C_{40}H_{56}O_2$), the yellow pigment of corn, just described by Paul Karrer in Zurich. But how to prove this hypothesis?

Fortunately Kuhn had received from Willstätter the manuscript of a German translation of Tswett's book [5]. There I read all the details and decided to try an experiment; a large glass tube was filled with calcium carbonate, and a solution of egg yolk lutein in carbon disulfide (a rather inappropriate solvent!) was filtered through; after washing with CS_2 two coloured zones were scraped out of the column; the pigments were then eluted and *crystallized*: one was identical to leaf xanthophyll (m.p. 193°C, $[\alpha]_{Cd}+145°$), the other to zeaxanthin (m.p. 207°C, $[\alpha]_{Cd}-70°$) [6].

This proved that Kuhn's hypothesis was right and — more importantly — that Tswett's method, when applied properly, was very useful for the separation and subsequent crystallisation of rather unstable pigments.

*This work as well as the following with carotene, was greatly facilitated by the availability of a new polarimeter with a cadmium lamp giving a strong red light $[\alpha]_{Cd644}$.

Then came the problem of carotene, the pigment of carrots which had been first crystallized in 1831 by Wackenröder. Paul Karrer had just published the structure of carotene based on the identification of various oxidation products and on a great deal of intuition. His formula has no asymmetric carbon atom; however, on examining solutions of carotene in the polarimeter I constantly found values varying from +15° to +60°; so either Karrer's formula was wrong, or the carotene of carrots was a mixture. I first tried to purify carotene by precipitating it as an iodide and then decomposing the iodide with thiosulfate; this then gave an optically inactive preparation. The mother liquor of the carotene iodide was also treated with thiosulfate and the solution left standing overnight; to my great delight, when returning in the morning, I found beautiful, glittering orange-red crystals in the flask. They had an optical rotation of more than +320°. α-Carotene was discovered!* [7, 8] (Figs. 44 and 45).

Later the use of chromatography on a special alumina ("Fasertonerde nach Wislicenus") gave better yields for the separation of the isomeric α- and β-carotenes [7, 8].

These results were published in a short note in *Naturwissenschaften*, dated February 17th, 1931 [7]; the results on lutein and the xanthophylls were published in the *Hoppe Seyler*, dated March 10th [6]; the detailed paper on the carotenes was sent to the *Berichte*, on March 18th, 1931 [8]. This paper is usually considered as marking the "birth of preparative chromatography"**.

---

*We proposed to call the optically active isomer α-carotene because it was supposed to contain at least one α-ionone ring and the inactive β-carotene (corresponding to the formula proposed by Karrer which has two β-ionone rings).

**In 1964, R. Kuhn, awarding me the A.W. Hoffman gold medal of the Chemische Gesellschaft said: "In Heidelberg, an meinem Institut haben Sie α und β Carotin getrennt und die Methode der Chromatographie zu neuem Leben erweckt" [9]. R.L.M. Synge, in *A retrospect on liquid chromatography* [10] wrote: "It is from the first chromatographic publication of these two (Kuhn and Lederer, 1931) that we can sharply fix the establishment of chromatography as a main branch of analytical endeavour".

Fig. 44. α-Carotene. ×225 [8].

Tswett's method was then very rapidly adopted by Paul Karrer's laboratory in Zürich and by Laszlo Zechmeister in Pécs (Hungary); in Kuhn's lab Alfred Winterstein and Hans Brockmann applied it to various problems.

Tswett's method became an essential tool in my further work, for the isolation of new natural compounds.

Another, very important event for my private life happened in Heidelberg. The city was a very romantic place and there is a famous song "Ich hab mein Herz in Heidelberg verloren . . ." That is what happened to me and to Mademoiselle Hélène Fréchet. She had come to Heidelberg to attend the lectures of Prof. Emil Gumbel, a mathematician and anti-fascist, who had to leave Germany even before Hitler came to power.

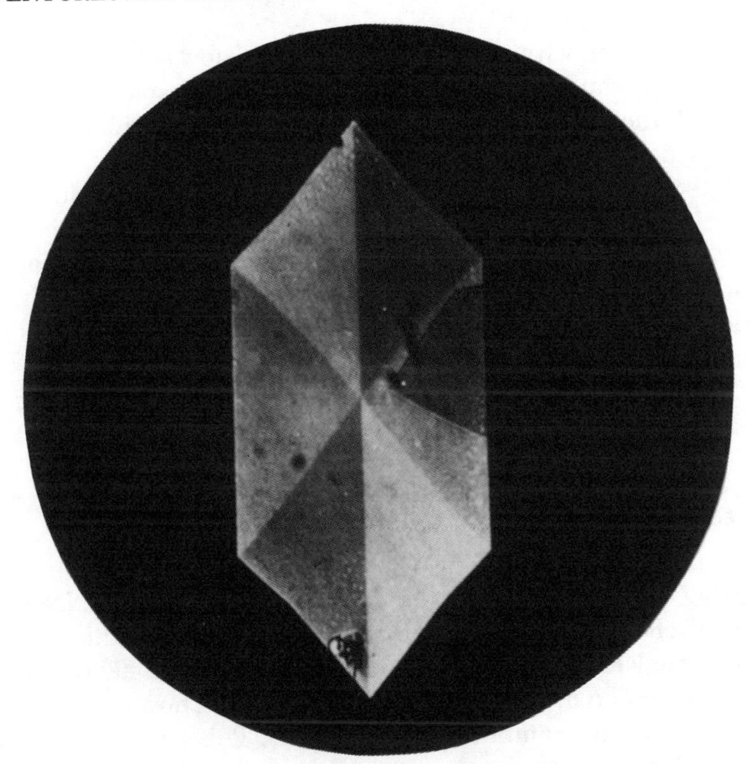

Fig. 45. β-Carotene. ×225 [8].

We met one Sunday afternoon, at a tea party and 6 months later, in June 1932, we were married in Paris, despite some resistance from my father-in-law, Maurice Fréchet (1878–1973), a well known mathematician and professor at the Sorbonne.

We spent a 3-day honeymoon near Fontainebleau, then I had to return to Heidelberg to work (this "sinister urge" has never left me . . .).

I had taken up another problem: the purification of "vitamin H". Paul György, at that time professor of paediatrics at the Heidelberg University, had found that rats fed a diet rich in raw egg-white got a severe eczematous disease, lost weight and finally died. He had asked Kuhn for his collaboration in isolating the liver factor which was named vitamin H. The first difficulty I encountered was that extraction with solvents gave only very low activity, as tested on rats by Paul György. Highly active aqueous solutions were at last obtained by digestion with papain; further purification steps included precipitation as a gold salt*.

After I had left Heidelberg it became clear that "vitamin H" was identical to biotin, isolated by F. Kögl in Utrecht. The toxic factor in raw egg-white had been purified by Theodor Wagner-Jauregg working in the same lab as I and was called avidin because it very avidly binds biotin. The strong affinity of biotin for avidin is now widely used in biochemical research.

My papers with Kuhn were usually written on Sundays in his home. Once he told me

> "I am so glad my wife has gone to Zürich to see her family, so that I can go to bed with my boots and the Zentralblatt".

During the war, Richard Kuhn became president of the Deutsche Chemische Gesellschaft and as such had to terminate his speeches with "Heil Hitler"; after the war he was rapidly cleared by the Americans and invited to Philadelphia. He has, however, been suspected of having committed "war crimes". This I doubt very much, I knew him quite well and am sure that he was not at all interested in politics and he was certainly not an antisemite. Richard Willstätter had been his teacher and he had always kept friendly relations with him. I was told he also protected several jews or half-jews during the war. I cannot believe that he took part in any criminal action although he was certainly a rather weak character and preferred to give in, when necessary.

At the Kaiser Wilhelm Institute I had many friends and good colleagues: the late Adam Deutsch, who later became professor in Lund, Max Hoffer, who found refuge at Hoffmann-La Roche in Nutley, NJ. Herman Blaschko of Otto Meyerhof's lab, who became professor in Oxford, the late Marcel Florkin from Liège, who later

---

*Paul Györgi left Heidelberg in 1934 and became professor at the University of Philadelphia. Our results were published in 1939 under the title: *Attemps to isolate the factor (vitamin H) curative of egg white injury* [11].

played a great role in the International Union of Biochemistry, etc.

We also met André and Marguerite Lwoff, who spent a year in Meyerhof's lab in 1932. André's friendship and support were most precious to me later on.

Following his suggestion, I started to examine the carotenoids of invertebrates. From the red, boiled shells of the lobster a crystalline pigment was obtained and named astacene*; the green eggs of the lobster yielded a related crystalline carotenoid astaxanthine which was the chromophore of a chromoprotein we called "ovoverdin" [12].

At the end of 1932 the political situation became more and more dangerous; on January 30th, 1933 Hitler became Reichskanzler and after a particularly virulent speech by the Führer, my wife and I left Heidelberg "à toute vitesse" on March 25th, 1933. I well remember our relief when we crossed the bridge over the Rhine at Kehl ("Ici commence le pays de la liberté"). Four days after we had left Heidelberg the Gestapo came to fetch me . . . (they missed me again 10 years later in Lyon).

Paris (April 1933–September 1935)

We settled in Paris, near the populous rue Mouffetard where I bought all sorts of sea food to isolate new carotenoids. Through a recommendation of Dr. W. Schoen of the Pasteur Institute, Prof. René Fabre, offered me a lab bench at the pharmacy of Hôpital Necker. There I had acetone, some glassware and a spectrophotometer. Astacene could be isolated from all crustacean species examined and also from the skin of the goldfish and two other red-coloured fishes, indicating that astaxanthine* is, indeed, a very widespread animal pigment. Several new carotenoids were crystallized; a hydrocarbon, torulene from red yeast [14], pectenoxanthin

---

*It was shown later by R. Kuhn and N.A. Sörensen [13] that astacene, a tetraketo β-carotene was in reality an artefact formed in alkaline medium from astaxanthine, a diketo-dihydroxy β-carotene.

from the gonads of *Pecten maximus* [15] which was later shown to be the first acetylenic carotenoid [16], echinenone, a provitamin A [17], pentaxanthine from the eggs of sea-urchins [18], etc... I started publishing short notes in the *Comptes Rendus de l'Académie des Sciences* and in *Comptes Rendus de la Société de Biologie*. For me at that time it was really "publish or perish", in the sense that I had to prove to myself that I was still capable of continuing research, despite the difficult circumstances; I also wrote two short books on carotenoids*; fortunately my wife corrected my still primitive French. I presented my manuscripts to professor Gabriel Bertrand (1867–1962) the "grand old man" of French Biochemistry whose lab was at the Pasteur Institute. He had the only chair of Biochemistry existing at that time in France; it had been created for Louis Pasteur, who died before occupying it; then it went to Emile Duclaux, then to Gabriel Bertrand. I never dreamt of course that the responsibility of this same chair, via Maurice Javillier (1875–1955), Eugène Aubel (1885–1975) and Claude Fromageot (1900–1958) would land on my frail shoulders in 1958.

At that time my financial situation was difficult; I could not get any fellowship; I had approached the Rockefeller Foundation, but they only gave fellowships to candidates who would return to their previous position: this, of course, was not my case!

Fortunately, I got to know Dr. Harry Plotz, a Boston doctor who had a lab at the Pasteur Institute; there I tried to purify poliovirus by chromatography, without success, but I finally obtained a modest fellowship from the Ella Sachs Plotz Foundation set up in memory of his wife. Harry Plotz also had a lab which he did not need and which he let me use at the Institut de Biologie Physicochimique (Fondation Edmond de Rothschild). There Dr. Yvonne Khouvine gave motherly help to arriving refugees.

In March 1935 I learnt from an international organisation that a position was offered in Odessa. I took the train, full of hope, but also full of anguish: would the "paradise" I had dreamt of really turn out to be one?

**Les Caroténoïdes des Plantes*, 1934, 83 pp., and *Les Caroténoïdes des Animaux*, 1935, 62 pp., Hermann, Paris.

My first contact with the Soviet Union was very discouraging: I was the only passenger crossing the frontier at Podvolotschisk, on the Bessarabian boarder; the train was first meticulously searched for hidden passengers (or saboteurs?), on the roof, under the seats, under the floor; then I was admitted to the restaurant where a dozen dirty and miserable children were waiting for some residues from the plates of the customers: the food was so bad that I could not touch it; I was glad to leave it to the "besprisorni" around me. Then I was allowed some hours rest in a room where wonderful posters promised heaven on earth. A sleeper car took me to Odessa; I could not sleep because a herd of bed bugs tried to feast on me. This was my first encounter with these disagreeable parasites; the second was a year later in Murmansk.

Odessa was in a miserable state and the position was not tempting so I went to Kiev where Prof. A. Palladin promised me a position as "Dozent" at the University. This however, never materialized.

On returning from Kiev, I applied for a fellowship of the "Caisse de Recherches", the precursor of the CNRS (Centre National de la Recherche Scientifique); I still remember the day when Jean Perrin, the famous physicist, came into my little lab and said:

"Your application has been examined; we are willing to give you a fellowship as attaché de recherches but as you are an Austrian citizen we can pay you only one third of the fellowship".

I then got a recommendation to the cultural attaché of the Soviet Embassy through Paul Langevin (also a renowned physicist). Fortunately, just at that time the Vitamin Institute in Leningrad needed a chemist; I went there, first by plane to Berlin, then with another small plane to Leningrad. I still remember the flight over the foggy baltic forests and the impression that the next little wind would bring us down hitting the top of the trees.

Leningrad (October 1935–December 1937)

The Vitamin Institute belonged to the Ministry of Food Industry, the head of which was Anastase Mikoyan*. I was offered a good salary (1500 rubles/month), a nice new 3-room flat overlooking the Fontanka and a rather well-equipped lab with 5 or 6 coworkers. Transportation of my family, furniture etc. from Paris to Leningrad and back would be paid for by the Institute. The 3-year contract was quickly signed and in October 1935 I took the train to Moscow with my wife and our two daughters, the youngest just 4 months old.

It is difficult to summarize our impressions of that heroic period. Leningrad is one of the most beautiful cities in the world; after the war we visited it several times and we were glad to see that all sequels of the terrible blockade during the last war (over 500 000 people died of starvation) had disappeared and the palaces restored to their former splendor.

I had been asked to help with the production of various vitamin preparations; we started some synthetic studies on vitamin C, following the fundamental work of Th. Reichstein; later a synthesis of vitamin B_1 was abandoned when Alexander Todd published his first results. My interest in vitamin A brought me once to Murmansk, where I visited a brand-new installation for the production of fish liver oil; then I wondered if the livers of the big Volga fishes (sturgeons, etc . . .) could not be used as source of vitamin A; I received tinned samples of such livers from all over the USSR and analyzed them by the well-known Carr Price reaction, using a pocket spectrometer I had bought for £ 25 in London; to my astonishment, all extracts gave a green-blue colour (λ_{max} 690) instead of the blue colour typical for vitamin A (λ_{max} 620); this led me, with Valentina Rosanova (who was later killed by a bomb during the siege of Leningrad) to publish a paper on "an abnormal Carr Price reaction" [19]. I got into contact with Sir Ian Heilbron (at that time

*I later met him at the French Embassy in Moscow on July 14th, 1954; he still remembered me and told me that Stalin had always insisted that vitamins were very important!

at the Chemistry Department of Manchester University) for further characterization of the new compound, which was active as vitamin A [19], and typical for freshwater fish [20]. This "vitamin A_2" had been simultaneously discovered by A.R. Morton (Liverpool) in goldfish livers and by George Wald (Harvard) in the eyes of the pike.

I spent two most interesting years in Leningrad, but the political situation then became very dangerous, especially for foreigners; this was the period sadly known as "Yeshowchina" during which many intellectuals were arrested, to be liberated a few months later, when Yeshow, the head of the GPU was found to be a "saboteur" and "liquidated".

Paris again (1938–1939)

On returning to Paris from Russia I at last got a position at the CNRS as Attaché de Recherche.

I first worked in one of Prof. Robert Lévy's labs in a basement of the Ecole Normale Supérieure where I isolated pigments of marine invertebrates: echinochrome from the eggs of the sea urchin *Arbacia pustulosa*, the related spinochromes from the spines of sea urchins [21], bonellin, the green pigment of the worm *Bonellia viridis* [22] etc.

Later Prof. Eugène Aubel and Prof. René Wurmser kindly offered me a laboratory at the Institut de Biologie Physicochimique and in April 1938 I got the degree of docteur ès Sciences at the Sorbonne with a thesis *Sur les caroténoïdes des animaux inférieurs et des cryptogames* [23], the greatest part of which I had previously written in Russian in view of a Russian thesis.

On March 12th, 1938, Austria was occupied by the nazis and I became "ex-Autrichien". In December 1938 I was very happy to become a French citizen.

In March 1939 I was contacted by Max Roger, Director of a perfume factory near Paris, who suggested a collaboration in the field of natural perfumes. I signed a contract and chose "animal perfumes" and especially ambergris and castoreum of which nearly nothing was known. This choice had most beneficial consequences for me a few years later as we shall see.

The cruel and dangerous years of war and occupation (1940–1945)

When the war broke out in September 1939 I was drafted into the army. Those who had a driving licence were put with the horses, those who had no licence were put on trucks. So I slept in the stables, behind a row of horses and I learnt that the gases produced by vegetarians have quite an acceptable odor in comparison to those of non-vegetarians.

The final defeat of the French army was to me quite understandable: even big guns were still drawn by horses and during our training we never fired a real shot: after a few months I was glad to be sent to a batallion of chemists in Paris, but we were not employed for any serious work. To my great relief I was dismissed on March 22nd, 1940 after our fourth child (and first son!) was born.

I immediately started work again at the Institut de Biologie Physicochimique. We began to extract 20 kg of castoreum (the dried scent glands of the Canadian beaver, which are highly appreciated in perfumery). But on May 10th, 1940 the "real war" started for France, and soon the German armies were approaching Paris. I put a 1-kg box of ambergris and extracts of castoreum in the trunk of an old car and moved southwards with my family until the armistice was signed on June 22nd.

We had arrived in a little village at the foot of the Pyrenees, exhausted and discouraged; the parents of my wife joined us, but my own parents were still in the occupied zone, near Paris. France was indeed cut in two: the occupied and the so-called "non-occupied" zone. No news from Paris, no news from the CNRS, food was rationed, the future was non-existent; but we had four children to feed. In this desperate situation I wrote to Claude Fromageot, professor of biochemistry at Lyon University. The answer came quickly "yes, you can come to work in my laboratory". So we put the children and all our belongings into the old car again; fortunately I also had in the trunk 20 litres of benzene and acetone (which were meant for the extraction of castoreum and ambergris); I used these solvents in a mixture with the 5-litre batches of gasoline we got on the way; once we had to wait overnight until the gasoline arrived at the station (we were No. 47 in the queue). But finally we arrived in Lyon and found a very nice villa with a big garden in Collonges au Mont d'Or, a few kilometers north of Lyon. I signed the lease but had not even enough money for the first rent. Fortunately Max Roger turned up when all funds were exhausted (I had been excluded from the CNRS in application of the Nuremberg laws) and renewed our contract. I was able to engage a very bright young technician, Daniel Mercier and an excellent research chemist Dr. Judith Polonsky who both stayed with me for nearly 40 years.

Space does not permit the description of all the personal problems of those four difficult war years: the battle for food, dodging the police etc . . .

However, those were still the golden years of "test tube chemistry": the day I arrived in Fromageot's lab, I unpacked my odorous treasures and started to fractionate the castoreum extracts made in Paris. In 2 weeks I had isolated two new crystalline pigments; their structure was described in 1942 in the *Travaux des Membres de la Société de Chimie Biologique* [24] which Prof. Jean Roche courageously edited in Marseille in the "free zone" (Jews were not allowed to publish in the occupied zone!).

Many other aromatic compounds were isolated; their structure brought me to the conclusion that the beaver accumulates in its scent gland (the secretion of which is a trail marker and pheromone) all sorts of aromatic compounds from the bark and buds of poplar trees and others on which it feeds. Some compounds are deposited as such, some are oxidized, some reduced, etc. [25].

Our work on ambergris showed that most odorous constituents were oxidation products of the triterpene ambrein which we formulated as a cyclized squalene [25, 26].

In 1943 Max Roger stopped his subsidies, but I was lucky in obtaining a contract with Henri Pénau and later Léon Velluz, directors at Roussel-Uclaf.

The last months before the liberation (May–August 1944)

We had anxiously followed the events in Russia, listening to the BBC. It is difficult to overestimate the role the BBC played in France during the German occupation in keeping up our morale. The surrender of Marshall Paulus at Stalingrad was clearly the beginning of the end and we now waited impatiently for the Allies to open the "second front" so insistently demanded by the Russians. I used to transcribe the good news from the Russian front, heard over the BBC, on the walls of the lavatory in the basement of the Institute ... In the meantime I was contacted several times by a mysterious person for performing some chemical manipulations for the resistance movement.

In May 1944 it became evident that important and dangerous events were soon to happen, so we decided to put our four children into safety; an aunt of my wife brought two of them to a little village south of Vichy, where food was still plentiful and where we found two rooms to rent; on May 24th I finally took our two elder daughters to the railway station Lyon-Perrache to bring them to the same little village, Les Amouillaux; an air-raid alarm — the first since the beginning of the war — delayed our departure, but we finally left (allied planes had hit a bridge on the Rhône that day). The next day I heard on the radio, that Lyon had been bombed; 1500 people had been killed. I immediately returned to Collonges au Mont d'Or where we lived and found our house intact, as well as my wife who was expecting our 5th child, but the Institute had been hit by a bomb which had dropped on the office of Claude Fromageot; five students had been killed in the courtyard; in my own lab no harm at all had been done.

On June 1st our 5th child (and second son) was born; a few days later a colleague came running into the lab informing me that the Allies had landed in Normandy. I immediately went home, took some belongings and left to join our children at the "Amouillaux". My wife could not yet travel, but joined us 10 days later.

We thought we were at last, personally, out of danger, but on August 1st we heard some shooting and the little village was occupied by German troops and serched for "enemies". I myself hid in the backroom, whereas my wife stayed with the five children in the front room. A soldier with a gun came in; he said "ah, enfants", my wife replied 'oui, cinq"; and he left . . . Our luck that day was due to the fact that the soldiers were mostly elderly men drafted into the regular army and not the ferocious Waffen SS.

On August 15th, walking in the woods, looking for mushrooms and berries I found a leaflet, dropped by a plane, announcing the landing of allied troops at the Mediterranean coast of France; this wonderful news clearly meant a rapid end of the Occupation. Lyon was indeed liberated soon and at the end of August we returned to Collonges. My mother who had remained there had suffered some maltreatment from French militia men, the day before they left, following the retreating German army*.

The first two years after the war (still in Lyon)

One memorable event, a few weeks after the liberation was the visit of a captain of the U.S. Second Armored Division, in full battle dress: it was my brother Wolfgang, who had been a soldier in the Austrian army when the nazis occupied Austria; he had been immediately dismissed as a Jew and we were able to get him a visa to France; he then left for the United States just before the war broke out and later volunteered into the U.S. Army; he landed in Normandy on June 9th and served as Commanding Officer of a Tank Destroyer Company through France, Belgium, The Netherlands and the Battle of the Bulge, until he was wounded in action shortly thereafter.**

In October 1946 the Swiss Chemical Society invited French chemists to a joint meeting in Basle. I submitted an abstract on the chemistry of ambrein; a few days later I received a cable from Prof. Leopold Ruzicka asking me to come to Zurich with my samples; I

*My father had died in March 1943 in Collonges from a heart attack.

**Born in Vienna in 1919, he is now Clinical Professor in the Department of Psychiatry of the University of California Medical School in San Francisco (author of two books and numerous clinical and popular articles).

learnt that he had a paper on the structure of ambrein and ambergris odorants unpublished since 1939, because of its importance for the perfume industry. My compounds were meticulously compared with the most distinguished ETH samples and were found to be identical [26]. Since then I have had many friendly and most stimulating contacts with the colleagues of the ETH: Vlado Prelog, Duilio Arigoni, Albert Eschenmoser, Oskar Jeger and others.

At the Basle meeting* I was contacted by Dr. Max Stoll, Head of Research of Firmenich, Geneva, with whom Ruzicka had been working and I have ever since enjoyed a privileged relationship with the firm, especially with Dr. Roger Firmenich and Dr. Max Stoll and in recent years with Dr. Günther Ohloff, the successor of Max Stoll. My contract with Firmenich, signed with the CNRS in 1950 has been essential over the past 30 years for the development of my laboratory in Paris and later in Gif, in procuring technical help, research workers, laboratory funds and even interesting research topics.

Paris once more (1947–1960)

In March 1947 we moved back to Paris and I started again in the same lab I had left in June 1940 at the Institut de Biologie Physicochimique. There I met Dr. Nine Choucroun who in 1939 had described a biologically active paraffin oil extract of tubercle bacilli. Her "PmKO" gave the impetus for a detailed study of the chemistry of Mycobacteria which started with the excellent doctor's thesis of Jean Asselineau** and led to more than 200 papers on the

---

*This meeting in Basle was also memorable because it was my first contact with the world outside France after 6 years of hardship. The shops were full of the most delicious things. Can you imagine, no food stamps were needed, you could buy as much butter, cheese, chocolate or anything you wished!

**Since 1960 Professor of Biochemistry at the University of Toulouse, author of a book *The Bacterial Lipids* (Hermann, Paris/Holden Day, San Francisco, 1962, 372 pp.).

chemistry of mycobacterial lipids, glycolipids, peptidolipids etc. and, in recent years, to the development of synthetic immunostimulants, which will be mentioned below.

From 1947 to 1960 my group at the Institut de Biologie Physico-chimique expanded rapidly, working on a large variety of problems.

Constituents of medicinal plants. With Judith Polonsky we were lucky to have the expert advice of Pierre Boiteau, an excellent specialist of the flora of Madagascar; whenever he drew our attention to a particular plant, we isolated new and interesting compounds: asiaticoside from *Centella asiatica* (a triterpene glycoside which is still used in France for treating ulcers) [27], calophyllolide and other analogous phenylcumarins from the nuts of *Calophyllum inophyllum* [28], darutoside, a diterpene triol glycoside from *Siegesbeckia orientalis* [29], etc.

The pheromone of the queen bee. In collaboration with Dr. Janine Pain, of the apicultural station of Bures-sur-Yvette, Dr. Michel Barbier succeeded in isolating the "perfume" of the queen bee (an important pheromone regulating the collective life in the beehive) and to determine its structure as 9-keto-2-decenoic acid [30], a result confirmed simultaneously by R.K. Callow in England.

Natural perfumes and aromas. My collaboration with Firmenich, Geneva, was first pursued in the field of ambrein-derived terpenoids; later a detailed analysis of the constituents of jasmin essence by Edouard Demole led to the isolation, structural elucidation and synthesis of an important new odorous compound: methyl jasmonate [31]. As a result of this work, Firmenich could produce the first synthetic jasmin essence containing most of the natural components. The volatile constituents of roasted cocoa were analyzed by Paul Dietrich [32], but the typical cocoa bitterness was very elusive; much later Willi Pickenhagen found that it was due to a complex between theobromine and various diketopiperazines formed during roasting [33].

Lysopine. Dr. C. Lioret of the Institute of Plant Physiology at Orsay had isolated a new amino acid from crown gall; with the help of Klaus Biemann, the well-known mass spectrometry specialist at MIT (Cambridge, MA) its structure (1) was determined. As it is the lysine analogue of octopine (2) (from octopus muscle), the first known "amino-imino acid", we called it lysopine; the stereochemistry, as shown below, was confirmed by synthesis from L-lysine [34]; the "opine" family was later expanded by other authors, most of the opines being isolated from crown gall tissue.

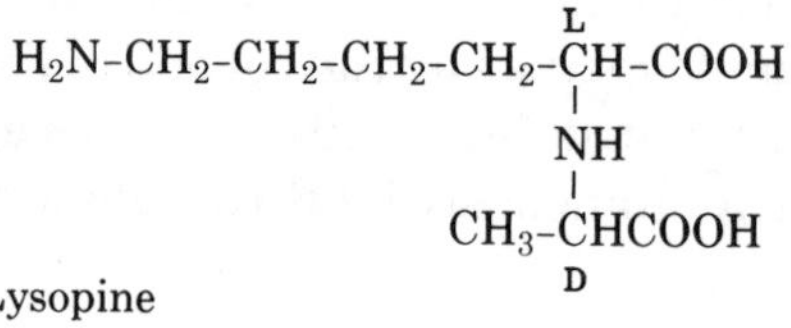

$$H_2N\text{-}CH_2\text{-}CH_2\text{-}CH_2\text{-}CH_2\text{-}\overset{L}{C}H(\text{-}NH\text{-}\underset{D}{C}H(CH_3)COOH)\text{-}COOH$$

1 Lysopine

$$H_2N\text{-}C(=NH)\text{-}NH\text{-}CH_2\text{-}CH_2\text{-}CH_2\text{-}\overset{L}{C}H(\text{-}NH\text{-}\underset{D}{C}H(CH_3)\text{-}COOH)\text{-}COOH$$

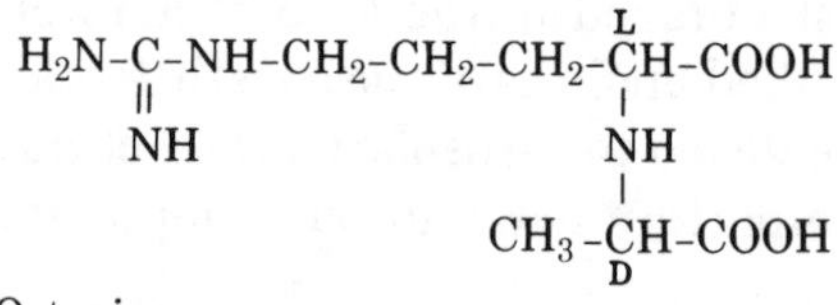

2 Octopine

Ascaryl alcohol. The embryologist E. Fauré-Frémiet (1883–1971) had isolated in 1913 from the eggs of the parasitic nematode *Ascar-*

is equi a lipid which he called "alcool ascarylique". I received a sample of the original material and we prepared larger quantities by extracting *Ascaris equi* from the Paris slaughterhouse. Dr. Judith Polonsky with Claudine Fouquey separated the crude unsaponifiable extract into three "alcools ascaryliques" A, B, and C. All three were characterized as glycosides of long-chain fatty alcohols or diols containing a new sugar, ascarylose, which was shown to be a 3,6-dideoxy hexose [35]. Two isomers had been previously described by Otto Westphal and Otto Lüderitz in Freiburg: abequose and tyvelose, immunodeterminant constituents of endotoxins of *Salmonella* species. A branched-chain structure had been proposed for tyvelose, but after a discussion with Otto Westphal and Otto Lüderitz in Freiburg, it was clear that ascarylose, abequose, and tyvelose (as well as paratose and colitose, discovered in the meantime) were all stereoisomers of a linear 3,6-dideoxyhexose. Abequose, tyvelose and paratose were synthesized in a joint effort; ascarylose was shown to be the optical antipode of tyvelose, and colitose the antipode of abequose [36, 37]. The fortuitous isolation of ascarylose thus led to the structural determination of the immunodeterminant constituents of the endotoxins of Gram-negative bacteria. Our friendly relations with Otto Westphal and his group brought me several times to Freiburg in later years.

A few words about my career

After the war, Prof. Eugène Aubel had proposed to the CNRS my promotion to "Maître de Recherche" and in 1952 I became "Directeur de Recherche" (a position equivalent to that of full professor).

In 1954, Claude Fromageot asked me to participate in the Biochemistry course of the Faculty of Science and to accept a position of Maître de Conférence at the Sorbonne. I agreed after some hesitation, because it meant less time for research and all sorts of administrative duties; I later realized how important it is to teach students; my course was on chemistry and biochemistry of lipids, a topic which I was quite familiar with. I tried to renew the course every year and finally published a little book* which was later improved with the help of my former pupil and assistant Robert Azerad**.

I was always rather nervous before the lectures and much preferred to give research lectures to graduate students, or even better at international meetings. There I had some problems, however, because our government urged us to speak in French. Once or twice I followed the official recommendation; for instance at the 5th International Congress of Biochemistry in New York in 1964, there were 6 congress lecturers; five of them, even the Russian Severin spoke in English, I was the only one to give my lecture *Excursion dans le monde mystérieux des Mycobacteries* in French; before I had even started half the auditorium had left (as expected!).

In 1958, after the untimely death of Claude Fromageot, Pierre Desnuelle, from Marseille (see his autobiography in Comprehensive Biochemistry, Vol. 35, pp. 283–331) became his successor, but soon realized that he preferred to stay in Marseille, so the chair was vacant again. My friend Jacques Monod at the Institut Pasteur did not want to take over the lab and coworkers of Fromageot, so I became Professor of Biochemistry.

Gif and Orsay (1960–1978)

In the late fifties Prof. Georges Champetier (1905–1979), vice director of the CNRS, had recognized the necessity of expanding the research facilities of Natural Products Chemistry in France. Early 1957 the CNRS decided to build an Institute for the Chemistry of Natural Products on its campus at Gif-sur-Yvette*** and I was to be the director; a few months later Prof. Paul Lebeau (1868–

Cours de Biochimie: Lipides, Ediscience, Paris, 1970, 124 pp.

**R. Azerad and E. Lederer, *Biochimie des Lipides*, Ediscience/McGraw-Hill, Paris, 1974, 246 pp.

***The little town of Gif-sur-Yvette, 30 km south of Paris, can be reached in 30 min by metro. The Chateau de Gif and its large park belonged to a Swiss banker, whose son was a schoolfriend of Frédéric Joliot-Curie. After the war, when Joliot-Curie was director of the CNRS, the estate was donated to the CNRS. The chateau is now

1959) of the Faculty of Pharmacy obtained a similar decision in favour of Prof. Maurice-Marie Janot (1903–1978), whose team had not enough space at the Paris Faculty of Pharmacy; at that time funds were still available, the initial sum was doubled and we both became directors of the Institut de Chimie des Substances Naturelles.

On December 1st, 1960 I went to Heidelberg to attend the ceremony in honour of the 60th birthday of Richard Kuhn and the same day my small office of the rue Pierre Curie was transferred to my large directorial office at Gif.

M.-M. Janot and I each had one wing of the building (mine was of course the eastern!) and we shared the responsibility of the general facilities (mass spectrometry, NMR, micro-analysis, library, pilot plant, later also X-ray cristallography).

Our research interests were complementary: his group working mainly on indole- and steroidal-alkaloids, mine on several more biochemical topics described below. On "my side" there was also a strong group of theoretical organic chemists with Bianca Tchoubar, Hugh Felkin and Irène Elphimoff-Felkin.

In a few years our Institute became an internationally known research center and many distinguished foreign colleagues were invited to give lectures there and to stay some time with us.

Konrad Bloch was one of our first guests in the Chateau in 1961, then Paul de Mayo (London, Ontario) and Ernest Wenkert (at that time Bloomington, IN) came every year and enlivened the atmosphere at the Institute.

I wish to mention especially one memorable event: in 1968 professor Wang Yu from Shanghai came with four of his colleagues and gave us a 2-h lecture on his great achievement: the synthesis of insulin; it was the first time he had lectured outside the Eastern block and his foreign competitors from Denmark, Germany and the USA had been invited. He visited Gif again in 1980 and I was

---

a guest-house for foreign visitors. On the campus of Gif there are now, besides the Institut de Chimie des Substances Naturelles, research laboratories for genetics, enzymology, neurophysiology, photosynthesis, peptide hormones, a phytotron, etc.

glad to meet him in his Institute in Shanghai in 1983.

Several foreign colleagues worked for some months with us: Clint Ballou (Berkeley), Patrick Brennan (at that time Dublin), Mayer Goren (Denver), John Law (at that time Harvard), etc.

In 1961 Prof. André Guinier, dean of the Orsay campus (4 km from Gif, now belonging to the Université Paris-Sud) asked my help for setting up an Institute of Biochemistry at Orsay. With the active participation of Hubert Clauser and others the new Institute was ready in 1963 and I obtained the transfer of my chair of Biochemistry from Paris to Orsay. Several assistants and my colleague and friend Hubert Clauser followed me to Orsay. From then on until my retirement my main research activities were at Gif, my teaching obligations at Orsay.

Let us now turn to some of the main topics of research at Gif and Orsay.

Biosynthetic studies

In Gif, we at last had the necessary equipment for biosynthetic studies with ^{14}C. Dr. Mireille Gastambide-Odier had shown in 1959 that the diphtheria bacillus (*C. diphtheriae*) synthesized the C_{32} corynomycolic acid (3)* by condensation of two molecules of $[^{14}]C$palmitic acid [38], thus laying the foundation for further studies of the biosynthesis of the larger mycobacterial mycolic acids. These showed that indeed the "mycolic acid condensation" was a general pathway for the CMN group (Corynebacteria, Mycobacteria, Nocardiae) producing the typical α-branched β-hydroxy acids.

$$CH_3(CH_2)_{14}CO_2H + \underset{\displaystyle C_{14}H_{29}}{\underset{|}{CH_2}}CO_2H \rightarrow CH_3(CH_2)_{14}\text{-}\overset{\displaystyle OH}{\overset{|}{\underset{\displaystyle H}{\underset{|}{C}}}}\text{-}\underset{\displaystyle C_{14}H_{29}}{\underset{|}{CH}}CO_2H$$

3 corynomycolic acid

*Discovered by Julio Pudles in 1951, synthesized in a racemic form by Judith Polonsky in 1954.

The first detailed structure (4) of a mycobacterial mycolic acid was elaborated by A.H. Etémadi [39]. He proved that tetracosanoic acid is incorporated as such and that the methyl group was due to a C-methylation reaction [40].

$$CH_3(CH_2)n_1CH{=}CH(CH_2)n_2CH{=}CHCH(CH_3)\text{-}(CH_2)_{17}\text{-}CH(OH)CH(C_{22}H_{45})COOH$$

4 α-Smegma mycolic acid ($n_1{=}15\text{–}19$; $n_2{=}12\text{–}16$)

Propionic acid was shown to be incorporated into phthiocerol (see below) and into the C_{32} mycocerosic acid (5) of mycobacteria [41].

$$CH_3(CH_2)_n\text{-}CH_2\text{-}CH(CH_3)\text{-}CH_2\text{-}CH(CH_3)\text{-}CH_2\text{-}CH(CH_3)\text{-}CH_2\text{-}CH(CH_3)\text{-}COOH$$

5 C_{32} mycocerosic acid n=18
6 preen gland fatty acids n=1,2

One day an American doctor came running into my office with a 3-month fellowship "to get a paper"; I suggested he inject the preen gland of a goose with ^{14}C-labelled propionic acid and then extract the methyl-branched acids (6) described previously by Murray [42]; the acid he obtained was indeed highly radioactive and degradation confirmed that it was formed in the preen gland from propionic acid. Dr. Noble got his paper [43]!*.

Further investigations with methyl branched fatty acids and phytosterols are mentioned in the chapter "the methyl group".

Phthiocerol and the mycosides

The structure of phthiocerol, a typical constituent of *M.tuberculosis* and *M.bovis*, had been studied in the forties by Einar Stenhagen

*Showing for the first time that vertebrates are capable of using propionate for the biosynthesis of such polymethyl-branched (low-melting) fatty acids.

and his wife Stina Ställberg-Stenhagen, using the quite new methods of monomolecular films, X-ray diffraction and mass spectrometry.

We had also got interested in phthiocerol and had isolated three "companions" and possible biosynthetic precursors, phthiocerolone, a methoxy-ketol, phthiodiolone, a keto-diol and phthiotriol [44].

A trip to Göteborg in 1958 gave me the opportunity to discuss the structure of phthiocerol with the Stenhagens and thus to make the close acquaintance of these two remarkable scientists, who unfortunately both died prematurely of tuberculosis. A joint letter to *Nature* [45] reported that phthiocerol was a mixture of two homologous methoxy-glycols of the structure (7). Both OH groups are usually esterified by mycocerosic acids (5).

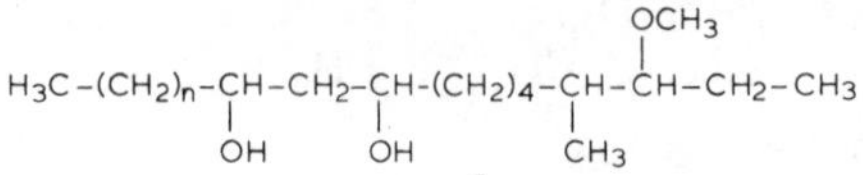

$$H_3C-(CH_2)_n-\underset{OH}{CH}-CH_2-\underset{OH}{CH}-(CH_2)_4-\underset{CH_3}{CH}-\overset{OCH_3}{CH}-CH_2-CH_3$$

7

n = 20,22 Phthiocerol

Later we became interested in some new "type-specific" mycobacterial compounds described by H.M. Randall, Donald W. Smith and A.P. MacLennan; as they are mycobacterial glycosides we proposed to call them "mycosides" [46].

The mycosides A (typical for *M. kansasii*) and B (typical for bovine strains) contain a phenolic derivative of phthiocerol, which we called "phenol-glycol" (8); this is glycosidized on the phenolic hydroxyl in both mycosides A and B [47]. Quite recently this "phenol-phthiocerol" has gained more interest as the aglycone of the principal antigen of the leprosy bacillus* [48]; this "*M. leprae* Phen-GL-1" induces suppression of mitogenic responses of lepromatous patients. Antibodies against the terminal disaccharide can be used for the detection of infection by *M. leprae*.

*Which is composed of 3,6-di-*O*-methylglucose, 2,3-di-*O*-methylrhamnose, 3-*O*-methylrhamnose linked as a trisaccharide to dimycocerosyl-phenol-phthiocerol.

The mycosides C, specific for strains of *M. avium* turned out to be glycosides of curious acyl oligopeptides, containing D-alanine, D-phenylalanine and D-*allo*-threonine (the latter had not yet been found in nature) [49].

The main structural features of the C-mycosides are as follows:
— a β-methoxylated C_{28} acid is amide-linked to the N-terminal nitrogen of a peptide chain having the structure D-phe-(D-*allo*-thr-D-ala)$_n$ where n is 1, 2 or 3;
— the C-terminal end of the oligopeptide is usually amidated by L-alaninol or ethanolamine;
— the terminal residue of the peptide chain is linked glycosidically to a 6-deoxyhexose; other 6-deoxyhexoses are attached to one or the other of the hydroxyl groups of D-*allo*-thr [50, 50a].

What may be the biological role of these curious compounds? I once suggested that they may be "in charge of the public relations of the cell" [50b] and, indeed, some C-mycosides are receptors for mycobacteriophages.

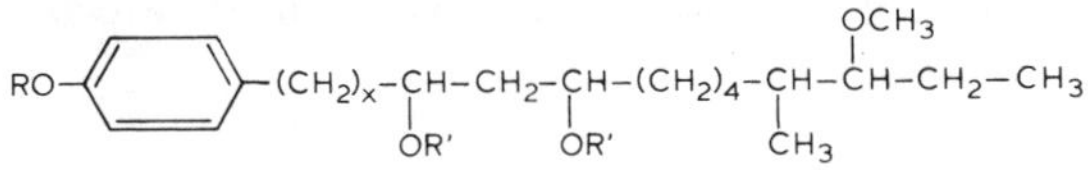

$$RO-C_6H_4-(CH_2)_x-\underset{OR'}{CH}-CH_2-\underset{OR'}{CH}-(CH_2)_4-\underset{CH_3}{CH}-\overset{OCH_3}{CH}-CH_2-CH_3$$

8

Phenolglycol A and B R=R'=H

Mycoside A R=trisaccharide (2-*O*-methyl-fucose, 2-*O*-methylrhamnose, 2,4 di-*O*-methyl-rhamnose).
R'=mostly mycocerosic acids
x=16, 18, 20

Mycoside B R=2-*O*-methyl-D-rhamnose
R'=mostly mycocerosic acids
x=14, 16, 18

From fortuitine to mass spectrometry of permethylated peptides

Our "Excursion into the mysterious world of Mycobacteria" [51] was extended to Corynebacteria and Nocardiae and brought many unexpected fruits. One of these can be described under the above heading.

During an analysis of the lipids of *Mycobacterium fortuitum* Dr. Erna Vilkas noted a precipitate gathering at the interface of ether and an aqueous layer. It turned out to be a nearly pure peptidolipid and was called "fortuitine". Chemical degradation gave a preliminary structure [52], not quite in agreement with analytical data. At that time Dr. Michael Barber of AEI, Manchester, was looking for well-defined high-molecular weight compounds for testing the capacity of the new MS9 mass spectrometer. I sent him a few mg of fortuitine and received 2 weeks later a letter giving the correct, full structure (9) with a molecular ion at 1359 [53]:

OAc OAc (on the two Thr residues)

$CH_3(CH_2)_{20}CO$-Val-MeLeu-Val-Val-MeLeu-Thr - Thr-Ala-Pro-OCH_3

9 Fortuitine

Here we had the proof that mass spectrometry not only could give molecular ions above 1300 but, more importantly, that the "sequence ions" obtained could easily lead to the discovery of unsuspected amino acid residues (*N*-methyl leucine in this case) and to the complete structure of acylated peptide methyl esters; other examples followed [54–56].

The two *N*-methylated amino acids and the one proline of fortuitine seemed to be the reason for its relative volatility. This led to the idea (first proposed by Jean van Heijenoort at Orsay) of permethylating acyl-oligopeptides to increase volatility for mass spectrometry. With Stephen D. Géro, Bhupesh C. Das and later Erna Vilkas and David Thomas a permethylation method was developed, allowing the sequencing of oligopeptides containing up to 10 or 12 amino acid residues [57–59]. The fortuitous isolation of fortuitine thus led to a new and useful method of peptide sequencing.

This method has, however, now lost some of its interest due to the development of fast atom bombardment mass spectrometry which allows sequence determination of peptides without derivatization.

Mass spectrometry was also essential for the determination of the structure of kitol (a vitamin A dimer from whale liver) [60], of the plastoquinones B and C [61], of the vitamin K_2 of Mycobacteria and Corynebacteria [62], of the mycobacterial peptidoglycan [63] and for the analysis of the mechanism of C-methylation reactions using CD_3-methionine (see below).

The methyl group: from biosynthetic reaction mechanisms to potential antiviral and antiparasitic drugs

Our interest in the biosynthesis of methyl-branched bacterial fatty acids led us to a more detailed study of the mechanism of the transfer of methyl groups to unsaturated fatty acids. Konrad Bloch at Harvard had shown that tuberculostearic acid (10-methyl stearic acid) was synthesized in Mycobacteria by the transfer of the methionine methyl group to the double bond of oleic acid; at the same time Klaus Hoffmann in Pittsburg had shown that the methylene group of lactobacillic acid (11, 12-methylene stearic acid) was synthesized by lactobacilli from vaccenic acid and the methyl group of methionine. So here was a problem: was a cyclopropane acid the precursor of tuberculostearic acid? We decided to study this question by the use of CD_3-labelled methionine, added to cultures of *Mycobacterium smegmatis* and examine the reaction products by mass spectrometry. A parallel study was carried out with yeast to examine the mechanism of the C-methylation leading to ergosterol.

We found that in the biosynthesis of tuberculostearic acid and of the ergosterol side chain the carbon atom of CD_3 methionine is transferred with only 2 of its deuterium atoms to a double bond of the precursor [65]. The loss of one deuterium atom of the CD_3 group of methionine was explained by the formation of a methylene and not a cyclopropane derivative [66]. Mass spectrometry confirmed the expected hydride shift [67].

Most other cases of C-methylation, for instance those leading to mycophenolic acid or to vitamin K_2 [68], were later found to occur with the transfer of the whole methyl group.

We also showed (more or less simultaneously with Duilio Arigoni at the ETH in Zurich) that the ethyl side chain of phytosterols was the result of two successive C-methylations of the 24–25 unsaturated precursors [69].

After having tackled this very academic problem we then looked for some "useful applications".

It was known that in some tumours hypermethylation of nucleic acids occurred, so why not try to inhibit these?

In 1971 Jean Hildesheim started in Gif to synthesize analogues of *S*-adenosyl homocysteine (SAH; 10), the universal inhibitor of biological transmethylations; in his in vitro experiments, however, SAH was always more active than the synthetic compounds [69a], so we were disappointed. Luckily, in September 1974 Dr. Malka Robert-Géro came to work with us; she had great experience, in particular with chick embryo fibroblasts infected with Rous sarcoma virus (RSV); very soon she discovered that the synthetic SAH analogues prepared by J. Hildesheim, especially 5′-deoxy-5′-*S*-isobutyl-thioadenosine (SIBA; 11) [69a] are strong inhibitors of RSV-induced cell transformation, whereas SAH is only weakly active [70]. This finding was rapidly extended to other oncogenic RNA as well as DNA viruses [71]. More recently, an antifungal antibiotic Sinefungin (12), a "carba-analogue of SAH" isolated by Lilly Research (Indianapolis) was shown by Malka Géro (and simultaneously by R.T. Borchardt, University of Kansas) to be a potent inhibitor of methyl transferases and of viral transformation [72]; it also has a strong antiparasitic activity against *Leishmania* species [73] in vitro, against *Trypanosoma brucei* in vivo in mice [74], and against *Entamoeba histolytica* in vitro [75].

Will any of these compounds be of future clinical or veterinary use? Clearly, more active and less toxic SIBA- and Sinefungin-analogues should be developed.

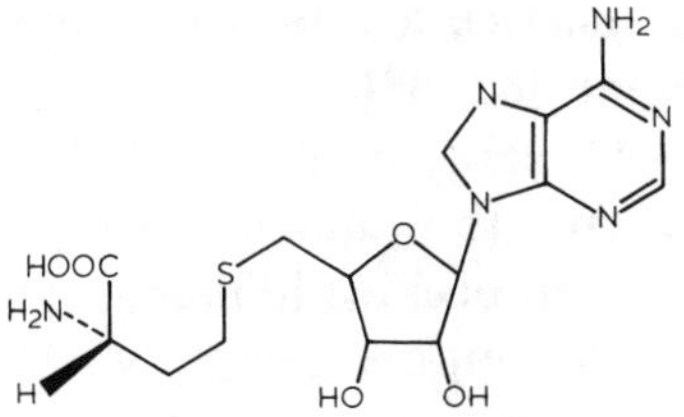

10 S-Adenosyl-homocysteine (SAH)

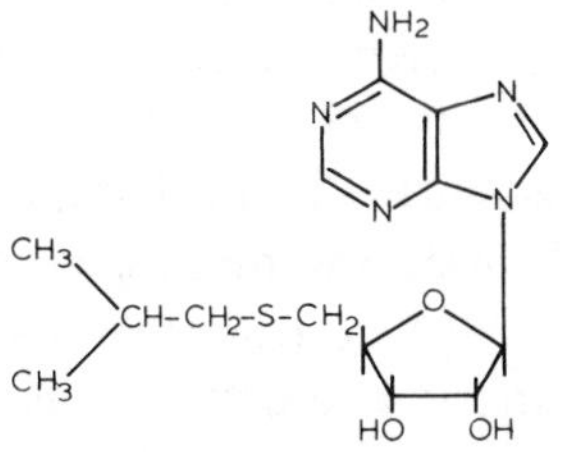

11 SIBA

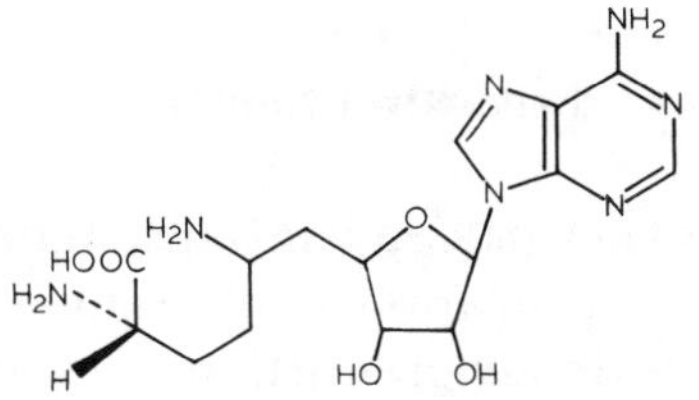

12 Sinefungine

Cord factor (trehalose dimycolate)

Cord factor, a "toxic lipid" extracted in 1950 by Hubert Bloch in René Dubos's laboratory in New York, from living, virulent, cord-forming Mycobacteria was found to be trehalose 6,6′-dimycolate by Hans Noll and Jean Asselineau [76]. Several years later, Adam Bekierkunst in Jerusalem revived interest in cord factor by showing its strong immunostimulant activities [77, 78]. Independently, Edgar Ribi in Montana isolated a biologically active glycolipid "P_3" which was finally shown to be identical to cord factor; today natural and synthetic trehalose diesters are recognized as strong immunomodulators and, in particular, as activators of macrophages [79, 80]. (For reviews see [81–83].)

Trehalose dimycolate (TDM) can cure mice after one injection into an established fibrosarcoma [84] and protects mice against various bacterial and parasitic infections [85, 86]; lower homologues of TDM can be synthesized, but are generally less active [81–83]. More recently the use of aqueous suspensions of TDM has been developed (by J.F. Petit at Orsay, with P. Lefrancier at Institut Choay) which are very stable, much less toxic and just as active as the TDM-oil preparations used previously, in particular for activating macrophages [79–80].

Mycobacterial cell walls coated with TDM ("Ribigen")* are used for treating ocular carcinoma of cattle and horses. TDM can also be applied in emulsion with *N*-acetyl-muramyl-L-alanyl-D-isoglutamine (MDP) for inducing tumour necrosis and as adjuvant for vaccines against influenza, *Brucella* etc. [81–83]

From Freund's adjuvant to MDP

Freund's adjuvant, an emulsion of dead mycobacterial cells in paraffin oil + Tween, containing an antigen in the aqueous phase, is widely used for increasing antibody production in animal experiments.

The first *soluble* mycobacterial fraction, which Robert White in London had found adjuvant-active, was Jean Asselineau's wax D [87]; when the chemical similarity of active wax D with cell walls became evident we got interested in cell-wall chemistry. Here I was very lucky to have excellent junior colleagues at the Institut de Biochimie at Orsay. Jean-François Petit had learnt (with J.M. Ghuysen in Liège and with J. Strominger in St. Louis) the delicate art of enzymatically dissecting bacterial cell walls and showed that the immunoadjuvant activities of whole mycobacterial cells used in Freund's complete adjuvant, are an intrinsic property of the cell wall; then, in 1972, Arlette Adam treated purified cell walls of

*Ribi Immunochem. Research Inc., Hamilton, Montana 59840 (U.S.A.).

Mycobacterium smegmatis with lysozyme and isolated the first water-soluble adjuvant (WSA) [88]; careful enzymatic and chemical degradations led then to the conclusion that a muramyl-dipeptide is the minimum active structure of the whole mycobacterial cell, in Freund's complete adjuvant. Fortunately I had started, some time before, a collaboration with Pierre Sinaÿ, professor of biochemistry at the University of Orléans, who had great experience with the synthesis of muramic acid derivatives.

Finally, C. Merser and P. Sinaÿ succeeded in synthesizing the "ideal molecule" MDP (13) by combining a suitably protected muramyl derivative with a protected L-ala-D-glu derivative prepared by Pierre Lefrancier at the Institut Choay [89]. The first few mg of synthetic MDP arrived at Orsay in May 1974, and, as expected, were found to be strongly active in the guinea pig for stimulating antibody production and producing delayed hypersensitivity [90].

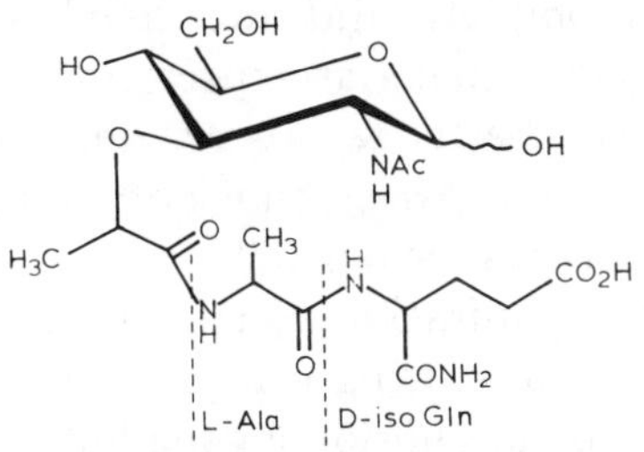

13 MDP

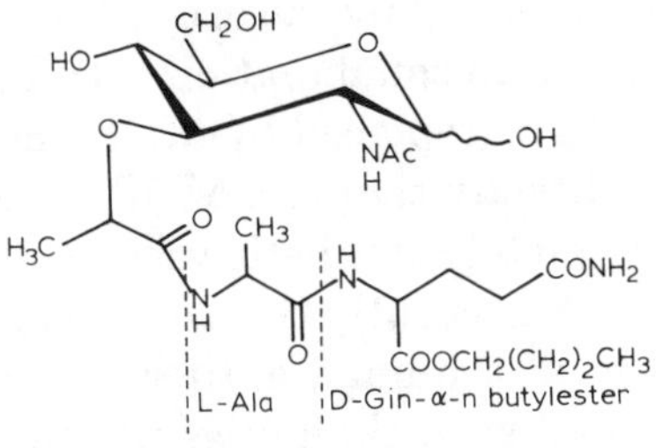

14 Murabutide

Since this first paper on MDP several hundred analogues and derivatives have been prepared and many large pharmaceutical firms have filed patents in this field. Our own work developed rapidly due to the chemical expertise of Pierre Lefrancier at the Institut Choay and the enthusiasm and biological competence of Louis Chedid and his group at the Pasteur Institute. MDP and many of its derivatives not only stimulate antibody production but also non-specific resistance to bacterial and parasitic infections (for reviews see [91–93]).

MDP has, however, some untoward effects, such as pyrogenicity;

fortunately a series of glutamine derivatives synthesized by P. Lefrancier were found by L. Chedid and coworkers to be as adjuvant-active as MDP and not at all pyrogenic [94, 95]; one of these "murabutide", *N*-acetyl-muramyl-D-alanyl-D-glutaminyl-*n*-butyl ester (14), successfully passed Phase I and II trials; an experiment with over 300 volunteers in Egypt organized by Louis Chedid, has shown that murabutide is well tolerated and that it increases significantly antibody production against tetanus toxoid. It is now a good candidate for clinical use as adjuvant for classical vaccines.

Another major prospect for muramylpeptides is their use as adjuvants for the new generation of synthetic vaccines now in development in several laboratories.

The pioneering work of Michael Sela and colleagues (Rehovot) has shown that synthetic peptides corresponding to antigenic sites of viruses or bacteria, when coupled to synthetic polypeptide carriers could produce antibodies against the corresponding natural proteins; the immune reaction is, however, rather weak, so it has to be increased by the use of Freund's complete adjuvant: then it was shown (in collaboration with L. Chedid) that - as expected - FCA could be replaced by MDP; moreover, better results are obtained by directly coupling MDP to the antigen-carrier, thus obtaining a completely synthetic vaccine. For reviews see [96, 97].

More recently the World Health Organisation (WHO), Geneva, started a human antifertility programme using a peptide corresponding to the antigenic site of human chorionic gonadotropin with nor-MDP in a squalene-arlacel emulsion as adjuvant*.

---

*P. Dukor, lecture at 8th Int. Symp. Medical Chem., Uppsala, August 1984 (in press). The general importance of adjuvants for vaccines is stressed in a recent report of WHO, Geneva, stating "The research for a clinically acceptable adjuvant for use in immunization against a number of diseases is generally recognized as a high-priority objective of the Special Programme of WHO as a whole" (UNDP/WORLD BANK/WHO Special programme for research and training in tropical diseases TDR/LEISH-SWG(5)/83.3.

Muramyl peptides as sleep factors. John Krueger, John Pappenheimer and Manfred Karnovsky at Harvard Medical School have shown that a "sleep factor" isolated from brain or human urine was a mixture of muramyl peptides [98]; MDP itself also prolongs slow-wave sleep in rabbits and monkeys when applied intracerebrally or even per os [99]. We were glad to participate in a study of structure–activity relationship in this field [100]; the structure of the main urinary sleep factor has been recently established as a disaccharide tetrapeptide (*N*-acetyl-glucosaminyl-*N*-acetyl-anhydro-muramyl-L-alanyl-D-glutamyl-*meso*-diaminopimelyl-D-alanine) [101].

Efforts are under way to produce synthetic muramyl peptides to be used as oral sleeping pills.

Muramyl peptides as vitamins. The natural role of peptidoglycan fragments for immunostimulation explains why "germ-free" animals and, to a lesser extent, newborn animals are highly susceptible to infection; their immune system needs stimulation by bacterial peptidoglycan products to become fully operative. In agreement with this hypothesis is the fact that newborns are not only very susceptible to infection, but they do not have slow-wave sleep! (M. Karnovsky, personal communication.) We are thus led to consider muramyl peptides as vitamins: organic trace compounds, derived from the food (or the intestinal flora) and indispensable for the normal health (immune status and sleep) of our organism [93].

In conclusion: the chemistry and biology of muramyl peptides is expanding rapidly; new active derivatives with specific properties are still discovered in nature and produced in the laboratory. We can foresee that soon certain muramyl peptides will be proposed as drugs for veterinary and clinical use, not only as adjuvants for natural and synthetic vaccines but also for stimulating nonspecific resistance in immunodepressed, malnourished or aged patients and — last but not least — for combined chemo- and immunotherapy.

However, despite all the favourable experimental evidence quoted above one should not forget the potential hazards of the use of immunomodulators [102].

Writing reviews

When I was about 10 years old my father told me to write short summaries of the books I had read, a very useful exercise, which later became most important to me.

When I was 16 and an amateur of studying microscopic organisms, I wrote a "digest" on *Amöboide Zellen im Tier- und Pflanzenreich* which was published a year later, to my great joy, in *Mikrokosmos*, a popular scientific magazine [103]*.

Until now, I have kept the habit of using my free time to write review articles; over the years they ranged from chromatography to natural pigments, animal perfumes, branched-chain fatty acids, terpenoids, bacterial lipids, glycolipids, biosynthetic problems, biochemistry of the methyl group, mass spectrometry of complex lipids, sequence determination of oligopeptides, the mycobacterial cell wall, inhibitors of methyl-transferases to immunomodulators; over 100 of these have been published, some of them in French, some in German, some in Russian, most in English.

My "rage d'écrire" also resulted in some books, first on carotenoids (see p. 452) then on chromatography; the original French text [104] was later rewritten in English and considerably expanded, with the help of my cousin Michael Lederer** who came from Australia to participate in this undertaking; a few years later (1957) a second edition was necessary and seems to have had quite some success [105].

My last literary exercise in chromatography was a collective work published in two volumes [106].

---

*It is amusing and perhaps not quite fortuitous that I am co-author 55 years later, of a paper describing the activation of macrophages [80] and, more recently, of a paper on the activity of the antifungal antibiotic Sinefungin and of the macrophage activator TDM against *Entamoeba histolytica* [75, 107].

**Born in Vienna 1924, he emigrated to Australia in 1938 and was one of the first to use paper chromatography for the separation of inorganic ions. From 1951 to 1960 in Paris, at the Institut du Radium, from 1960 to 1979 in Rome, head of the Laboratorio di Cromatografia del CNR, since 1979 at the Institut de Chimie Minérale et Analytique of the University of Lausanne; editor of the Journal of Chromatography (Elsevier, Amsterdam).

Life with industry

When I married in June 1932 I was still living at the expense of my father. I then obtained, through Kuhn, a monthly fellowship of 200 DM from I.G. Farben, for the work on "vitamin H" mentioned above.

On arriving in Paris, I did some consultations on flavins and other vitamins for the firm Byla; this brought me in touch with Dr. Henri Pénau who later helped me again during the war; he was an extremely amiable and helpful man.

My contacts with Max Roger have been mentioned above; during the German occupation he became Mayor of Neuilly; through him I received a false identity card: I was to be Edouard Lefèvre, born in Abbeville (where the archives of the City Hall had been destroyed). I never had the occasion of using this document, but have kept it preciously; one never knows . . .

My very fertile collaboration with Firmenich, Geneva, which started in 1947 has also been mentioned above.

After my first publications on mycolic acids and the lipids of Mycobacteria I was invited by Dr. Albert Wettstein, director of research of the Pharma Abteilung of Ciba, to come to Basle to discuss a collaboration. A contract was signed with the CNRS in 1950 and initiated an intensive collaboration which lasted for nearly 20 years. My frequent visits to Basle brought me in contact with Prof. Thaddeus Reichstein whose friendly and helpful attitude I will never forget.

Hubert Bloch, with whom we had a pleasant collaboration since 1951 for determining the structure of his "cord factor", later became director of biological research at Ciba-Geigy. In 1970, however, he terminated the collaboration for some obscure reason, considering perhaps that no useful application would develop.

I then obtained a contract with Carter-Wallace (New Jersey) where Frank Berger, the inventor of meprobamate, a friend of Louis Chedid, was director of research. Two years later, however, Carter-Wallace was reorganized and stopped most of its research projects.

Finally, with the help of Agence Nationale de Valorisation de la

Recherche (ANVAR) and the active interest of Jacques Monod Louis Chedid and I obtained a new contract with the Pasteur Institute and the Laboratoires Choay. This contract has been renewed ever since and has allowed me to continue some research activity after retiring.

I am greatly indebted to Dr. Jean Choay, scientific director and to Pierre Willaime, president of the Laboratoires Choay for their comprehensive attitude over the last years.

I always considered that a collaboration with industry could be interesting and useful for the laboratory, but only if it concerned a scientifically valid project and if it left sufficient liberty for research and publication; it also should guarantee enough funds for coworkers, technical help, equipment, etc.

In France collaboration with industry was for a long time considered to be a "maladie honteuse"; only recently the CNRS has started to encourage such collaboration.

Politics

By birth I belong to a more or less persecuted (or let us say disadvantaged) minority. This has had very beneficial effects on my activities: to succeed, I always had to try to do better than others. It also gave me a strong sense of solidarity with persecuted social or political minorities.

As a youth my sympathies were polarized by the programme of the "Kommunistische Manifest" of Marx and Engels and the success of the Russian Revolution.

It took me a very long time (in fact until Krouchtchev's report denouncing Stalin's crimes) to wake up to reality.

Later, a new model seemed to be more promising: maoism; the old Chinese culture and wisdom would surely be applied in the best interest of the people; was it not a great success that after some 20 years of maoism there were no more famines, no more internal wars, etc.? Alas, I was not quite right here again.

In the late sixties I attended several meetings of the "Tribunal Russell" which at that time was principally staging various actions against the war in Vietnam.

In 1969 I organized an International Meeting at Orsay against chemical warfare in Vietnam. Amongst the distinguished foreigners attending were Dorothy Hodgkin and Mrs. Nguyen Thi-Binh*, at that time Foreign Minister of the GRP (Gouvernement de la République Provisoire du Sud-Vietnam); it was an impressive moment when she arrived in a big black car, preceded and followed by policemen on motorcycles.

Another admirable Vietnamese woman should be mentioned here: Dr. Duong Quynh-Hoa. She once visited Orsay as Health Minister of the GRP and we had the great pleasure of having her to lunch at home. In 1982 we met her again in Hochiminhville (ex-Saïgon) where she now is in charge of a Research Centre for Paediatrics in a children's hospital.

Quite often I took part in street demonstrations in Paris in defense of human rights along with my friends and colleagues Alfred Kastler, Laurent Schwartz, Henri Cartan and others.

We also organized various boycott actions; against scientific meetings in Greece (during the military regime), in Spain (before the death of Franco), etc.

We also tried to help the research efforts of developing countries. Students from Africa, India, South-East Asia and others were always welcome. I wish to mention especially Radhouane Ellouz from Sfax (Tunisia) who got his doctors's degree in 1970 and who then founded the now flourishing "Institut des Sciences et Techniques" in Sfax; his wife Farielle Ellouz participated at Orsay in the final work on MDP (see above) got a doctorate in Pharmacy and is now teaching biochemistry at the University of Monastir.

Amongst our Vietnamese students, Duong Tan-Phuoc became director of an Institute of Oils and Fats in Hochiminhville, Nguyen Huu-Khôi is director of a laboratory of the Chemistry of Natural Substances in Dalat, etc.**.

*Mrs. Nguyen Thi-Binh later participated in the Peace Conference with H. Kissinger at Gif. We met her again in Hanoi in 1982 when she was Minister of Education.

**We also sent some pieces of equipment to Vietnam; this did not always turn out well: our first mass spectrometer which we had in Gif since 1961 was sent to Hanoi in 1975 but on unloading in Haiphong there was no strong enough crane and the box fell into the sea; one can imagine the state of the electronics after three months under water!

Having been a refugee myself, several times in my life, I quite naturally understood the plight of others and helped them whenever I could: they came from Brasil, Chile, Argentine, Spain, from Prague, Hungary, Rumania, etc. I am glad to say that I always found some funds for them.

Fortunately all those who started work in one of my laboratories have either returned to their native country or mostly stayed in France and obtained stable positions with industry, or the CNRS.

Retirement

Since I retired in October 1978, I have kept a little office and a secretary on the campus of Gif and I also share a small office at the Institut de Biochimie at Orsay with my colleagues there.

My departure from the Institut de Chimie des Substances Naturelles was overcompensated by the arrival of Sir Derek Barton as director of the whole Institute, assisted by Pierre Potier as "co-director".

Looking back, I can say that I owe the few successes I had in my research to a number of lucky circumstances.

Emil Fischer once said "wer Methode hat, der hat Erfolg". The methods which gave me a start in the early thirties were the micromanipulations* learnt in Vienna and then of course preparative chromatography started in Heidelberg.

Then, from 1934 to 1945 I had great luck in surviving with my family. These years were of course very unfavourable for my scientific activity; they might have been the most productive in my life.

From 1947 to 1960 I was lucky again to be able to develop a research group at the Institut de Biologie Physicochimique in Paris, in a warm atmosphere of friendship; I have already mentioned Eugène Aubel and André Wurmser as well as Yvonne Khouvine,

*2–3 mg were sufficient for a C, H or N-determination instead of the 200 mg samples burnt in the classical apparatus which I still found in use in France, in 1933.

who were always ready to help. Boris Ephrussi (1901–1979), the founder of Genetics in France and a most fascinating personality, also worked there*.

Among the friends and colleagues of the younger generation, let me mention Marianne Manago, now President-elect of the International Union of Biochemistry, Mike Michelson, Bernard Pullman, now the very successful Administrator of the Institut de Biologie Physico-Chimique and also Alberte Pullman, his wife.

From 1961 on, I had the great luck to move into the brand-new Research Institute at Gif; there again new methods and instruments accelerated further progress: mass spectrometry, NMR, isotopes for biosynthetic work, a pilot plant for large-scale extractions, etc...

Luck I also had in the choice of several research topics as described above; several times some chance observation or minor problem led to important results, such as "from ascaryl alcohol to immunodeterminant sugars of endotoxins", "from fortuitine to mass spectrometry of peptides", "from the methyl group to potential antiviral and antiparasitic drugs", "from Freund's adjuvant to MDP", etc...

Can this repeated luck be rationalized? Louis Pasteur said: "Le hasard ne favorise que les esprits préparés" and Albert Einstein: "Imagination is more important than knowledge"; I suppose I had some sort of "esprit préparé" and some imagination, but I am sure that my knowledge was always insufficient.

And last but not least, I was incredibly lucky to have married Hélène Fréchet in 1932; in 1982, at our golden wedding, we were

*He had discovered the "petite" mutant yeast, which was red; I had started extracting the dark red, hydrosoluble pigment and showed its solution once to Sir Alexander (now Lord) Todd when he came to my lab; just at that time vitamin B_{12} had been described and Todd thought that the yeast pigment could well be the same compound; preliminary biological experiments seemed to confirm the identity, but then came the news from Basle, that it was quite inactive; it also had no cobalt and could not be crystallized; that was a great disappointment.

surrounded, in our garden, by our seven children with two sons-in-law, four daughters-in-law and eight grandchildren (a ninth was born later). My private life has proceeded smoothly and allowed me to concentrate on scientific occupations.

George Bernard Shaw once wrote:

"All autobiographies are lies. I do not mean unconscious, unintentional lies: I mean deliberate lies. No man is thought to be good enough to tell the truth about himself during his lifetime, involving, as it means, the truth about his family his friends and his colleagues".

The story I told here is of course intended to be true, in all details, but the reader will perhaps notice that I have said only a minimum about my family, my friends and my colleagues. I think this was a wise precaution.

If I were asked where I would prefer living, I would answer: I cannot think of living anywhere else than in France, where human rights and democracy are still respected, certainly near Paris, to be able to participate in and enjoy major cultural and scientific events; more particularly in a little town south of Paris, called Sceaux, with its huge public parc, with its boulevard Colbert where there is a small, old house with a small garden where I grow flowers, fruits, vegetables and where on weekends the children and grand-children come to enliven our quiet life*.

---

*My children were of course conditioned for higher education; one year at the Lycée Français in London gave them the English polish considered essential. They have mostly obtained positions in the University or at the CNRS. Our eldest daughter has a Doctorat ès Lettres and is professor of applied linguistics at the University of Paris; the other six have inherited mathematical talents from their grandfather Maurice Fréchet. Together they have one Dr. ès Sc. in mathematics, one in physics and one in chemistry; one is a cardiologist after having been assistant in mathematics; one has an M.Sc. in mathematics, one in physics. They all have suffered from their father's leitmotiv at home "only scientific research is worthwhile and makes one happy". If I had to start again, would I be able to say something else?

Acknowledgements

Over the years my work was supported by CNRS and, in part, by grants from DGRST; INSERM; Fondation Wakssman; Fondation de la Recherche Médicale; Ligue Nationale Française contre le Cancer; The National Institutes of Health, Bethesda; WHO; The Cancer Research Institute, New York, and a contract with ANVAR; Laboratoires Choay; Institut Pasteur, and Sanofi.

This autobiography is based, in part, on a previous text *Fifty Years of Scientific Research (the Fool's luck)*, in M. Kageyama et al. (Eds.), Science and Scientists, Essays by Biochemists, Biologists and Chemists, 1981, pp. 315–322. I thank the Publishers for giving me the copyright.

REFERENCES

1 C.V. Wurzbach, Biographisches Lexikon des Kaiserthums Oesterreich, Wien, Hof- und Staatsdruckerei, 1872, pp. 234-246.

2 K. Przibram, Neue österreichische Biographie ab 1815. Grosse Österreicher. Amalthea-Verlag, Zürich, pp. 184-191.

3 E. Späth and E. Lederer, Ber. D. Chem. Ges., 63 (1930) 120-125; 743-748; 2102-2111.

4 E. Lederer, J. Chromatogr., 73 (1972) 361-366; Ouspechi Chromatographii; isdatelstvo Naouka Moscow (1972) 26-30.

5 M. Tswett, Kromophilli v rastitel'nom i żivotnom mire (Chromophylls in the plant and animal kingdom), Warsaw, 1910.

6 R. Kuhn, A. Winterstein and E. Lederer, Z. Physiol. Chem., 197 (1931) 141-160.

7 R. Kuhn and E. Lederer, Naturwissenschaften, 19 (1931) 306.

8 R. Kuhn and E. Lederer, Ber. D. Chem. Ges., 64 (1931) 1349-1357.

9 R. Kuhn, Nachr. Chem. Techn., 12 (1964) No. 14, 287.

10 R.L.M. Synge, Biochem. Soc. Symp. No. 20, 1970, p. 176.

11 P. György, R. Kuhn and E. Lederer, J. Biol. Chem., 131 (1939) 745-759.

12 R. Kuhn and E. Lederer, Ber. D. Chem. Ges., 66 (1933) 488-495.

13 R. Kuhn and N.A. Sörensen, Ber. D. Chem. Ges., 71 (1938) 1879-1888.

14 E. Lederer, C. R. Acad. Sci., 197 (1933) 1694-1695.

15 E. Lederer, C. R. Soc. Biol., 116 (1934) 150-152; 117, 411-412.

16 G. Galasko, J. Hora, T.P. Toube, B.C.L. Weedon, D. André, M. Barbier, E. Lederer and V.T. Villanueva, J. Chem. Soc. Ser. C., (1969) 1264-1265.

17 T. Moore and E. Lederer, Nature, 137 (1935) 996.

18 E. Lederer, C.R. Acad. Sci., 201 (1935) 300-302.

19 E. Lederer and V.A. Rosanova, Biokhymia, 2 (1935) 293-303.

20 A.E. Gillam, I.M. Heilbron, W.E. Jones and E. Lederer, Biochem. J., 32 (1938) 405-416.

21 E. Lederer and R. Glaser, C. R. Acad. Sci., 207 (1938) 454-456.

22 E. Lederer, C. R. Acad. Sci., 209 (1939) 528-530.

23 E. Lederer, Bull. Soc. Chim. Biol., 20 (1938) 554-566; 567-610; 611-634.

24 E. Lederer, Bull. Soc. Chim. Biol., 24 (1942) 1155-1162.

25 E. Lederer, Nature, 157 (1946) 231; Centenary Lecture J. Chem. Soc., (1949) 2115-2125.

26 E. Lederer, F. Marx, D. Mercier and G. Pérot, Helv. Chim. Acta, 29 (1946) 1354-1365.

27 P. Boiteau, A. Buzas, E. Lederer and J. Polonsky, Nature, 163 (1949) 258-259.

28 J. Polonsky and E. Lederer, Bull. Soc. Chim. France, (1954) 924-932.

29 A. Diara, E. Lederer and J. Pudles, Bull. Soc. Chim. France, (1959) 693-700.

30 M. Barbier, E. Lederer et T. Nomura, C. R. Acad. Sci., 251 (1960) 1133-1135.

31 E. Demole, E. Lederer and D. Mercier, Helv. Chim. Acta, 45 (1962) 672-685; 685-692.
32 P. Dietrich, E. Lederer, M. Winter and M. Stoll, Helv. Chim. Acta, 47 (1964) 1581-1590.
33 W. Pickenhagen, P. Dietrich, B. Keil, J. Polonsky, F. Nouaille and E. Lederer, Helv. Chim. Acta, 58 (1975) 1078-1086.
34 K. Biemann, C. Lioret, J. Asselineau, E. Lederer and J. Polonsky, Bull. Soc. Chim. Biol., 42 (1960) 979-991.
35 C. Fouquey, E. Lederer and J. Polonsky, Bull. Soc. Chim. Biol., 39 (1957) 101-132.
36 C. Fouquey, E. Lederer, O. Lüderitz, J. Polonsky, S. Stirm, R. Tinelli and O. Westphal, C. R. Acad. Sci., 246 (1958) 2417-2420.
37 C. Fouquey, O. Lüderitz, E. Lederer, J. Polonsky and O. Westphal, Nature, 182 (1958) 944.
38 M. Gastambide and E. Lederer, Nature, 184 (1959) 1563; Biochem. Z., 333 (1960) 285-295.
39 A.H. Etémadi, R. Okuda and E. Lederer, Bull. Soc. Chim. France, (1964) 868-870.
40 A.H. Etémadi and E. Lederer, Biochim. Biophys. Acta, 98 (1965) 160-167.
41 M. Gastambide, J.M. Delaumény and E. Lederer, Chem. Ind., (1963) 1285-1286; Biochim. Biophys. Acta, 70 (1963) 670-678.
42 K.R. Murray, Austral. J. Chem., 15 (1962) 510-515.
43 R.E. Noble, R.L. Stjernholm, D. Mercier and E. Lederer, Nature, 199 (1963) 1600-1601.
44 H. Demarteau-Ginsburg, A. Ginsburg and E. Lederer, Biochim. Biophys. Acta, 12 (1953) 587-588.
45 H. Demarteau-Ginsburg, E. Lederer, R. Ryhage, S. Ställberg-Stenhagen and E. Stenhagen, Nature, 183 (1959) 1117-1119.
46 E. Lederer, A.P. MacLennan, H.M. Randall and D.W. Smith, Nature, 186 (1960) 887-888.
47 M. Gastambide-Odier, P. Sarda and E. Lederer, Tetrahedron Lett., (1965) 3135-3143.
48 S.-N. Cho, D.L. Yanagihara, S.W. Hunter, R.H. Gelber and P.J. Brennan, Infect. Immun., 41 (1983) 1077-1083.
49 M. Ikawa, E.E. Snell and E. Lederer, Nature, 188 (1960) 558-560.
50 E. Vilkas and E. Lederer, Tetrahedron Lett., (1968) 3089-3092.
50a M. Chaput, G. Michel and E. Lederer, Biochim. Biophys. Acta, 63 (1962) 310-326.
50b E. Lederer, Pure Appl. Chem., 7 (1963) 247-268.
51 E. Lederer, Sixth Int. Congr. Biochemistry, Proc. Plen. Sessions, IUB, 33 (1964) 63-78.
52 E. Vilkas, A.M. Miquel and E. Lederer, Biochim. Biophys. Acta, 70 (1963) 217-218.
53 M. Barber, P. Jollès, E. Vilkas and E. Lederer, Biochem. Biophys. Res. Commun., 18 (1965) 469-473.

54 M. Barber, W.A. Wolstenholme, M. Guinand, G. Michel, B.C. Das and E. Lederer, Tetrahedron Lett., (1965) 1331-1336.
55 F. Lanéelle, J. Asselineau, W.A. Wolstenholme and E. Lederer, Bull. Soc. Chim. Fr., (1965) 2133-2134.
56 E. Vilkas, A. Rojas, B.C. Das, W.A. Wolstenholme and E. Lederer, Tetrahedron, 22 (1966) 2809-2821.
57 B.C. Das, S.D. Géro and E. Lederer, Biochem. Biophys. Res. Commun., 29 (1967) 211-215.
58 E. Vilkas and E. Lederer, Tetrahedron Lett., (1968) 3089-3092.
59 D.W. Thomas, E. Lederer, M. Bodansky, J. Izdebski and I. Muramatsu, Nature, 220 (1968) 580-582.
60 C. Giannotti, B.C. Das and E. Lederer, Chem. Commun., (1966) 28-29.
61 B.C. Das, M. Lounasmaa, C. Tendille and E. Lederer, Biochem. Biophys. Res. Commun., 21 (1965) 318-322.
62 S. Beau, R. Azerad and E. Lederer, Bull. Soc. Chim. Biol., 48 (1966) 569-581.
63 A. Adam, J.F. Petit, J. Wietzerbin-Falszpan, P. Sinaÿ, D.W. Thomas and E. Lederer, FEBS Lett., 4 (1969) 87-91.
64 J. Wietzerbin-Falszpan, B.C. Das, I. Azuma, A. Adam, J.F. Petit and E. Lederer, Biochem. Biophys. Res. Commun., 40 (1970) 57-63.
65 G. Jauréguiberry, J.H. Law, J.A. McCloskey and E. Lederer, Biochemistry, 4 (1965) 347-353.
66 G. Jauréguiberry, M. Lenfant, B.C. Das and E. Lederer, Tetrahedron Supp. 8, part I, (1966) 27-32.
67 M. Lenfant, H. Audier and E. Lederer, Bull. Soc. Chim. Fr., (1966) 2775-2777.
68 M. Guérin, R. Azerad and E. Lederer, Bull. Soc. Chim. Biol., 47 (1965) 2105-2114.
69 E. Lederer, Quart. Rev., 23 (1969) 453-481.
69a J. Hildesheim, R. Hildesheim and E. Lederer, Biochimie, 53 (1971) 1067-1071.
70 M. Robert-Géro, F. Lawrence, G. Farrugia, A. Berneman, P. Blanchard, P. Vigier and E. Lederer, Biochem. Biophys. Res. Commun., 65 (1975) 1242-1249.
71 M. Robert-Géro, P. Blanchard, F. Lawrence, A. Pierre, M. Vedel, M. Vuilhorgne and E. Lederer, in E. Usdin, R.T. Borchardt and C.R. Creveling (Eds.), Transmethylation, Elsevier, Amsterdam, 1979, pp. 204-214.
72 M. Vedel, F. Lawrence, M. Robert-Géro and E. Lederer, Biochem. Biophys. Res. Commun., 85 (1978) 371-376.
73 U. Bachrach, L.F. Schnur, J. El-On, C.L. Greenblatt, E. Pearlman, M. Robert-Géro and E. Lederer, FEBS Lett., 121 (1980) 289-291.
74 D.K. Dube, G. Mpimbaza, A.C. Allison, E. Lederer and L. Rovis, J. Trop. Med. Hyg., 32 (1983) 31-33.
75 A. Ferrante, I. Ljungström, G. Huldt and E. Lederer, Trans. Roy. Soc. Trop. Med. Hyg., 78 (1984) 837-838.

76 H. Noll, H. Bloch, J. Asselineau and E. Lederer, Biochim. Biophys. Acta, 20 (1956) 299-309.
77 A. Bekierkunst, I.S. Levij, E. Yarkoni, E. Vilkas, A. Adam and E. Lederer, J. Bacteriol., 100 (1969) 95-102.
78 A. Bekierkunst, I.S. Levij, E. Yarkoni, E. Vilkas and E. Lederer, Infect. Immun., 4 (1971) 245-255.
79 M. Lepoivre, J.P. Tenu, G. Lemaire and J.F. Petit, J. Immunol., 129 (1982) 860-866.
80 J.P. Tenu, E. Lederer and J.F. Petit, Eur. J. Immunol., 10 (1980) 647-653.
81 E. Lederer, Springer Semin. Immunopathol., 2 (1979) 133-148.
82 E. Lederer, J. Med. Chem., 23 (1980) 819-825.
83 E. Lederer and L. Chedid, in E. Mihich (Ed.), Immunological Approaches to Cancer Therapeutics, Wiley, New York, 1982, pp. 107-135.
84 E. Yarkoni, H.J. Rapp, J. Polonsky and E. Lederer, Int. J. Cancer Res., 22 (1978) 564-569.
85 I.A. Clark, Parasite Immunol., 1 (1979) 179-196.
86 G.R. Olds, L. Chedid, E. Lederer and Adel A.F. Mahmoud, J. Infect. Dis., 141 (1980) 473-478.
87 R.G.S. Johns, E. Lederer and R.G. White, Immunology, 1 (1958) 54-56.
88 A. Adam, R. Ciorbaru, J.F. Petit and E. Lederer, Proc. Natl. Acad. Sci. USA, 69 (1972) 851-854.
89 C. Merser, P. Sinaÿ and A. Adam, Biochem. Biophys. Res. Commun., 66 (1975) 1316-1322.
90 F. Ellouz, A. Adam, R. Ciorbaru and E. Lederer, Biochem. Biophys. Res. Commun., 59 (1974) 1317-1325.
91 M. Parant, Springer Semin. Immunopathol., 2 (1979) 101-118.
92 A. Adam, J.F. Petit, P. Lefrancier and E. Lederer, Mol. Cell. Biochem., 41 (1981) 127-147.
93 A. Adam and E. Lederer, Med. Res. Rev., 4 (1984) 111-152.
94 P. Lefrancier, M. Derrien, X. Jamet, J. Choay, E. Lederer, F. Audibert, M. Parant, F. Parant and L. Chedid, J. Med. Chem., 25 (1982) 87-90.
95 L. Chedid, M. Parant, F. Audibert, G. Riveau, F. Parant, E. Lederer, J. Choay and P. Lefrancier, Infect. Immun., 35 (1982) 417-424.
96 M. Sela, in Progress in Immunology V, Academic Press Japan, Tokyo, 1983, pp. 13-21.
97 T.M. Shinnik, J.G. Sutcliffe, N. Green and R.A. Lerner, Annu. Rev. Microbiol., 37 (1983) 425-446.
98 J.M. Krueger, J.R. Pappenheimer and M.L. Karnovsky, J. Biol. Chem., 257 (1982) 1664-1669.
99 J.M. Krueger, J.R. Pappenheimer and M.L. Karnovsky, Proc. Natl. Acad. Sci. USA, 79 (1982) 6102-6106.
100 J.M. Krueger, J. Walter, M.L. Karnovsky, L. Chedid, J.P. Choay, P. Lefrancier and E. Lederer, J. Exp. Med., 159 (1984) 68-76.
101 S.A. Martin, M.L. Karnovsky, J.M. Krueger, J.R. Pappenheimer and K. Biemann, J. Biol. Chem., 259 (1984) 12652-12658.

102 R.H. Gisler, F.M. Dietrich, G. Baschang, A. Brownbill, G. Schumann, F.G. Staber, L. Tarcsay, E.D. Wachsmuth and P. Dukor, in Drugs and Immune responsiveness, J.L. Turk and D. Parker (Eds.), McMillan, New York, 1979, pp. 133–160.
103 E. Lederer, Mikrokosmos, 19 (1925) 109–112.
104 E. Lederer, Progrès Récents de la Chromatographie, I. Chimie Organique et Biologique, Actualités Scientifiques et Industrielles, No. 1079, Hermann, Paris, 1949, 146 pp.
105 E. Lederer and M. Lederer, Chromatography, a Review of Principles and Applications, Elsevier, Amsterdam, 1953, 465 pp; 2nd ed., 1957, 711 pp.
106 E. Lederer et al., La Chromatographie en Chimie Organique et Biologique. Masson, Paris, 1959, I, 671 pp; 1960, II, 874 pp.
107 A. Sharma, A. Haq, S. Ahmad and E. Lederer, Infect. Immun., 48 (1985) 634–637.

G. Semenza (Ed.) Selected Topics in the History of Biochemistry: Personal Recollections (Comprehensive Biochemistry Vol. 36) © 1986 Elsevier Science Publishers

Chapter 10

Recurrent Luck in Research

N.W. PIRIE

Rothamsted Experimental Station, Harpenden, Herts. AL5 2JQ (U.K.)

Throughout life I have been consistently lucky. First, by being born in 1907 and so to have lived through a period of unusual technological change. For example: when a boy, I could bicycle freely and safely on country roads, cars were rare and noisy; when an adolescent, cars were reliable but so simple that an amateur could repair them; when an adult, the sky was nearly empty, I could fly as I wished without planning or radio control. Heredity gave me a robust body and enough sense to treat it reasonably. A liking for verbal precision makes me doubt the need for the word **tired**. We have the words **bored**, **exhausted** and **sleepy**: different remedies suit each of these states, and choosing the right remedy may promote extra activity. My skill in doing that has excited comment from those less able to make accurate diagnoses. Thus, a listing in *Brighter Biochemistry* [1], in the style of the Old Testament, of the official and unofficial activities of the inmates of the Cambridge Biochemistry laboratory, included "Pir, who sleepeth not neither doth he rest" and the article about our survey of the middle of Spitsbergen (Svalbard) [2] described me as "nearly indefatigable when going over rough country".

My father was a painter. He observed animal behaviour keenly and accurately. During the time when I knew him, he usually painted cats, ducks and hens; earlier he had painted larger animals. There are some lively paintings of cattle done while visiting a

Plate 16. N.W.Pirie in 1976 with an early version of the juice extractor (Butler and Pirie, 1981).

brother in Texas. His paintings fetch good prices now, but they were only a minor source of income in his lifetime. Because we were all frugal, we lived comfortably on my mother's inherited income. I have never been debarred from any desired activity or purchase by inability to pay: that convenient situation is as much a consequence of an unaquisitive and economical, or parsimonious, nature as of an adequate income.

An elder brother and sister were the perfect stimulus to an intellectually aggressive child: another piece of luck. My father was given a subscription to the Oxford English Dictionary as a wedding present. My youth was therefore punctuated by the arrival of its successive volumes. At first, my main interest was in the bits of gold leaf which careless book binders left in the tissue paper packs used to protect the Morocco leather binding. Gradually I began to share my father's enthusiasm for etymology.

Astigmatism and starting to stammer when about 7 could be regarded as bad luck. However, the former may explain my complete lack of interest in every form of game, and the latter kept me away from school for about half the time between 7 and 14. Avoidance of conventional education is a common prelude to a research career. Continuous schooling from 14 to 18, coupled with my mother's income, got me into Cambridge. School mathematics, especially geometry, had suited me. I therefore started by studying Chemistry, Physics and Mathematics. After a month I realised I was making nothing of University Mathematics and switched to Physiology in its place. When a boy, I had kept mice and rabbits and was familiar with the anatomy and reproductive processes of animals. My mother had trained as a nurse; I read her old Physiology books avidly but, before getting to Cambridge, I was unaware that a living could be earned in any aspect of Biology except Medicine. My father urged me to be an engineer because, from about 10, I saw more quickly than he did where a lever should be put when we were moving rocks in the garden, a support when he rearranged the lofts in his studio, or the chimney pipe from his coke stove for maximum heat conservation.

I feel genuinely indebted to only two of the people who were paid, formally, to teach me: Hopkins and G.F.C. Searle. After graduation, Haldane was influential, but in my last year as a student of Biochemistry he was on sabbatical leave. Searle was a queer character: vegetarian, antivivisectionist and with some other bees-in-his-bonnet. He had a marvellous understanding of what could be learnt from class experiments. He taught real Physics - electricity, optics, properties of matter etc. - not ephemeral stuff about the inside of atoms. He got angry if he found you taking trouble to make one measurement, out of the set needed in an experiment, significantly more precisely than the others. To be a little slapdash was reprehensible laziness: to take useless trouble was far worse, it showed you were not thinking.

Sulfur compounds and animal nutrition

During my second year in Cambridge I discovered Biochemistry and revelled in it. This was partly because of Hopkins' personality, and partly because his laboratory seemed livelier than any other laboratory I had contact with. The article on him in this series [3] is by me, although a change made after proof may obscure that fact.

For reasons which now elude me, I got interested in titration curves during the final year of Biochemistry and, with Kathleen Pinhey [4], plotted the titration curve of glutathione. One reason for the collaboration was that she was an early user of the glass electrode: those who showed any interest in our paper were as often interested in the electrode as in the glutathione. More perhaps by good luck than by strict logic, we deduced that it is γ-glutamyl-cysteinyl-glycine. Trying to get evidence for our assignment of pK values to the 4 titratable groups, we plotted curves in the presence of formaldehyde and found that the apparent pK which was 8.7 in water, diminishes as would be expected with an $-NH_2$ group, whereas the 9.1 pK increases. Cysteine shows the same phenomenon and, in the presence of formaldehyde and at pH 9.3, behaves as a negative buffer or like Buridan's Ass [5]. Minute amounts of acid or alkali, by shifting the formaldehyde from $-NH_2$ to $-SH$, have a greater effect on pH than they have on the pH of an equal volume of water.

When glutathione is extracted by boiling yeast or liver, separations are tedious because colloidal material is also extracted. Extracts made by adding a mixture of ethanol, ether and sulfuric acid are cleaner, presumably because colloids do not diffuse through the osmotically damaged but unbroken cell walls. From these extracts, without further purification, cuprous glutathione can easily be separated [6]. Glutathione suddenly became abundant - at least in Cambridge. Methods similar to this are regularly used in commercial production. Surprisingly, the method was patented [7] many years later. Hopkins' characteristically perceptive recognition of the merits of the cuprous derivative for separating glutathione led me to use the technique for preparing several other cysteine derivatives, and to study the soluble compounds formed when excess cuprous copper is added. As a side-line on that, I confirmed Haldane's observation that 100% carbon monoxide smells of onions. I was using carbon monoxide to sequestrate cuprous copper.

Another by-product of that work was the introduction of freeze-drying into Biochemistry [8]. The technique had been used in Microbiology, and it was the theme of About's romantic novel *L'homme a l'Oreille Cassé* published in 1862. When working in a scarcely heated lab, a solution of the –SS– form of glutathione, in a well arranged desiccator, froze spontaneously. I was so pleased with the open feathery texture of the dry product that I used the technique consistently thereafter with non-crystalline substances. I also showed it to Keilin who showed it to his many visitors.

Glutathione had, at that time, no clearly defined function in vivo. Some people suggested that it was a by-product of protein digestion which resisted further digestion. Hele and I [9] therefore fed dogs on it and found that it yielded sulfate as readily as cyst(e)ine whether given by mouth or injected. We tried some other cyst(e)ine peptides: some are less readily digested. An interesting by-product of that work was the observation that chloroacetylcysteine, when neutralised for injection, titrates as a dibasic acid and liberates chloride ions.

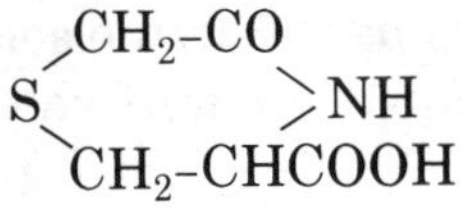

$$S\begin{cases} CH_2\text{-}CO \\ \\ CH_2\text{-}CHCOOH \end{cases}\! > NH$$

was probably formed. It is not readily metabolised to sulfate: acetyl cysteine is also poorly metabolised. The most interesting result from a set of measurements of the oxidation of various S compounds by tissues and tissue slices in vitro was the observation that rat chorion, like liver and kidney but unlike some other tissues, oxidises thiosulfate to sulfate.

In 1932, methionine+cystine accounted for less than half the S present in casein. My improved method of methionine isolation increased the yield fivefold. That still left some S unaccounted for although it made methionine the dominant form in which S was present. N.R. Lawrie and I tried to find out whether a growing rat needed both cystine and methionine. The problem, at that time, was to get vitamin sources which did not contain enough S to vitiate the experiment. We failed. While purifying what was then called the "B complex" we noticed that our best preparations always had a green fluorescence and joked that that was probably the vitamin. Perhaps foolishly we abandoned the project. Others isolated riboflavin later.

Cysteine is slightly, and cystine strongly, levorotatory in acid solution. Using this rotation change I found [10] that thiocarbamides, and thioglyoxalines such as ergothionine and thiohistidine, catalyse the oxidation of cysteine. The method was also used to follow the establishment of equilibria between cyst(e)ine and optically inactive substances such as thioglycollic acid.

Krebs' urea cycle, and the stimulating presence of Krebs himself in the lab for 2 years, made me think about the old problem of hippuric acid formation. This is a puzzle because more glycine can be excreted in hippuric acid than is present in the diet, and an equivalent amount of urea is no longer excreted. Birds conjugate benzoic acid with ornithine, and we and chimpanzees conjugate phenylacetic acid with glutamine. An early suggestion that the first step was conjugation with an amino acid which was then whittled down to glycine, was held to have been disproved by the results of feeding various benzoylated amino acids. However, benzoylated arginine and ornithine had not been tried. Because they are part of the Krebs' cycle, they seemed to be the amino acids most likely to be involved. So I fed them to dogs. Dogs were chosen because they conjugate benzoic acid partly with glycine and partly with glucuronic acid. From the ratio of the two products excreted it should be possible to distinguish benzoic acid freed from the amino acid by hydrolysis, and therefore available for conjugation with glucuronic acid, from benzoic acid which had not been detached from the larger amino acids. I got no conclusive results. It is now known that tissues can conjugate benzoic acid with glycine, but I still think the more complicated mechanism may exist. It would bring birds into line with most mammals, though it would still leave us and chimpanzees anomalous in our manner of handling phenylacetic acid. With isotopically labelled arginine and ornithine it should be easy to find out.

Brucella antigens and related topics

By 1932 I was getting bored with S metabolism - hence the excursion into hippuric acid formation. Therefore, when Prof. H.R. Dean in the adjacent Pathology laboratory asked Hopkins to suggest someone to collaborate in a study of the antigenic relationship between *Brucella melitensis* and *Br. abortus*, I readily took up the new line of work. After 7 years, 5 papers [11] were published. They do not clarify the relationship, but they describe some interesting properties of *Brucella* antigens.

Some bacterial antigens are released spontaneously into the culture medium. Others were, at that date, often released by brutal methods such as boiling or grinding the dried bacteria. We decided to use gentle methods. It seemed to me likely that the carbohydrate present in antigen preparations made from bacteria which, because they do not grow well except on surfaces, had been grown on agar, often were agar. An early by-product of the main work was a study of some widely used colour reactions for detecting agar [12]. Agar differs from most other carbohydrates more in the rapidity of colour development on heating with orcin or diphenylamine in acid solution, than in the tint of the colour. Hepta-acetyl DL-galactose was made by acetolysis of agar. Because this gives colour reactions similar to those of agar, I suggested that, in agar, galactose is in the open chain, or aldehyde, rather than in the usual pyranose form. That suggestion was withdrawn when evidence was published that there is 3,6-anhydro galactose in agar and which gives similar colour reactions. Hyperacetylated sugars were then a novelty, so I made a few others by acetolysing the mercaptals of some penta-acetyl hexoses and tetra-acetyl pentoses.

Antigen, in what we regarded as essentially the native state, separates from *Brucella* when suspensions in 2% phenol are left at 0°C for a few weeks. Centrifuges running at 15 000 rev./min (16 000 × *g*) held only 35 ml at that date, so the antigen was usually concentrated by precipitation with ammonium sulfate. That was a useful purification step because, although agar breakdown products precipitate along with the antigen at room temperature, unlike the antigen they do not redissolve at -5°C. After 2 or 3 cycles of differential precipitation and centrifugal sedimentation, the centrifuge pellets are colourless, slightly birefringent and give solutions which become birefringent when stirred. That suggested the presence of either rods or plates. We argued in favour of plates because, under dark-ground illumination, precipitates with antisera seemed to show edge-to-edge adherence of plates. *Br. melitensis* is 0.6 × 1.6 μ; from that size and the yield of antigen we calculated that it could be a brittle shell, 10 to 20 mμ thick, on the living organism. That would be invisible by ordinary microscopy. Electron microscopy was still in the future. Our preparations were not merely fragments of disintegrated bacteria because these contain 12% N, whereas the antigen contains 4 to 5% N.

Sodium dodecyl sulfate (SDS) was introduced into Biochemistry as a means for disintegrating tobacco mosaic virus [13]. It was not at that time on sale but was one of a group of new wetting agents which had been sent to a golfing enthusiast in the lab for testing as herbicide-spreading agents. SDS was the most powerful and convenient. As with freeze-drying, I showed its merits to Keilin and he spread the news around. It is interesting that, after nearly 50 years, it is still widely used. It converts *Brucella* antigen into material which is no longer sedimentable at 16 000 × *g* but still precipitates with antiserum though in a narrower precipitation zone.

We dismantled the "native" antigen progressively. By repeated precipitation with neutral ethanol, or by a single precipitation with a mixture of ethanol and ether containing 0.5% HCl, 20% of the dry matter, and half the P, separates as a lipid fraction. After this, the antigen is no longer birefringent, it sediments more slowly at $16\,000 \times g$ and, when injected intradermally, is more toxic than "native" antigen or the original bacteria. Some material giving colour reactions for amino acids can be removed by precipitation with ethanol from solution in 80% acetic acid. Brief acid hydrolysis releases an insoluble substance with some phospholipoidal properties, and inorganic phosphate. Antigenicity and toxicity are then lost, but the soluble material, which accounts for half the mass of the "native" antigen, is retained on dialysis and contains a single component sedimenting in the analytical ultracentrifuge as a particle of mass 3300. On further acid hydrolysis, formic acid is liberated and an $-NH_2$ group appears. *N*-formyl compounds were not known to occur in nature at that time: several have been found since. The material gives colour reactions similar to, but not identical with, those of glucosamine, and the $-NH_2$ pK is 6.8 whereas that of glucosamine is 7.8.

Our picture of the "native" antigen was therefore:- a backbone of formylamino carbohydrate and possibly another carbohydrate, combined with phosphate, amino acids and at least two phospholipids. So far as I am aware, no part of our picture has been contradicted — nor has it, in detail, been confirmed. After a long and enjoyable collaboration, Miles' interests and mine began to diverge. Nevertheless, we briefly collaborated again in editing the first symposium, on "*The Nature of the Bacterial Surface*", of the Society for General Microbiology in 1949.

By the time our publications appeared, *Br. abortus* infection in cattle was being successfully controlled with the live vaccine S 19. There are always more problems with a live than a killed vaccine. Miles and I had made material from *Br. abortus* similar to our *Br. melitensis* antigen, so I was asked to make it in bulk to test as a killed vaccine. The antigen protected guinea pigs as effectively as the live vaccine [14]. In cattle, though it increased the level of circulating antibody in the blood more than the live vaccine, its protective effect was doubtful [15].

After 11 years in Cambridge I began to want a change of scene and so got Hopkins to write to Landsteiner, in the Rockefeller Institute (New York), asking him to let me work there for 2 or 3 months. In spite of his politically reactionary outlook (he called Roosevelt "that communist in the White House") we got on very well. This may have been because I, not being dependent on his good-will, argued with him more vigorously than his other colleagues thought prudent. He probably found their complete subservience boring. Conditions might have been different if I had accepted the job he offered me. We diazotised 3-amino pyridine and two of its carboxylic acids, coupled them to horse serum, and studied the antigenic specificity of the products [16]. Nothing unexpected came of this work, but I enjoyed doing some straightforward organic chemistry.

Plant viruses

While working with Miles, I was often in the Pathology Department and there met Bawden who was learning serological techniques. We were members of the same College so I already knew him slightly. He showed me precipitin reactions with sap from plants infected with potato virus X. When I found that he did not know what the end point was, expressed as a concentration of indiffusible dry matter rather than dilution, I took away, to make the measurements, some of the fluids he was using. That started a collaboration which lasted until his death 38 years later. Our work did not contribute directly to the control of any virus infection. Nevertheless, it was probably useful in dispelling the aura of mystery which surrounded viruses in the early 1930s. We gave them describable physical properties.

This is not the place to write about Bawden: I have done that elsewhere [17]. It may, however, be in order to comment on collaboration because I have been lucky in having had two distinguished collaborators. One became Foreign Secretary and the other Treasurer of the Royal Society. Hopkins taught me that someone's time is being wasted when people with similar capabilities work together. He recognised the need for periods of tuition, but thought that people should set out on their own as soon as possible. He would have been horrified, or at least mystified, by the vast army that now sometimes seems to claim responsibility for one publication. Prolonged collaboration is permissible only between people with contiguous rather than overlapping skills. It is greatly improved if collaborators like each other's company enough to go to a pub to discuss progress. By good luck, Miles and Bawden knew much more Physics and Biochemistry than was known by most microbiologists or plant pathologists, and so understood what I was doing. I never got an equal understanding of their expertise.

Viruses had interested me when I was a student: that interest was reinforced by marriage to a biochemist who worked first on Rous sarcoma virus and then on bacteriophage. As a result of that, I had many illuminating conversations with W.E. Gye who was a premature advocate of the viral origin of some cancers. Potato virus X can be assayed both by infectivity and serologically. It is destroyed by proteolytic enzymes [18], but we were careful to point out that "there is no evidence that other equally important substances may not also be present". When teaching, I constantly stressed that it is a common biochemical fallacy to assume that, because a preparation *contains* so-and-so, it *is* so-and-so.

Potato virus X was, and still is, troublesome material to work with. In 1936 Bawden moved from the Potato Virus Research Station in Cambridge where, as he put it, "the most sophisticated piece of apparatus (was) a recalcitrant Primus stove", to Rothamsted. Facilities there were much better, and it was no longer necessary to work only on viruses which infect potatoes. Like most plant virus workers we therefore then concentrated on tobacco mosaic virus. In a few weeks, using techniques which had been standard in protein chemistry for half a century, I made preparations from which a liquid crystalline layer separated. Luckily I had been fascinated by the idea of liquid crystallinity when a student, so I recognised what had happened. Liquid crystallinity and the curious "herring bone" pattern given by a TMV solution when it dries, made me think that we had something which would interest my friend Bernal. His X-ray measurements gave the width of the particles, but left their length uncertain. On the assumption that a suspension of rods becomes liquid crystalline at a concentration at which each rod has no longer a sphere, or a squat cylinder, to rotate in, I calculated a length for the TMV rods and hence the approximate mass of each particle. From this it was clear that many thousands of particles were present in the most dilute inoculum that could cause infection. It was therefore illegitimate to assert that the properties of the dominant components of our preparations, in spite of their novelty and interest, were those of the virus itself. Because there was so much of this anomalous protein in sap from infected plants, we wondered whether it could be a by-product of infection, somewhat like Bence-Jones protein, rather than the virus itself.

Several claims that TMV had been isolated were made in the early 1930s. We thought them all baseless. Our scepticism included Stanley's claim for a "crystalline globulin" in 1935. The infectivity of his preparation was nearly irrelevant. Sap from infected plants is infective at such great dilutions that it would be surprising if anyting made from it, unless it had actually been boiled, were not infective. Evidence for crystallinity was poor, and crystals, though beautiful, need not be uncontaminated. The general principles of protein chemistry made me sceptical about a supposed globulin containing 20% N: to reach that value it would have to be made largely of arginine and glycine. A year later, Stanley brought the N content down to a more reasonable level. At the same time however, he increased our scepticism by claiming that his product contained 0.00% P and S. Such complete separation from other sap components containing P and S seemed unlikely. Scepticism increased still more when Eriksson-Quensel and Svedberg [19] found that the specific volume of the material was 0.646. They commented that that was a surprising value for a protein, and so presumably checked their measurement, but offered no explanation. We were puzzled that so many people lacked, and still lack, any awareness of biochemical improbabilities and accept(ed) Stanley's claims.

Our material contained 0.5% P [20,21] and we separated RNA from the protein in various ways. Luck intruded again. As a student

I had been interested in nucleic acids to such an extent that I think I was familiar with all the literature of what was then an unfashionable subject. In one session of the final Biochemistry examination I answered only one question: it was on them (I was reproved for this but passed the examination). Bawden and I did not think of viruses as very small bacteria; we dimly recognised the significance of their inability to multiply except in a host. Nevertheless, we expected an infective agent to be fairly complex. I searched carefully for carbohydrate and lipid components in our preparations. We accepted the presence of nucleic acid as a fact — not as a matter of philosophical significance. However, when teaching students, I used to ridicule the "tetranucleotide hypothesis". While working with Landsteiner, I met Levene from time to time (Beilstein was kept in his lab rather than the library) and teased him on the matter. He replied: "After all: I only called it a hypothesis". Recognising that nucleic acid was very much larger than a tetranucleotide, we argued (e.g. [22]) that it was just as likely as protein to be the vehicle of biological specificity.

Scepticism about the size of nucleic acids depended partly on the absence of positive evidence for the number 4. Four nucleotides were known, and Levene saw no reason to postulate unnecessary complexity. Scepticism also depended on an ingrained dislike for orthodoxy. This showed in other ways. For example: when a student, continental drift was ridiculed in Cambridge, so I advocated it; the "fundamental particles" were electrons and a nucleus, so I argued that the nucleus also had parts. I may add in parentheses that I dislike the "big bang" hypothesis. Perhaps that will go the way of other orthodoxies.

The physical properties of our preparation of TMV excited considerable attention, partly from their intrinsic interest, and partly because we used a goldfish and sea-horse to stir a dilute solution in a tank between crossed polarisers so as to demonstrate flow birefringence at a soirée. We then supplied material to others for the scientific study of fish movements. The presence of 5% RNA excited no interest whatever. Not unexpectedly, Stanley denied its importance and claimed, for a time, that it could be removed without loss of infectivity. Gradually he incorporated most of our description into his own. This hastened progress in virus research — it leaves unanswered the puzzle of what he made originally. Whatever it was, it can have contained little TMV unless an improbably large number of analytical and arithmetical mistakes were made.

Bawden had a prodigious memory. He made few notes, but carried all the relevant virus literature in his head. This enabled him to choose, for study, viruses which are durable and easily transmitted. We made a few other virus preparations, including the troublesome potato virus X, as liquid crystalline nucleoproteins, and then made tomato bushy stunt virus as rhombic dodecahedral crystals. The flood of metaphysical balderdash released by the early claim that TMV was crystalline, would probably have been even greater if the genuinely crystalline preparation had come first. Those who wanted to classify viruses as living organisms, but sensed some incompatibility between the living and crystalline states, had found solace in our argument [21] that TMV is no more crystalline than a shoal of fish. Its properties were those necessarily shown by any collection of rods of fairly uniform width. With bushy stunt, no similar escape from a careful analysis of baseless prejudices was possible — assuming (we repeatedly emphasised that it was an assumption) that out preparations were the virus and not by-products of the infection.

When working with the group of tobacco necrosis viruses, we thought we had evidence justifying our cautious attitude. Six of them yielded crystalline preparations. One crystallises so readily that crystals are visible with a hand lens along the edge of the pellet sedimented by centrifuging. If the pellet is suspended in a little water and centrifuged at low speed before the crystals have time to dissolve, the supernatant fluid is more infective than the crystalline fraction. A few years later, B. Kassanis isolated serologically unrelated viruses from the culture. The one which is crystallisable is, when alone, not infective. The one which has not been crystallised is infective and, in its presence, the crystallisable one can multiply too. That was the beginning of the now fairly widely recognised phenomenon of "satellitism".

Searching for evidence that the main component of other preparations actually was the virus, we studied how infectivity, and serological and physical properties, were affected by various disruptive agents. The most definite result was that infectivity is lost after treatment with agents too mild to affect the other properties. There were a few surprising results. For example; bushy stunt loses infectivity, without other obvious change, when frozen in the absence of salt at its isoelectric point. And TMV is readily split in strontium nitrate solution [23].

We were remiss in regarding the loss of 99% of infectivity as the same as total loss. Others discovered that there is residual infectivity in disrupted TMV and that it is associated predominantly with the fraction containing RNA. This residual infectivity is destroyed by many agents — some not known to affect nucleic acids. We therefore doubted the attribution of infectivity to nucleic acid if that term is used so narrowly [24] as to include only structures made from the conventional nucleotides. Thiaminase + thiamine is one of the more interesting inactivating systems. Perhaps this mysterious enzyme is concerned in vivo with an aspect of nucleic acid metabolism. It is hard to believe that its normal function is to destroy a vitamin. Some of these destructive agents, e.g. spermine and interferon, are now known to affect other nucleic acids. Our work, and that of others, made us doubt the validity of the "dogma" that "information" invariably goes by the route DNA to RNA to protein. Elsewhere [25] I have discussed the manner in which Avery's evidence, in 1944, that DNA could be responsible for an infective process, was incorporated into general thinking. By that date, our evidence made it probable that 12 viruses or virus strains, with widely different biological properties, contained RNA. Because of this historical sequence, it is odd that DNA was immediately accepted as the normal carrier of "information", and that viruses containing RNA are categorised as retroviruses.

We did not study the conditions in which RNA from viruses other than TMV loses infectivity, but we did some comparable work on one of the necrosis viruses [26]. It is inactivated by preparations containing mitochondria from both healthy and virus-infected leaves, and by products of their metabolism. Inactivation by mitochondrial preparations can be prevented by inhibitors such as azide, or by excluding air. We invoked such phenomena to explain the varied durability of infectivity in crude preparations, and some differences in the infectability of test plants in different physiological states. RNA separated from TMV is inactivated in a similar manner [24].

Several infective fractions, with differing physical properties, can be separated from the sap of plants infected with TMV. Although that could be the result of contamination with TMV with the usual physical properties, we studied the components of preparations at various stages during their purification. A basic difficulty in work such as this is that leaf sap is not a physiological fluid; therefore, substances in it may associate with each other only during the process of extraction. This difficulty arises with many other viruses [27]. In spite of that reservation about the original source of components which could be removed from TMV preparations without loss of infectivity, preparations made by centrifuging sap soon after expression were compared with preparations made from aged sap or by precipitation with acid or ammonium sulfate [28]. TMV is so robust that it is easily separated from many components of normal leaves. Less stable viruses have to be kept cold and handled quickly: contamination is then probable. Leaf microsomes (later called ribosomes) are one contaminant. They can be made [29] from tomato, tobacco and bean leaves, and contain RNA, ribonuclease and other enzymes. I commented on their tendency to aggregate linearly, as shown in electronmicrograms, but my main concern was to get rid of them without destroying the accompanying viruses.

Fragments of ribosomes and similar material accompany, or may be combined with, TMV made by gentle methods. Various treatments, e.g. incubation with citrate, remove them: the product, though still infective, then has significantly different properties from gently made TMV. The processes which strip them away could be called "purification". On the other hand, these Procrustian techniques make an *écorché* rather than native virus. Unfortunately, little attention is paid to these phenomena, and many analyses and physicochemical studies are still made which, if some components are picked up during the extraction, are contaminated; if, by contrast, these are genuine components of native virus, the complex deserves study.

Proteins other than TMV, which are present in infective sap but not sap from healthy plants, deserve study if, as we argued [22], virus infection should be thought of as a general derangement of host metabolism. That review surveyed most of the processes which might be involved, commented on many unwarranted assumptions made by others, and suggested that the study of virus multiplication would probably shed more light on the mechanism of protein synthesis than the latter study would shed on virus multiplication. In it we also commented that

> "the specificity of protein synthesis is as likely to result from specific nucleotide arrangements as from any other type of specific structure",

and that virus multiplication could as logically be considered an aberration of nucleic acid as of protein metabolism. In 1952, such comments were taken as further examples of our notorious nucleic acid mania.

No one primarily interested in virus multiplication as a special and peculiar subject, would choose plants as research material. Those plant viruses that have been carefully studied do not penetrate an undamaged leaf surface. The damage done during inoculation confuses observations made during the first stages of multiplication. Thereafter, viruses are not extruded from plants: damage during extraction introduces further opportunities for artefact formation. It can be claimed that plants offer exceptional experimental opportunities because of their tolerance of extensive ranges of temperature and nutrition. A counterpart to that possible advantage is that leaves in different positions on a plant differ in infectability. Furthermore, because we regarded virus infection as a general derangement, or hijacking, of the synthesising systems of a plant, we foresaw that, in the limit, infection should be unobservable. Many examples show no correlation between the severity of an infection, assessed by stunting or early death, and the amount of virus, or at any rate of anomalous nucleoprotein, extractable from the plant. TMV is an extreme example. An infected plant can survive although about half the protein in its sap, and 80% of the protein in the fibre left after expressing the sap, can be TMV. When an infection leads to local lesions rather than to a systemic infection, there seems to be a correlation between physiological conditions which produce few lesions, and those which produce small lesions [30]. That suggests the possibility that the limiting case of severity is an infection which destroys an individual cell so abruptly that the infection does not spread to neighbouring cells and therefore remains unobserved by ordinary techniques.

By 1960 it seemed probable that viruses are disassembled soon after entering the host and that, although the complete structure is probably the normal means of dissemination in the field, something approximating to nucleic acid probably spreads infection between cells. We therefore started a prolonged study of the behaviour of TMV RNA in leaves and leaf extracts, hoping that what we knew by then about conditions in which RNA lost infectivity would explain the effects of different physiological states, and positions on the plant, on the infectability of leaves. In spite of many interruptions, we had got the results into some sort of order by 1972 when the work ended with Bawden's untimely death. If anyone finds the two resulting papers [31] hard to understand, I may add by way of apology that they were even harder to write. They would have been clearer had Bawden lived because his capacity to hold all aspects of a subject in his mind simultaneously was greater than mine.

Among the treatments examined for their effects on the amount of infective RNA in the final extract were: variation in the sequence of extraction events, extraction with exclusion of air or in the presence of yeast RNA, and removal of small molecules by freezing the intact leaves and allowing them to thaw in 5 to 10 vols. of water. That was a logical reversal of the technique I had used when making glutathione 40 years before. Similar experiments were made with uninfected leaves to which TMV RNA, labelled with ^{32}P, was added at various stages so that the location of RNA or its derivatives could be followed.

From these experiments we concluded that destruction by leaf ribonuclease removes less RNA from circulation than fixation by leaf fibre, and that sequestration on calcium phosphate, if that is allowed to precipitate during the extraction, can remove all the RNA. Adsorption of nucleic acid by calcium phosphate was well known: the novel feature was coprecipitation. With coprecipitation, calcium phosphate can carry out with it its own mass of RNA [32]. The technique has been used to get fibroblast DNA into a form which is taken in by mouse cells. It is possible that sequestrations of this type play a part in both calcium and nucleic acid deposition and metabolism.

Leaf protein as a human food

Purification is the separation of one type of molecule from another. While purifying viruses present in leaf extracts, I therefore learnt much about the properties of proteins in uninfected leaves. As I discarded unwanted protein precipitates, I often ate pieces if they were reasonably free from ammonium sulfate. When the war started in September 1939, various groups of scientists in Cambridge met to discuss how their knowledge and skill should be used. Those with whom I was associated were critical of the inertia of the government: we had already published two books condemning the prewar proposals for air-raid precautions as wholly unrealistic. Remembering the food shortage during the 1914–18 war, I thought that food needed as vigorous attention as armaments, and suggested that grassland would yield more human food if used to produce leaf protein (LP) than if used to produce fodder for cattle. Nothing was done about most of our proposals immediately. With the "fall of France" in 1940, things began to be taken more seriously and I was asked to cooperate with the Food Investigation Board and Imperial Chemical Industries in studying LP production. There was some conflict from the start. I said I already knew enough about the properties of LP; the only problem was to devise practical methods for extracting leaf juice on a large scale. After 40 years this still seems to be the main problem.

For as long as suitably lush grass was available during 1940 and 1941, I toured the country testing different types of large-scale mill and press. This showed me that, for this purpose, rubbing the leaf assiduously is more important than chopping it finely, and that there is no need for intense pressure to separate juice from fibre. If rubbing has been adequate, the pressure exerted between finger and thumb gets out nearly as much juice as comes out at any pressure. But it is important to maintain the pressure for the few seconds that the juice needs to run away from the pressing surfaces. These historical and mechanical points are described in greater detail elsewhere[33].

Those who manufactured or used the pieces of large-scale equipment I tested were keenly interested in what I was trying to do, and were surprisingly tolerant of the messes I made in their works. The officials who sponsored the work were ambivalent. Sometimes I was told that it was so important that I must be more secretive and not explain, when apologising for the mess, just what was proposed, lest the Germans got hold of the idea. At other times I was told it was a lot of nonsense and could never be practical. I did not then know that similar ideas had been patented by Ereky in 1927 and Goodall in 1936.

When "Lend-Lease" food began to come to Britain from the U.S.A. there was clearly no more need for projects such as LP production. After the war, a circular from the Agricultural Research Council asked the institutes with it was concerned about their research projects which could benefit agriculture and the food supply in the Less Developed Countries (LDCs). My proposal that work on LP should be revived was accepted and we got a grant with which equipment, of a type which the earlier work suggested would be effective, was installed. This could handle a ton of greenstuff per hour. Some critics argued that LP should first be made on a lab scale for testing on rats. When there are two equally logical possible courses of action, critics invariably argue that, whichever course is being followed, it is the wrong one. I argued that amino acid analyses, and in vitro digestibility studies, showed that extracted LP would be nutritionally valuable: the important step was to demonstrate that extraction was a practical farm operation.

After a brief "honeymoon", the ARC lost interest in the project and withdrew support. The original concern with conditions in the

LDCs seemed to have been forgotten, and it was argued that LP could be of no value to British agriculture. I disagreed. In 1942 [34] I had stressed that the partly dewatered leaf residue was a winter feed for ruminants which could be economically conserved and that, because leafy crops give a greater yield of protein and dry matter than any other type of crop, far from robbing ruminants of their fodder, LP production could increase the available cattle fodder. My interest in the welfare of the LDCs is undiminished. But the general principle of fodder fractionation has a role in wealthy countries also. Most work on the subject in Australia, France, Ireland, Japan, New Zealand, Poland, Spain, U.S.A. and U.S.S.R. is designed to make winter fodder for cattle and feed for pigs and poultry.

Work proceeded slowly for the next few years. Nevertheless, by 1957 we were able to make enough LP for a successful pig feeding trial at the National Institute for Research in Dairying. Then the whole situation improved. Bawden, who had been made Director of Rothamsted in 1958, persuaded the Rockefeller Foundation to give us a 5-year grant. He could not persuade the ARC to make work on LP an integral part of the Rothamsted programme, but in other ways he gave us unstinting support. The grant allowed us to make a series of improved pulpers and presses. These, as before, could process about a ton of crop per hour and, after testing at Rothamsted, were given to other institutes, e.g. in India and Uganda. With increased staff, large scale production became easy and several tons of LP were sent to the Rowett Research Institute for a very successful pig-feeding trial. That, coupled with a poultry feeding trial by J. Bibby & Sons, and various trials on rats, demonstrated that LP had the expected good feeding value.

Having established the nutritive value of LP, the feasibility of large-scale production, and the possibility of getting 2 tons of dry, 100% protein from a hectare in a year, I reverted to my original concern with the food supply in the LDCs. It seemed obvious that a foodstuff would be most useful if it could be made on a small scale, from local crops, for local use, by relatively unskilled people. We therefore made a series of what were called "Village Units" which, after testing, were sent to India, Nigeria and Papua-New Guinea.

This was Intermediate, or Appropriate, Technology. Unfortunately, the concept had not yet been enunciated and the Rockefeller Foundation had not become interested in that line of work. However, it generously allowed the remainder of the grant to be used in making "Village Units" and on employing a succession of cooks to develop methods for presenting LP on the table. But it did not renew the grant.

Progress would then have been slow if the International Biological Program (IBP) had not been formally established in 1964 [35]. By a combination of luck and chicanery, I managed to get on to both the U.K. National Committee for the IBP and the international committee concerned with Use and Management of Resources. Consequently, we got grants to make equipment suitable for agronomic work on the yields of LP attainable with different crops in different climates, and to send the equipment to Aurangabad, Calcutta and Ibadan. The most valuable and sustained work on LP, especially on the relevant agronomy, still comes from India. This is partly because, among the countries which need a new source of protein, India is best able to do research. Another reason is that, as a result of a lucky conversation in a pub, which was passed on to Jayaprakash Narayan, I met him, and he interested the Indian government in this line of work.

Agronomic work in countries where LP would be useful was now well established: work at Rothamsted on the process of extraction was in danger of languishing. An application to the Wolfson Foundation for support was making little headway when I met my old friend (later Lord) Zuckerman in the middle of Papua-New Guinea. The plane taking him and Lord Mountbatten to Australia got stuck in a bog and I helped to push it out. He was a Wolfson Trustee: this lucky meeting allowed me to promote the application on a personal basis. In 1966 we got a grant and, 12 years later, the Wolfson Foundation gave another grant with which the first satisfactory "Village Unit" [36] was made. Several copies and variants of it have been made since then.

As soon as LP was being made in conditions of reasonable hygiene, we supplied it to Waterlow who found [37] that it was a satisfactory replacement for half the milk in the diets of infants in

Jamaica. Soon after that, it proved equally satisfactory for infants in Nigeria and schoolboys in India. As a result of these successes, the Nuffield Foundation financed an attempt at LP production and use in Jamaica, and the Nestlé Foundation one in Uganda. Each collapsed, without achieving anything, because of local political events.

Carol Martin, the chairman of the international charity "Find Your Feet", became interested in LP through reading a popular book on the various novel foods which were being suggested. We discussed how a large-scale trial could be managed and decided that the Sri Avinashilingam Home Science College in Coimbatore (India) would be the ideal site. Through the IBP I had met Rajammal Devadas, the Principle of the College, and knew that a project would be efficiently and energetically managed by her. Also, in 1970, under the auspices of the IBP, she had hosted the first international conference on LP. Comparisons between diets containing LP, or other sources of protein, started in 1975 and still continue in Coimbatore. Results are published from time to time, and are so satisfactory that "Find Your Feet" is helping with trials elsewhere in India and in Ghana, Mexico and Sri Lanka. LP for all these trials is made locally with "Village Unit" type equipment.

In a comparison on adolescents in Pakistan, a supplement of LP promoted slightly better growth than an equal amount of protein given as milk. That was unexpected because the amino acid composition of LP, though nearer to the ideal for a human food than that of most seed proteins, is not as good as that of milk. Carotene in LP probably corrected a slight vitamin A deficiency in the home diet of these adolescents. Vitamin A deficiency is so common in many parts of the world that the carotene in LP may be nearly as important nutritionally as the protein. I am therefore now studying the conditions needed for carotene stability in preserved LP.

There have now been enough feeding trials. LP is acceptable and valuable, and should soon lose its position as a novelty and become a standard dietary component. Before this happens, more work is needed on the selection of crops suited to different climatic regions — especially dual purpose crops. We know something about LP from potato and sugar beet tops, but scarcely anything about LP from leaves which will become an abundant by-product if coppiced trees are used extensively in "energy plantations". More work is also needed, along the lines of the recent "Find Your Feet" projects. These are designed to find out how LP production can be integrated with normal rural customs. And much more work is needed on the design of equipment for extracting leaf juice: I find that aspect of work on LP particularly pleasant and interesting.

Food in general

No one who was taught by Hopkins could be unaware of the social and political importance of food. In the 1930s, the surveys made by Orr, later the first Director General of FAO, dramatised the poor state of nutrition in the LDCs, and even in relatively prosperous countries such as Britain. With encouragement from Hopkins and Orr, we formed the "Committee Against Malnutrition" which circulated a regular Bulletin bringing the facts to the attention of local authorities and the medical profession. In it, arguments about how adequately a family living on unemployment pay could feed itself was a perennial topic. The government produced figures suggesting that the unemployed could eat well if they made more use of cheap cuts of meat such as "scrag end of neck": we ridiculed these figures, and pointed out, incidentally, that there would not be enough of the cheap cuts to go round unless the meat supply depended on giraffes. Then the League of Nations took up the subject, and there was no further need for our amateur efforts. That apprenticeship directed my attention to the lamentable state of the world food supply, and to the need for more radical thinking about agriculture. War in 1939 gave the topic immediate practical importance. It seemed clear that, with extended use of wheat and potatoes, Britain could produce the food energy the population needed, but there would be a protein shortage. Hence my advocacy of LP.

Soon after the end of the war it was obvious that the situation was still much the same. Le Gros Clark and I [38] edited a collection of 12 essays called *Four Thousand Million Mouths* which argued that, with proper use of existing knowledge, enough food could be produced to feed the world adequately. Now that world population approaches 5000 million, it is amusing to recall that half the reviewers called our title, and the opening words of the preface

> "Within the lifetime of some of our children the world's population may be expected to reach 4000 millions"

alarmist. Since then, in many articles and a book [39], I have discussed the same general problem, and the specific problem of the many possible protein sources other than LP. This is a subject plagued by fashion and non-quantitative thinking. For example: I argue that such projects as the cultivation of algae in tanks, and yeast on hydrocarbons, will have a negligible effect on human welfare. The more recent articles (e.g. [40] narrow the issue to a discussion of the ratio between the amount of money and effort devoted to saving life through hygienic and medical services, and the amount devoted to agriculture. Saving life is obviously commendable, but planners should remember that when a child's life is saved, someone should be financed to produce the 20 to 30 tons of food that will be needed in a lifetime.

Contraception

Pincus worked in Cambridge in 1937–8 and we became good friends. As soon as the war ended, he invited me to work in the Worcester Foundation for Experimental Biology in Massachusetts. I went with no clear idea of what I would do there. On a brief visit it is best to fit into a line of work already in progress so, finding that problems had arisen in work on hyaluronic acid and hyaluronidase, I worked on them because my experience with agar and *Brucella* seemed relevant. I had a long-standing interest in the techniques and principles of contraception, knew that ova are released encased in cumulus containing hyaluronic acid, and knew that sperm carried hyaluronidase. Only a bigotted antiteleologist would fail to sense a connection. We modified hyaluronic acid in several ways and found that a nitric ester was an effective inhibitor of hyaluronidase. It prevented the dispersal of cumulus from rabbit ova in vitro and, when put into rabbits vaginas before normal copulation, prevented conception [41]. Our results interested the Searle Company, which had financed my visit to Worcester, enough for us to take out 5 joint patents on substances such as these. When I returned from Worcester, I was often asked what I had been doing there and replied that I had been designing an explosive contraceptive. This was, of course, a joke, there are only 2 nitratable groups in the hexose units of hyaluronic acid.

During my year in Worcester there were opportunities for many hours of conversation with Pincus about problems arising from increasing world population. He realised that his vast knowledge of the physiology of steroid action, till then mainly directed at stress and ageing, could be applied to contraception. As is well known, he was the "father of the pill". Because that joint paper on rabbits [41] is Pincus' first on contraception, I can perhaps claim to be one of its grandfathers, and to be the only scientist who has worked on both sides of the food/population equation. Pincus assumed that I would like conditions in his lab so much that I would stay there, but I preferred Rothamsted.

Contraception is too important to be left to a few specialised research workers. As I put it in 1952 [42]

> "Our knowledge of the enzyme processes in yeast and liver is now extensive. Most of this knowledge was gained by people not specifically interested in yeast or liver but interested in enzymic processes and using tissues such as these as convenient material. Had other tissues been as readily available they would have been used and somewhat different mechanisms would have been discovered. If by now as much work had been done on the gametes as on other tissues, the design of an efficient contraceptive would probably be a matter of development rather than fundamental research."

That is still the position. Now that there is anxiety (perhaps excessive) about long-term effects of the "pill" and IUD, more attention should be given to antihyaluronidases. They are likely to be more specific and less toxic than conventional spermicides.

Nature origins and distribution of life

People who came to talk to me about the properties of viruses used, in the 1930s, to ask whether they were alive or not. In self-protection, I wrote an article [43] explaining that the question was meaningless. Nevertheless, some systems obviously are alive: how they arose is a fundamental problem in Biology about which I have been writing articles (about 40 so far) ever since. They are extremely repetitive. There is little relevant new material to discuss, but requests for contributions to symposia come along, and old illusions, which deserve correction, reappear regularly.

Obviously, biochemical processes which are now widespread, or even universal, need have nothing to do with the *origins* of life. They are simply the most successful processes which have evolved so far. For various purposes, organisms use a diverse set of reactions — often involving unusual elements. Had ambient conditions been somewhat different, a different group of reactions could have become dominant. In a similarly sceptical vein, the articles point out that ambient conditions can vary more than is often assumed. It is not a coincidence that the boiling point of water in our atmosphere is a little greater than the thermal death point of almost all organisms. Organisms do not adapt to non-existent environments: they would probably adapt to hotter environments if these offered an exploitable energy flux at the same site for a few million years. Among my other heresies are: in a biologically interdependent world, commitment to molecules belonging to one stereoisomeric system is no odder than the general use of right-hand screws in engineering; in the beginning, as now, sunlight was the dominant source of low-entropy energy; tar-coated mineral surfaces were at first much more important than the "soup" of small molecules which was probably also present. The last point is now beginning to gain acceptance, but it is the clay minerals, rather than minerals containing the transition elements, which get attention.

One way of avoiding discussion about origins is to assume that life came from elsewhere [44]. There is no need to discuss that possibility until there is evidence for something resembling life elsewhere. Nevertheless, it remains a possibility. When journeys into space became possible, and when methods for getting information about conditions on other planets were vastly improved, the Royal Society made me chairman of a subcommittee on "Biological Experiments in Space". We organised 5 symposia on relevant basic Biology and on practical matters such as the Physiology of astronauts and the recycling of their excretions. Less interest is taken in these matters in Britain than in other countries. Our attitude may not be mistaken: there are more pressing problems. However, as a result of lack of interest, the subcommittee has been disbanded. It was interesting, and perhaps useful, while it lasted.

Jobs, offers of jobs, etcetera

Hopkins persuaded the Rockefeller Foundation to endow a second demonstrator in Biochemistry in 1932. I got the new post: Hopkins had noticed my liking for practical work and defined my duties as supervising the lab work of the final year students throughout the day for the first month of the course. That appointment was probably facilitated by a tentative offer of a job in the neighbouring department of Pathology. Congenial as the job was, after a few years I felt restive and asked about jobs in the Lister Institute (London) and the department of Pathology in Oxford. I had known Florey, the professor, when he worked in Cambridge: perhaps that is why he did not offer me a job. The story is flatteringly inverted in the Royal Society obituary notices on Lord Florey and Sir Ernst Chain. I take the opportunity of correcting it here because the inversions make the ludicrous suggestion that Hopkins refused to release me — an attitude which would have been completely out of character.

In 1937, Sir John Russell said he would try to make a job for me in Rothamsted, and asked me to design part of a lab on which work was about to start. He had problems with the proposed new appointment, but got round them by use of that ingenuity for which he was well known (cf. [45]). I was appointed "Virus Physiologist" in 1940. Although Biochemistry, had the term been invented at the time, was an important aspect of 19th century work at

Rothamsted, the subject had lapsed and my group was formally part of Plant Pathology. I became head of an independent Biochemistry department in 1947.

As already mentioned, I had not been tempted by two offers of jobs in U.S.A. The position at Rothamsted involved so little administration, and the department was so adequately financed, that I paid little attention to suggestions that I should be a candidate when a few professorships fell vacant. The position was different when Biochemistry was reorganised at Imperial College (London). The main reason for the difference was the personality of the Rector, Sir Roderick Hill. He did not just offer me a job: he preached sermons at me about my duty to instill scepticism among students and develop Biochemical Engineering as a subject. Planning for a move to Imperial College went ahead until, in 1954, Sir Roderick died. His successor was less interested in revitalising Biochemistry so, although I got a letter appointing me to the job, I thankfully gave up the idea of exchanging the comfort of Rothamsted for the hurly-burly of a professorship.

Being proposed for Fellowship of the Royal Society is an honour conferred by scientists who are familiar with the work you have done and with the possibility that you will do more. Getting the Fellowship is, to a large extent, a matter of luck. From the very large number of candidates, the Council chooses a list; few Council members are familiar with the work of more than one or two of the candidates. It may be added that there is no reason to think that the intrusion of any element of "democracy" would lessen the importance of luck. I was lucky. When I was proposed, there happened to be three people on the Council who knew my work — a most improbable state of affairs. So I got in without waiting. It may be that we attach too much importance in Britain to Fellowship of the Royal Society. However, it is the adhering body to the International Unions and similar organisations, e.g., the IBP, its committees are therefore influential. And it controls a considerable amount of money. I am indebted to it for paying for, or facilitating, more overseas visits for conferences or research, than I can conveniently count. And it awards prizes. Obviously, luck comes in here too. Bawden was Treasurer when I got the Copley medal. I do not know to whom I am indebted for the surprising award of the first Rank Prize for Nutrition and Agronomy. It has been extremely useful because Rothamsted used it to take out an annuity on me: this helps to finance work which continues after I was retired in 1972. If my luck holds, work may go on for another few years. LP production is now firmly established as a commercial process end a theme for many types of research. The international conference in Coimbatore in 1970 has been followed by conferences in Aurangabad in 1982 and in Nagoya in 1985. Activity on my part is no longer needed to keep the project alive: it is, however, enjoyable.

REFERENCES

1 Anon. (R.S.A.), Brighter Biochem., 8 (1931) 24.
2 R.M.Jackson, Geog. J., 78 (1931) 277.
3 N.W.Pirie, in G. Semenza (Ed.), Comprehensive Biochemistry, Vol., 35, Elsevier, Amsterdam, 1983, p. 103.
4 N.W.Pirie and K.G.Pinhey, J. Biol. Chem., 84 (1929) 321.
5 N.W.Pirie, Biochem. Soc. Symp., 17 (1959) 26.
6 N.W.Pirie, Biochem. J., 24 (1930) 51.
7 L.Rapkine, (1945) U.S. Pat., 2376186.
8 N.W.Pirie, Biochem. J., 25 (1931) 6.4.
9 T.S.Hele and N.W.Pirie, Biochem. J., 25 (1931) 1095.
10 N.W.Pirie, Biochem. J., 27 (1933) 1181.
11 A.A.Miles and N.W.Pirie, Brt. J. Exp. Pathol., 20 (1939) 83,109, 278; Biochem. J., 33 (1939) 1709, 1716.
12 N.W.Pirie., Brt. J. Exp. Pathol., 17 (1936) 269.
13 M.Sreenivasaya and N.W.Pirie, Biochem. J., 32 (1938) 1707.
14 J.S.Paterson, N.W.Pirie and A.W.Stableforth, Brt. J. Exp. Pathol., 28 (1947) 223.
15 J.S.Paterson and N.W.Pirie, J. Comp. Pathol. Ther., 58 (1948) 227.
16 K.Landsteiner and N.W.Pirie, J. Immunol., 33 (1937) 265.
17 N.W.Pirie, J. Gen. Microbiol., 72 (1972) 1; Biog. Mem. Roy. Soc., 19 (1973) 19.
18 F.C.Bawden and N.W.Pirie, Brt. J. Exp. Pathol., 17 (1936) 64.
19 I-B. Eriksson-Quensel and T.Svedberg, J. Am. Chem. Soc., 58 (1936) 1863.
20 F.C.Bawden, N.W.Pirie, J.D. Bernal and I.Fankuchen, Nature, 138 (1936) 1051.
21 F.C.Bawden and N.W.Pirie, Proc. Roy. Soc. B, 123 (1937) 274.
22 F.C.Bawden and N.W.Pirie, Symp. Soc. Gen. Microbiol., 2 (1953) 21.
23 N.W.Pirie, Biochem. J., 56 (1954) 83.
24 F.C.Bawden and N.W.Pirie, J. Gen. Microbiol., 21 (1959) 438.
25 N.W.Pirie, Nature, 240 (1972) 572.
26 F.C.Bawden and N.W.Pirie, J. Gen. Microbiol., 16 (1957) 696.
27 N.W.Pirie, In "Interaction of viruses and cells" pll. Int. Cong. Microbiol. 1953; Cold Spring Harbor Symp., 11 (1946) 184.
28 N.W.Pirie, Biochem. J., 63 (1956) 316.
29 N.W.Pirie, Biochem. J., 47 (1950) 614.
30 F.C.Bawden and N.W.Pirie, Proc. Roy. Soc. B, 182 (1972) 297.
31 F.C.Bawden and N.W.Pirie, Proc. Roy. Soc. B, 182 (1972) 319.
32 N.W.Pirie, Proc. Roy. Soc. B, 185 (1974) 343.
33 N.W.Pirie, Science, 152 (1966) 1701; Leaf Protein and Other Aspects of Fodder Fractionation, Cambridge University Press, London, 1978, 2nd ed. 1986.
34 N.W.Pirie, Chem. Ind., 61 (1942) 45.

35 E.B.Worthington (Ed.), The Evolution of IBP, Cambridge University Press, 1975.
36 J.B.Butler and N.W.Pirie, Expl. Agric., 17 (1981) 39.
37 J.C.Waterlow, Brt. J. Nutr., 16 (1962) 531.
38 F.LeGros Clark and N.W.Pirie (Eds.), Four Thousand Million Mouths, Oxford University Press, Oxford, 1951.
39 N.W.Pirie, Food Resources, Penguin Books, 1969; 1976.
40 N.W.Pirie, Africa Health, 3 (3) (1980) 20; Ecol. Food Nutr., 10 (1981) 257.
41 G.G.Pincus, N.W.Pirie and M.C.Chang, Arch. Biochem., 19 (1948) 388.
42 N.W.Pirie, Lancet, i (1952) 54.
43 N.W.Pirie, in Perspectives in Biochemistry, Cambridge University Press, London, 1937, p. 11.
44 N.W.Pirie, Phil. Trans. Roy. Soc. A, 303 (1981) 589.
45 N.W.Pirie, Dict. Natl. Biogr., 1961–70, (1981) 908.

Name Index